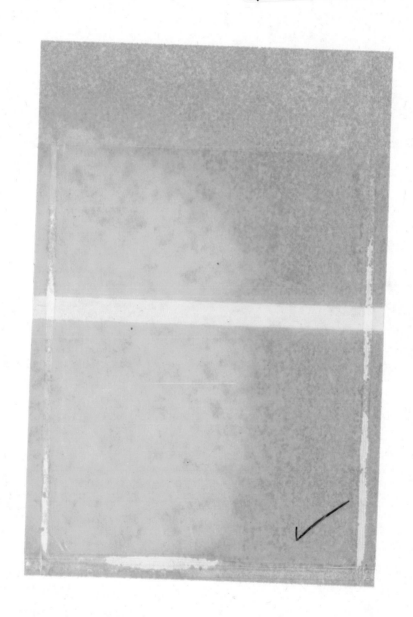

ENCYCLOPEDIA OF ENERGY-EFFICIENT BUILDING DESIGN:
391 practical case studies

by

Dr. Kaiman Lee
architect
landscape architect
planner

To my mother

Dr. Hazel Lee,

and Cathy, Rita & Birk

PREFACE

Energy, its sources, production, distribution, purchase, utilization and conservation are topics that will be important to our society for a long, long time. So many words have been spoken about the need for efficient utilization of energy resources. But so relatively little has been accomplished that it is imperative that the experience existing in the field be available in an easily used, broadly oriented and well organized reference work. From visionary Arcosanti to a family ski house in Colorado. From grand to modest, visionary to pragmatic, ethereal to nuts and bolts. This remarkable compilation of concise, verbal and graphic information in building design for energy efficiency is an important piece of research for our times. The energetic Dr. Lee has collected, sorted, distilled, organized, filed and cross filed a panoramic panoply of energy related efforts in the building industry.

What is very important to researchers and practicing professionals who will use these volumes, is that he has rendered no judgements, aesthetic, operational or economic. He has quite rightly recognized that buried in the ugliest form may be a clue to unique approaches as yet untried. Behind the glossiest facade may be a theoretical weakness to be improved. Each user of this data bank is free to draw on its currency at will and expend it in the furtherance of this important goal for mankind, survival.

Clifford Douglas Stewart, AIA
February, 1977

TABLE OF CONTENTS

INTRODUCTION

The subject of energy presupposes two issues: energy production and energy conservation. Great amounts of energy can be generated through the application of solar, wind, hydro, and bio-fuel energy production. Integration of these energy production methods into architectural design will not only conserve our existing energy resources, but will soon be mandatory in view of the diminishing and non-renewable supply of fossil fuels. Although the use of alternative energy production devices is still in an experimental stage, substantial advancements in this area are now in progress. By studying buildings and designs which incorporate these energy-gathering devices, additional application and effectiveness of such techniques become more apparent. Essential to the operation of these alternative energy production devices is the architect's awareness of the importance of climatic conditions. Also, passive design features such as building orientation, natural ventilation, and landscaping must be a part of every design solution. Accordingly, this publication examines most of the recent outstanding projects which exemplify the utilization of energy conscious design principles.

Included in these two volumes are abstracts of 391 building projects that apply innovative methods using either natural and renewable energy sources or concepts of energy conservation in building design. Each project is described in a concise abstract, and illustrations provided by these contributors are included to aid the reader. (Thus, the variation of the quality of graphics is inevitable.) The format of the abstracts is purposely simple so that all readers should feel at ease in referring to this book. The abstracts are listed alphabetically by title and followed by two indices: one pertaining to building type, and the other to energy-conserving or generating techniques. Project titles often appear more than once in the latter index as many of these buildings combine several techniques for maximum energy efficiency.

Some readers may be interested in simply examining the numerous and diverse practical applications possible for any building type. Home owners may find these abstracts educational; with minimal effort, one can acquire a workable language of architectural concepts relating to energy conservation techniques which will facilitate communication with both architects and builders. Furthermore, architects and builders should find this publication a helpful reference in adapting the various energy-efficient methods to their own designs. Engineers, as well, can adapt their own expertise to the new architectural forms and ideals. They may also find useful information concerning the manufactured hardware that is presently available for efficient energy production.

This publication is made up of very timely material. The immediate demand has never been greater for a comprehensive source where the vast amount of available information can be easily obtained. It is important that we examine early examples which have succeeded or failed in their intended purposes so that we may integrate and improve them for future use. Hopefully, these volumes will encourage the careful study of energy conscious innovations, and facilitate the invention of newer, more efficient methods than are commonly used. Nearly all of the abstracts have been edited and approved by an authority associated with each project. The degree of accuracy of each abstract is heavily reliant upon information given by these authorities and other source material that was utilized. No attempt has been made here to pass judgement on any single project, but rather it is hoped that a more objective and factual description has been presented. The analysis of these abstracts and recommendations could not have been accomplished without delaying the availability of this publication. The author has long felt the urgent need of a reference book such as this, and has made it accessible as soon as possible. This becomes even more apparent when the reader finds many in-between page numbers; these have been included even though the material was received just before the publication went to the printer.

The author wishes to thank all the contributing authors whose valuable time and effort was spent editing the written abstracts and supplying illustrations. In a sense, this book has hundreds of authors. Due credit is given to the Spring 1976 class at the Boston Architectural Center involved in the process of abstracting. Written abstracts were keypunched and listed at the Ark-2 computer facility, and many thanks are accorded to its developer, Mr. Clifford D. Stewart, AIA, for his generosity and concern in the author's work. Special thanks are extended to Ms. Ruth Eisenbud for her consistent technical aid in the compilation of these abstracts, and to Ms. Annelise Huber for her infinite patience in additional editing and finishing of this publication.

ALPHABETICAL LISTING
OF TITLES

TITLE LISTING ACCORDING
TO PROJECT TYPE

COMMUNITY CENTER, STUDENT CENTER AND CITY HALL

COMMUNITY COLLEGE

CONDOMINIUM

CONFERENCE CENTER

COTTAGE

DAY CARE CENTER

FARM HOUSE

FIRE STATION

GUEST HOUSE

GREENHOUSE

HEALTH FACILITY

HOTEL

HOUSE

HOUSING

INDUSTRIAL BUILDING AND FACTORY

OFFICE BUILDING - LOW RISE

POST OFFICE

POTTERY STUDIO

POULTRY SHELTER

PUMPING STATION

RESEARCH AND EDUCATION CENTER

SCHOOL

TITLE LISTING ACCORDING
TO TECHNIQUES

AIR TYPE SOLAR COLLECTOR

```
UNIV. OF ARIZONA ENVIRON. RESEARCH LAB. SOLAR TEST HOUSE    922
USDA ARS, MIKE, AND FLOYD GRANGER HOUSE                     944
WAGONER HOUSE                                               950
WASWANIPI PROTOTYPE HOUSE                                   953
WAVERLY HOMES                                               962
WEST PEARL CONDOMINIUM, SOUTH BUILDING                      971
WHITEWATER STATE PARK OFFICE/INFORMATION BUILDING           977
WILDER HOUSE                                                979
ZAUGG SOLAR HOUSE #3                                       1013
ZWILLINGER SOLAR HOUSE                                     1018
```

ARTIFICIAL LIGHTING

```
ALBANY HIGH SCHOOL                                          65
DALLAS WORLD TRADE CENTER                                  312
ENERGY COST CUTTER HOUSE                                   368
HOLLIS MIDDLE SCHOOL                                       450
HYATT REGENCY HOUSE HOTEL AND PLAZA                        460
MARKEM CORPORATION PLANT                                   532
NORTHERN ARIZONA UNIVERSITY EXECUTIVE CENTER BUILDING      623
SOLAR OFFICE BUILDING                                      820
SOUTHFIELD, MICHIGAN OFFICE BUILDING                       835
```

ATTRIUM

```
CHILDREN'S HOSPITAL OF PHILADELPHIA                        213
ECOLOGY HOUSE                                              352
UNDERGROUND OFFICE                                         917
WATSON SOLAR HOUSE                                         958
ZERO ENERGY HOUSE                                         1016
```

AUTOMATIC ENVIRONMENTAL CONTROL

```
DALLAS WORLD TRADE CENTER                                  312
JACKSON HOUSE                                              470
SEWARD'S SUCCESS                                           768
SOLARON SYSTEM FOR THE GUMP GLASS CO.                      831
SUBURBAN MINNEAPOLIS OFFICE BUILDING                       852
WAGONER HOUSE                                              950
```

BEAD WALL SOLAR COLLECTOR

```
ASPEN AIRPORT TERMINAL                                     80
BOSTON SOLAR HOME                                          126
ERIE-LAKAWANNA HOUSE                                       378
FREESE HOUSE                                               412
MONTE VISTA GREENHOUSE                                     584
NEW HAMPSHIRE FARMHOUSE                                    613
PATOKA STATE PARK NATURE CENTER                            652
SHORE HOUSE                                                784
TYRRELL HOUSE                                              914
```

CLIVUS MULTRUM

```
COTE NORD PROTOTYPE HOUSE                                  281
MAINE AUDUBON BUILDING                                     526
MINNEAPOLIS SOLAR HOME                                     562
POINTE BLEUE PROTOTYPE HOUSE                               679
PROJECT OUROBOROS                                          696
WASWANIPI PROTOTYPE HOUSE                                  953
```

CONVECTION

```
CHERRY CREEK SOLAR RESIDENCE - 419A ST. PAUL STREET        203
COTE NORD PROTOTYPE HOUSE                                  281
CROFT SOLAR HOUSE                                          300
DELAP HOUSE                                                325
```

GLAZING

GREENHOUSE

```
NIEHUSER HOUSE                                                    619
NORTH BRANFORD HOUSE                                              622
PEALE SCIENCE CENTER                                              653
PHILIPS SOLAR HOUSE                                               667
PRINCE EDWARD ISLAND ARK                                          687
PROJECT OUROBOROS                                                 696
R. & N. DAVIS HOUSE                                               706
RCA CORPORATE HEADQUARTERS                                        711
REGINA                                                            718
S.R. HOUSE HOUSE                                                  737
SETON VILLAGE HOUSE                                               767
SEWARD'S SUCCESS                                                  768
SOLAR ENERGY RESOURCE CENTER                                      810
SPRINBROOK LAKE HOUSE                                             842
UNIVERSITY OF MAINE SOLAR HOUSE                                   925
URBIS ET ORBIS                                                    942
WATSON SOLAR HOUSE                                                958
WINONA CITY, MINNESOTA                                            993
ZERO ENERGY HOUSE                                                1016
```

HEAT PUMP

```
ACCESS TEST HOUSE NO. 3                                            50
ALUMNI HOUSE CONFERENCE CENTER                                     68
ANNUAL CYCLE ENERGY SYSTEM HOME                                    75
ASSOCIATION FOR APPLIED SOLAR ENERGY HOUSE                         84
BLAKEWOOD ELEMENTARY SCHOOL                                       114
BRIDGERS & PAXTON OFFICE BUILDING                                 142
CAPE COD SOLAR HOUSE                                              159
CARPENTER HOUSE                                                   168
CONCORD NATIONAL BANK                                             269
CUSTOM LEATHER BOUTIQUE BUILDING                                  309
DENVER COMMUNITY COLLEGE                                          327
ELECTRA III HOME                                                  361
ENERGY COST CUTTER HOUSE                                          368
ESSEN HOUSE                                                       383
FEDERAL BUILDING, SAGINAW, MICHIGAN                               393
GOVERNMENT EMPLOYEES INSURANCE COMPANY BUILDING                   426
GRASSY BROOK VILLAGE                                              434
HILL HOUSE                                                        446
HOWARD BELL ENTERPRISE HOUSE                                      459
HYDE HOUSE                                                        461
MILLS HOUSE                                                       554
NEBRASKA HOUSE                                                    606
NEW CENTURY TOWN                                                  608
NORTHWESTERN FINANCIAL CENTER                                     626
PACIFIC NW BELL TELEPHONE CO. COMMUNITY DIAL OFFICE BUILDING      644
PHILIPS SOLAR HOUSE                                               667
PHOENIX COLORADO SPRINGS HOUSE                                    668
PROJECT OUROBOROS                                                 696
RAVENWOOD ELEMENTARY SCHOOL                                       708
RESEARCH AND ENVIRONMENTAL DESIGN INSTITUTE MILL                  721
ROGER STRAND RESIDENCE                                            732
SCOTTSDALE GIRLS' CLUB SOUTH                                      757
SOKA SOLAR HOUSE                                                  798
SOLAR 1                                                           800
SOLAR HOMES INC. HOUSE                                            817
SOUTH COUNTY HOSPITAL                                             833
SUNRAY I HOUSE                                                    871
SUNRAY II HOUSE                                                   872
TERA ONE                                                          881
UNIVERSITY OF ARIZONA SOLAR ENERGY LABORATORY                     923
UNIVERSITY OF MAINE SOLAR HOUSE                                   925
UNIVERSITY OF TENNESSEE SOLAR HOUSE 2                             940
WEINBERG HOUSE                                                    966
WESTMINSTER HIGH SCHOOL                                           973
WILTON WASTEWATER FACILITIES                                      983
WINEWOOD OFFICE PARK                                              990
XEROX TRAINING AND MANAGEMENT DEVELOPMENT CENTER                 1000
YANAGIMACHI SOLAR HOUSE II                                       1004
ZAUGG SOLAR HOUSE #2                                             1010
```

HEAT RECOVERY

HEATILATOR FIREPLACE

HYDRO

HYDROPONICS

ICE COOLING

INSULATION

SKYLIGHTS

SOD ROOF

SOLAR MEMBRANE

SPRAY WATER COOLING

STRUCTURAL STORAGE

SWIMMING POOL

THERMAL MASS

TRICKLING WATER TYPE SOLAR COLLECTOR

UNDERGROUND

WATER TURBINE

CASE STUDIES

TITLE	ACCESS TEST HOUSE NO. 3
CODE	
KEYWORDS	HOUSE ENERGY SOLAR HEATING GREENHOUSE AUXILIARY RECYCLING RETROFIT COLLECTOR WIND GENERATOR HEAT PUMP RECOVERY
AUTHOR	ACCESS PROGRAM, SCHOOL OF ARCHITECTURE, UNIVERSITY OF WISCONSIN – MILWAUKEE
DATE & STATUS	SOME ENERGY SAVING FEATURES WERE ADDED TO THE ACCESS TEST HOUSE NO. 3 IN 1976; SOME MORE WILL BE ADDED IN 1977.
LOCATION	2915 N. FIRST ST., MILWAUKEE, WISCONSIN
SCOPE	THE ACCESS TEST HOUSE NO. 3 WAS FUNDED BY H.U.D. FOR THE ACCESS PROGRAM TO DEVELOP AND DEMONSTRATE LOW COST SELF-HELP ENERGY AND RESOURCE CONSERVATION. (IT IS ONE OF THE ONLY PROGRAMS WHICH STRESSES THE RETROFITTING OF URBAN RESIDENCES FOR ENERGY CONSERVATION.) INITIAL PROJECTS IN THIS PROGRAM ARE: ROOF-MOUNTED SUPPLEMENTARY SOLAR HEATING SYSTEMS (AIR TRANSFER MEDIA), VERTICALLY MOUNTED SOLAR HEATING PANELS, REFLECTIVE INSULATING SHUTTERS, ZONE AND NIGHT SET BACK CONTROL OF AUXILIARY FORCED AIR HEATING SYSTEM, SOLAR HEATED GREENHOUSE, AND A WATER RECYCLING SYSTEM.
TECHNIQUE	DURING THE FIRST PHASE OF THE PROJECT, MORE INSULATION WAS ADDED. WEATHER-STRIPPING AND SEALING WERE ADDED AROUND THE DOORS AND WINDOWS. (THE WINDOWS WITH A SOUTHERN EXPOSURE WERE ENLARGED, AND REFLECTIVE INSULATED SHUTTERS WERE INSTALLED AROUND THE OTHER WINDOWS.) IN ADDITION, A 400 SQ. FT. SOLAR COLLECTOR AND A 100 SQ. FT. SOLAR DOMESTIC WATER HEATER WILL BE INSTALLED ON THE SOUTH WALL. A SOLAR HEATED GREENHOUSE WAS PROVIDED BOTH TO INSULATE THE EXTERIOR WALL OF THE GARAGE SHOP FROM DIRECT COLD AND TO PROVIDE FOR THE PRODUCTION OF VEGETABLES. A HYDROPONICS BASEMENT, TO PROVIDE FOR THE GROWTH OF ALGAE WHICH COULD SERVE AS A SUPPLEMENTAL PROTEIN SOURCE, IS RECOMMENDED IN A FUTURE PROPOSAL. A WIND GENERATOR WAS PROVIDED FOR THE PRODUCTION OF ELECTRICITY TO RUN MINOR HOUSEHOLD APPLIANCES.

DURING THE SECOND PHASE OF THE PROJECT, A HEAT PUMP WILL BE ADDED TO AMPLIFY THE COLLECTED SOLAR HEAT. THE HEAT STORAGE SYSTEM WILL BE INSTALLED IN THE BASEMENT. A HEAT RECOVERY SYSTEM, AND A ROOM FOR MONITORING THE SOLAR COLLECTION SYSTEM WILL ALSO BE INSTALLED IN THE BASEMENT AT THIS TIME. A ROOM WILL BE PROVIDED FOR STORAGE OF FRESH FRUITS, AND CANNED VEGETABLES AND FRUITS. |
MATERIAL	REFLECTIVE INSULATING SHUTTERS, GREENHOUSE, WEATHER-STRIPPING, SOLAR COLLECTOR, DOMESTIC HOT WATER HEATER, HEAT PUMP, WIND GENERATOR
AVAILABILITY	ADDITIONAL INFORMATION MAY BE OBTAINED FROM THE DEPT. OF ARCHITECTURE, THE UNIVERSITY OF WISCONSIN, MILWAUKEE, WISCONSIN 53201.
REFERENCE	ACCESS PROGRAM 2915 N. FIRST ST., ACCESS TEST HOUSE NO. 3 MILWAUKEE, WISCONSIN, DEPT. OF ARCHITECTURE, UNIVERSITY OF WISCONSIN, 1976, 3 P

before

after

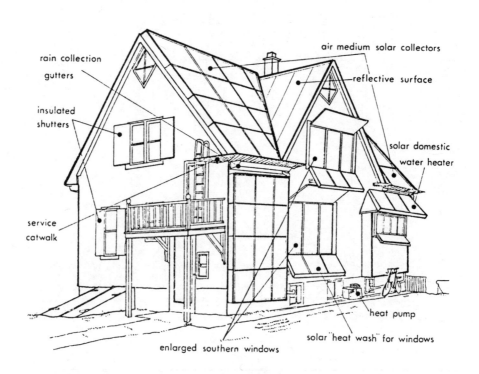

rain collection gutters

insulated shutters

service catwalk

air medium solar collectors

reflective surface

solar domestic water heater

heat pump

enlarged southern windows

solar "heat wash" for windows

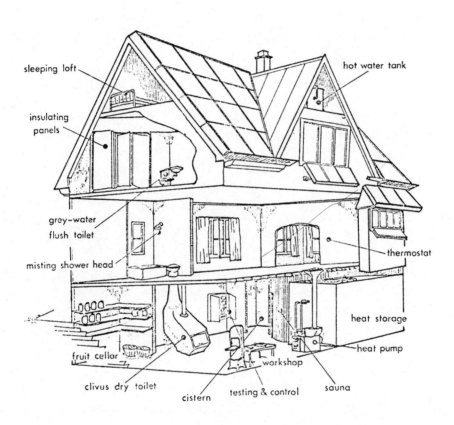

sleeping loft

hot water tank

insulating
panels

grey-water
flush toilet

misting shower head

thermostat

heat storage

fruit cellar

heat pump

clivus dry toilet

workshop

sauna

cistern

testing & control

TITLE	ACORN VILLAGE HOUSE
CODE	
KEYWORDS	HOUSE INSULATION WATER TYPE COLLECTOR
AUTHOR	ACORN STRUCTURES, INC.
DATE & STATUS	THE ACORN VILLAGE HOUSE WAS COMPLETED IN FEBRUARY 1975.
LOCATION	ACTON, MASSACHUSETTS
SCOPE	THE BUILDING IS A HOUSE-AND-GARAGE COMBINATION, WITH THE COLLECTOR ON THE GARAGE, AND THE STORAGE SYSTEM BENEATH THE HOUSE. THE HOUSE IS A 1,400 SQ. FT., SINGLE-STORY, 3-BEDROOM HOUSE, OF TRADITIONAL APPEARANCE, WITH UNHEATED ATTIC AND A CRAWL SPACE FOUNDATION.
TECHNIQUE	THERE IS 3-1/2" OF FIBERGLAS INSULATION IN THE WALLS, AND 6" IN THE ROOF. THE GARAGE IS 20' FROM THE HOUSE. IT IS 24' IN THE EAST-WEST DIMENSION AND THE SAME IN THE NORTH-SOUTH DIMENSION. THE SOUTH-FACING ROOF SLOPES 47-1/2 DEGREES FROM THE HORIZONTAL. THE GARAGE IS NEITHER INSULATED, NOR HEATED.
	THE COLLECTOR AREA OCCUPIES 480 SQ. FT. GROSS, 420 SQ. FT. NET. IT IS A WATER-TYPE COLLECTOR, MOUNTED ON THE 24' X 20' GARAGE ROOF, SLOPING 47-1/2 DEGREES FROM THE HORIZONTAL. IT CONSISTS OF SIX PANELS, OF SPECIAL DESIGN. EACH IS 20' X 4'. THE ABSORBING STRUCTURE, OF THE RAYTHEON DESIGN, EMPLOYS A FORMED, 0.015" ALUMINUM SHEET. COPPER TUBES, 3/8" IN DIAMETER, 19' LONG AND 5-1/2" APART ON CENTERS, ARE CLIPPED TO THIS SHEET. THE HEADERS ARE 1" IN DIAMETER. THE BLACK COATING IS NON-SELECTIVE. THE LIQUID IN THE COLLECTOR IS WATER WITH AN ADDITIVE TO KILL ALGAE (BARCLAY ALGAECIDE BM). NO ANTIFREEZE IS ADDED, AND BECAUSE THE LIQUID IS DRAINED BEFORE A FREEZE-UP CAN OCCUR, DRAINING OCCURS AUTOMATICALLY WHEN THE 300-W CENTRIFUGAL CIRCU-LATING PUMP STOPS. THE COLLECTOR IS SINGLE GLAZED, WITH A 20' X 4' SHEET OF 0.025" KALWALL SUN-LITE (POLYESTER AND FIBER-GLAS). THE SHEET IS HELD IN WAVY FORM BY WAVY EDGE STRIPS TO INCREASE STIFFNESS, AND ALLOW FOR THERMAL EXPANSION AND CON-TRACTION. THE PANEL FRAME IS MADE OF WOOD. THE BACKING IS 2" OF FIBERGLAS. THE PANEL SEALANT IS SILICONE.
	AN EARLIER COLLECTOR CONSISTED OF 18 PPG PANELS, EACH OF WHICH EMPLOYED AN OLIN BRASS CO. 6' X 3' ROLL-BOND ALUMINUM SHEET WITH INTEGRAL PASSAGES. GLAZING WAS OF THERMOPANE. ANTIFREEZE WAS USED, AND A HEAT EXCHANGER WAS USED. THE NEW COLLECTOR IS SIMPLER AND MORE COST EFFECTIVE.
	THE STORAGE SYSTEM CONTAINS 2,400 GALLONS OF WATER IN A WOOD-AND-CONCRETE-BLOCK TANK BENEATH THE FLOOR AT THE NORTH END OF THE HOUSE. THE TANK IS 10' X 9' X 3-1/2', IS FITTED WITH A NEOPRENE LINER, AND IS INSULATED WITH 1" OF URETHANE FOAM ON THE SIDES AND ON THE BOTTOM, AND 2" OF STYROFOAM ON THE TOP. THE HEAT IS DELIVERED FROM THE STORAGE SYSTEM TO THE ROOMS BY A FAN-COIL SYSTEM, EMPLOYING A LOW-POWER (380 WATT) FAN.
MATERIAL	FIBERGLAS INSULATION; 0.015" ALUMINUM SHEET; 3/8" DIAMETER COPPER TUBES; NON-SELECTIVE BLACK COATING; BARCLAY ALGAECIDE BM; 300-W CENTRIFUGAL CIRCULATING PUMP; 0.025" KALWALL SUN-LITE (POLYESTER AND FIBERGLAS); WOOD; 2" FIBERGLAS; SILICONE; PPG PANELS; OLIN BRASS CO. 6' X 3' ROLL-BOND ALUMINUM SHEET; THERMOPANE; ANTIFREEZE; HEAT EXCHANGER; WOOD-AND-CONCRETE-BLOCK TANK; NEOPRENE LINER; 1" OF URETHANE FOAM; 2" OF STYROFOAM; FAN-COIL SYSTEM; LOW-POWER (380 WATT) FAN
AVAILABILITY	ADDITIONAL INFORMATION MAY BE OBTAINED FROM ACORN STRUCTURES, INC., BOX 250, CONCORD, MASSACHUSETTS 01742.
REFERENCE	SHURCLIFF, W.A. SOLAR HEATED BUILDINGS, A BRIEF SURVEY CAMBRIDGE, MASSACHUSETTS, W.A. SHURCLIFF, MARCH 1976, 212 P

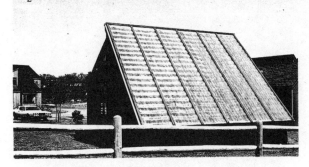

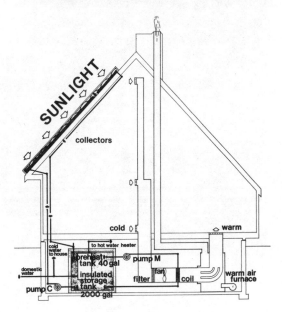

SUNLIGHT

collectors

cold

warm

cold water to house

to hot water heater

pump M

domestic water

preheat tank 40 gal

insulated storage tank 2000 gal

pump C

filter

fan

coil

warm air furnace

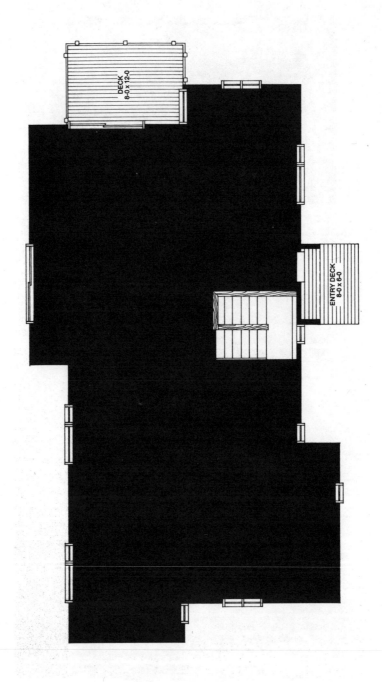

DECK
8-0 x 12-0

ENTRY DECK
8-0 x 6-0

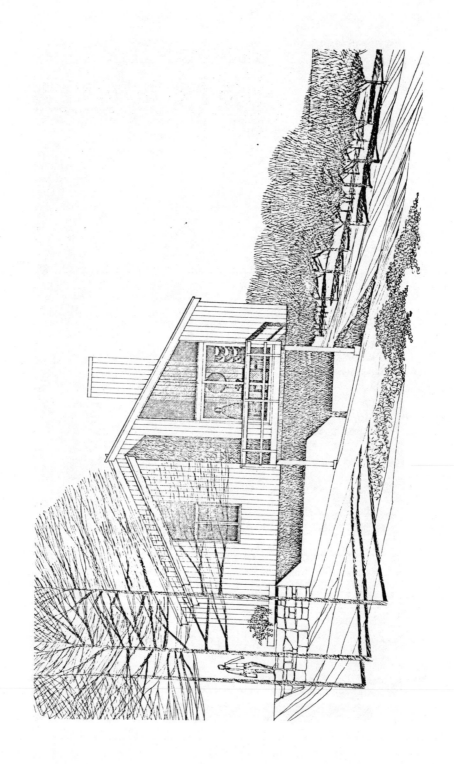

58

TITLE	ADMINISTRATION BUILDING FOR SCHENKER A.G.
CODE	
KEYWORDS	EXTERIOR VENETIAN BLIND SUN SENSITIVE ELECTRONIC CONTROL DOUBLE GLAZING OFFICE
AUTHOR	ALFONS BARTH, JAQUES AESCHIMANN, J. SCHLEUTERMAN, H. WEGMANN, WULLSCHLEGER AND RUETSCHI
DATE & STATUS	THE CONSTRUCTION OF THIS BUILDING WAS COMPLETED IN 1974.
LOCATION	SCHONENWERD, SWEDEN
SCOPE	THIS 3-STORY OFFICE BUILDING IS CONSTRUCTED OF CONCRETE STEEL AND GLASS. EXTENSIVE PROVISIONS HAVE BEEN MADE FOR FUTURE EXPANSION FROM THE PRESENT 3 LEVELS TO 5, AND FROM 3 TEN-METER BAYS TO 5. THE STAIRS, ELEVATORS AND MECHANICAL EQUIPMENT ARE ALREADY SIZED FOR SUCH EXPANSION, AND ALL INTERIOR PARTITIONS – EVEN AROUND THE TOILET ROOMS – CAN BE KNOCKED DOWN FOR FUTURE CHANGES.
TECHNIQUE	THIS BUILDING USES EXTERIOR VENETIAN BLINDS, WHICH ARE OPENED AND CLOSED AS NEEDED BY MEANS OF AUTOMATIC ELECTRONIC CONTROLS, SENSITIVE TO EXTERIOR LIGHT LEVELS. GUIDES HOLD THE VERTICALLY TELESCOPING BLINDS, WHICH SHADE THE WALLS OF DOUBLE GLAZING.
MATERIAL	DOUBLE GLAZING, ELECTRONICALLY CONTROLLED EXTERIOR VENETIAN BLINDS
AVAILABILITY	FOR ADDITIONAL INFORMATION SEE REFERENCE BELOW.
REFERENCE	ARCHITECTURE PLUS SWISS BLINDS: AN OFFICE BUILDING THAT REACTS AUTOMATICALLY TO THE SUN ARCHITECTURE PLUS, JULY/AUGUST 1974, P 99

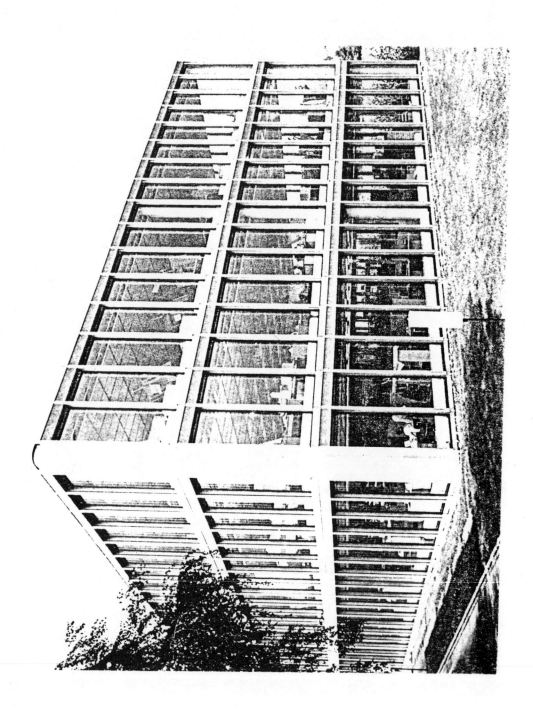

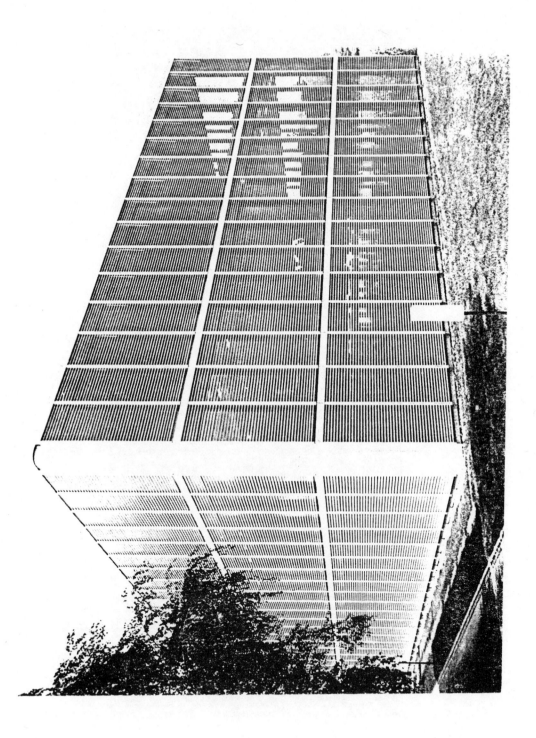

61

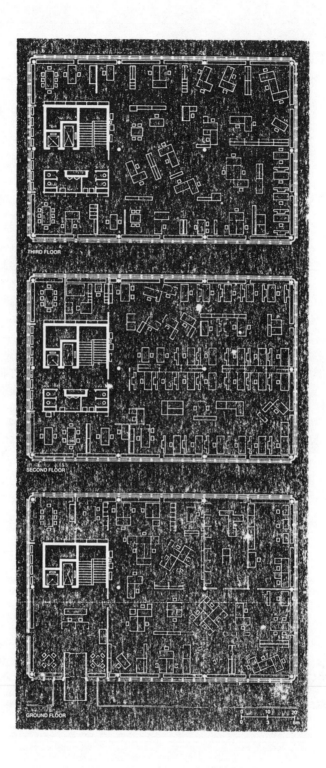

THIRD FLOOR

SECOND FLOOR

GROUND FLOOR

62

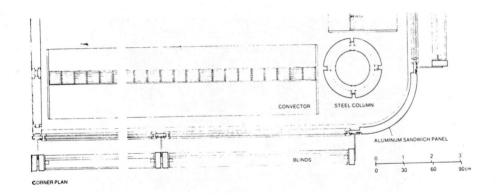

CONVECTOR   STEEL COLUMN

ALUMINUM SANDWICH PANEL

BLINDS

CORNER PLAN

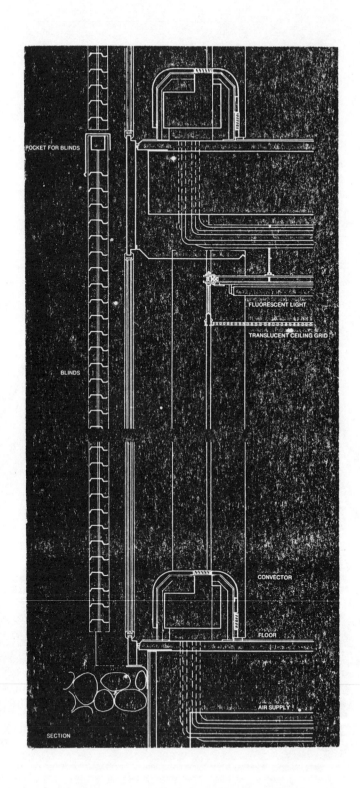

POCKET FOR BLINDS

FLUORESCENT LIGHT

TRANSLUCENT CEILING GRID

BLINDS

CONVECTOR

FLOOR

AIR SUPPLY

SECTION

TITLE	ALBANY HIGH SCHOOL
CODE	
KEYWORDS	MULTI TRACK PLANNING SCHOOL ENERGY BURIED INTERIOR COURTS MINIMUM GLAZED AREA LARGE WINDOWS INSULATING GLASS HEAT RECOVERY SYSTEM RECLAIMS NATURAL LIGHT
AUTHOR	RICHARD G. JACQUES ASSOCIATES / RIST-FROST ASSOCIATES
DATE & STATUS	ALBANY HIGH SCHOOL WAS COMPLETED IN 1974.
LOCATION	ALBANY, NEW YORK
SCOPE	IN 1971, THE CITY OF ALBANY DECIDED TO REACTIVATE A 1966 PROJECT TO CONSOLIDATE THE CITY'S HIGH SCHOOL FACILITIES ON ONE SITE. ALTHOUGH A DESIGN HAD BEEN COMPLETED IN 1966, BUDGET PROBLEMS HAD FORCED THE SCRAPPING OF THE PROJECT AFTER FOUNDATION PILES HAD BEEN PUT IN PLACE ON THE SITE. A NEW ARCHITECT WAS COMMISSIONED TO DESIGN A FLEXIBLE HIGH SCHOOL CAMPUS FOR 2,800 STUDENTS, USING AS MANY OF THE EXISTING PILES AS POSSIBLE. A PROGRAM BUDGET OF $17 MILLION WAS SET, CONSTRAINED BY A LIMITATION ON BONDED INDEBTEDNESS: COMPLETIONS OF THE SCHOOL WAS SET FOR SEPTEMBER 1974 (AN ABSOLUTE REQUIREMENT FOR POLITICAL REASONS).

THE PROJECT WAS BROKEN INTO SEPARATE BUT INTERCONNECTED BUILDINGS, ALLOWING CONSTRUCTION OF ONE WHILE THE NEXT WAS BEING DESIGNED.

THE SCHOOL WAS THE STATE'S FIRST EXAMPLE OF A PHASED MULTI-TRACK PLANNING/DESIGN/CONSTRUCTION PROGRAM APPLIED TO A MAJOR SCHOOL FACILITY. IT ALLOWED OCCUPANCY OF THE SCHOOL 18 TO 24 MONTHS SOONER THAN WOULD HAVE BEEN POSSIBLE WITH CONVENTIONAL PROCE-DURES.

ENERGY CONSERVATION WAS OF PRIME IMPORTANCE FROM THE BEGINNING OF THE PROJECT. THE SITE CHOSEN BY THE CITY WAS ONCE A CITY PARK IN A RESIDENTIAL NEIGHBORHOOD. THE MAIN ACADEMIC BUILDING, A THREE STORY STRUCTURE, IS SET INTO A NATURAL BOWL WITH THE FIRST LEVEL PARTIALLY BURIED. THIS BOWL AND THE TREES AROUND THE PARK COMBINE TO SHELTER THE SCHOOL FROM THE AREA'S STRONG WINTER WINDS.

TECHNIQUE	ARRANGING THE BUILDINGS AS A CAMPUS AROUND A SERIES OF INTERIOR COURTS ALLOWED FOR A MINIMUM GLAZED AREA ON THE SIDES OF THE BUILDINGS FACING OUTWARD FROM THE SITE. FACADES FACING THE COURTYARDS HAVE LARGE WINDOWS, GLAZED WITH INSULATING GLASS, WHILE SIDES FACING THE OUTSIDE ARE BRICK. BUILDINGS ARE CONNECTED BY GLASS AND STEEL BRIDGES AT THE SECOND LEVEL.

A MULTIPLE PIPE WATER MEDIUM HEAT RECOVERY SYSTEM RECLAIMS EXCESS HEAT FROM VARIOUS PARTS OF THE CAMPUS AND REDIRECTS IT TO WHERE HEAT IS NEEDED. THE MECHANICAL SYSTEMS ARE LOCATED IN TOWERS THAT ARE SEPARATE FROM THE ACADEMIC AREA OF THE SCHOOL.

FLUORESCENT LIGHTING IS USED THROUGHOUT MUCH OF THE SCHOOL. FIXTURES PRODUCING LITTLE GLARE WERE USED TO REDUCE THE REQUIRED LIGHTING INTENSITY. THE LARGE INWARD FACING WINDOWS BRING NATURAL LIGHT TO MANY CLASSROOMS.

MATERIAL	INSULATING GLASS, BRICK, STEEL BRIDGES, MULTIPLE PIPE WATER MEDIUM HEAT RECOVERY SYSTEM, FLUORESCENT LIGHTING
AVAILABILITY	ADDITIONAL INFORMATION MAY BE OBTAINED FROM RICHARD G. JACQUES ASSOCIATES, 41 STATE STREET, ALBANY, N.Y. 12207.
REFERENCE	AIA ENERGY NOTEBOOK ENERGY, AIA ENERGY NOTEBOOK AN INFORMATION SERVICE ON ENERGY AND THE BUILT ENVIRONMENT AMERICAN INSTITUTE OF ARCHITECTS, WASHINGTON, D.C., 1975

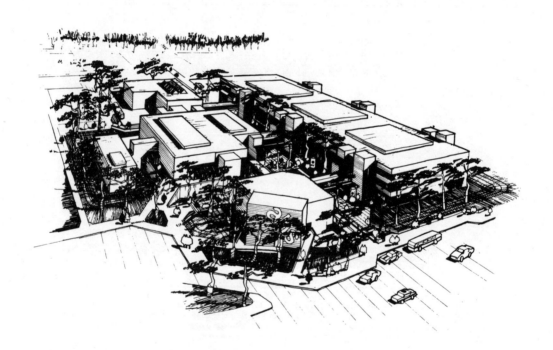

TITLE	ALLERS HOUSE
CODE	
KEYWORDS	LOG CABIN AIR TYPE COLLECTOR STORAGE
AUTHOR	SUN MOUNTAIN DESIGN
DATE & STATUS	THE ALLERS HOUSE WAS UPDATED FOR SOLAR HEATING IN 1974.
LOCATION	SANTA FE, NEW MEXICO
SCOPE	THE BUILDING IS A LOG-CABIN TYPE BUILDING. IT HAS ONE STORY, TWO BEDROOMS, AND A LOFT. THERE IS NO BASEMENT.
TECHNIQUE	THE CHINKS BETWEEN THE LOGS ARE FILLED WITH FIBERGLAS.

THE COLLECTOR HAS 500 SQ. FT. OF GROSS AREA. IT IS AN AIR TYPE COLLECTOR, MOUNTED ON THE SOUTH ROOF, AND SLOPING ABOUT 30 DEGREES. RADIATION IS ABSORBED BY THE NON-SELECTIVE BLACK COATING ON THE ALUMINUM SHEET. THE GLAZING IS DOUBLE; THE OUTER LAYER IS OF KALWALL SUN-LITE FIBERGLAS-REINFORCED POLY-ESTER, AND THE INNER LAYER IS OF TEMPERED GLASS. THE AIRSPACE BETWEEN THE ALUMINUM SHEET AND THE GLASS SHEET IS 1/2". BEHIND THE ALUMINUM SHEET THERE IS A 4" AIRSPACE FOR THE AIRSTREAM, DRIVEN BY A 3/4 HP BLOWER, THAT CARRIES ENERGY TO THE STORAGE SYSTEM. THE INSULATED BACKING IS 6" OF FIBERGLAS.

THE STORAGE SYSTEM HAS 35 TONS OF STONES IN A SHALLOW, 2-1/2' HIGH, BIN BELOW THE FLOOR. THE STONE DIAMETER IS 1" TO 3". THE DIMENSIONS OF THE STONE-FILLED REGION ARE 2' X 8' X 36'. ALONG EACH 36' LONG SIDE OF THIS REGION, THERE IS A HALF-CULVERT, 2' IN DIAMETER, FOR AIR INPUT OR OUTPUT, WITH GRILLS OF EXPANDED METAL, TO EXCLUDE STONES FROM THE CULVERT. THE INSULATION OF THE BIN IS 4" PUMICE BELOW, AND SOME INSULATION ABOVE. WHEN THE ROOMS NEED HEAT, THE ROOM AIR IS CIRCULATED THROUGH THE BIN OF STONES.

MATERIAL	FIBERGLAS; NON-SELECTIVE BLACK COATING; ALUMINUM SHEET; KALWALL SUN-LITE FIBERGLAS-REINFORCED POLYESTER; TEMPERED GLASS; 3/4 HP BLOWER; 6" FIBERGLAS; 35 TONS OF STONES; SHALLOW, 2-1/2' HIGH BIN; 4" PUMICE
AVAILABILITY	ADDITIONAL INFORMATION MAY BE OBTAINED FROM SUN MOUNTAIN DESIGN, 107 CIENGA ST., SANTA FE, NEW MEXICO 87501.
REFERENCE	SHURCLIFF, W.A. SOLAR HEATED BUILDINGS, A BRIEF SURVEY CAMBRIDGE, MASSACHUSETTS, W.A. SHURCLIFF, MARCH 1976, 212 P

TITLE	ALUMNI HOUSE CONFERENCE CENTER
CODE	
KEYWORDS	SOLAR PANEL COLLECTOR STORAGE TANK
AUTHOR	RICHARD JAQUES ASSOCIATES
DATE & STATUS	THIS BUILDING WAS FINISHED IN 1976.
LOCATION	STATE UNIVERSITY OF NEW YORK AT ALBANY
SCOPE	THIS BUILDING CONTAINS 7,000 SQ. FT. OF FLOOR SPACE.
TECHNIQUE	THERE ARE APPROXIMATELY 2,200 SQ. FT. OF SOLAR PANELS MOUNTED ON THE ROOF OF THIS STRUCTURE, SLANTED AT A 45 DEGREES ABOVE THE HORIZONTAL AND AN AZIMUTH 17.5 DEGREES WEST OF SOUTH. HEAT STORAGE IS FURNISHED BY TWO 8,000 GALLON WATER TANKS BURIED NEAR THE CENTER. A HEAT RECLAIM UNIT IS USED TO BOOST THE OUTPUT OF THE SOLAR COLLECTORS. WHEN THE SUN IS STRONG AND THE WATER IN THE STORAGE TANK IS 100 DEGREES OR OVER, THE SYSTEM OPERATES PURELY THROUGH THE SOLAR PANELS. WHEN THE TEMPERATURE OF THE WATER DROPS TO NO LOWER THAN 50 DEGREES, THE WARM WATER IS EVAPORATED AND COMPRESSED BEFORE USED FOR HEATING. THE SYSTEM ALSO HAS A CONVENTIONAL BACKUP HEATING SYSTEM. IN SUMMER, THE HEAT PUMP SYSTEM CONVERTS TO A COOLING SYSTEM.
MATERIAL	SOLAR COLLECTOR PANELS, STORAGE TANKS, HEAT RECLAIM UNIT, EVAPORATOR, CONDENSOR
AVAILABILITY	ADDITIONAL INFORMATION MAY BE OBTAINED FROM DONALD STEWART, ASSISTANT DIRECTOR, ATMOSPHERIC SCIENCES RESEARCH CENTER, UNIVERSITY OF NEW YORK, 1400 WASHINGTON AVENUE, ALBANY, NEW YORK 12222.
REFERENCE	HEALEY, JAMES TECHNICAL REPORT ON THE ALUMNI HOUSE CONFERENCE CENTER, A ALBANY, NEW YORK, ATMOSPHERIC SCIENCES RESEARCH CENTER PAMPHLET, 1976, 20 P

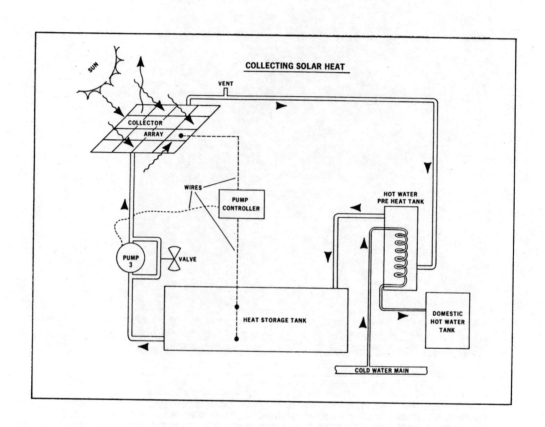

COLLECTING SOLAR HEAT

SUN

VENT

COLLECTOR
ARRAY

WIRES

PUMP
CONTROLLER

HOT WATER
PRE HEAT TANK

PUMP
3

VALVE

HEAT STORAGE TANK

DOMESTIC
HOT WATER
TANK

COLD WATER MAIN

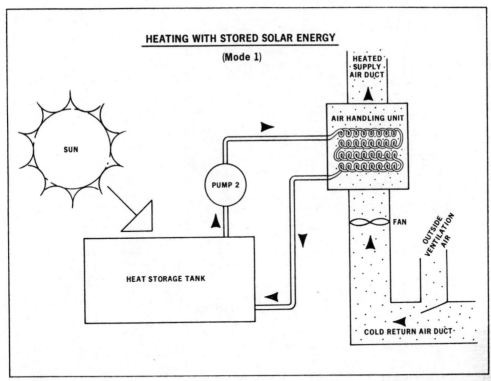

HEATING WITH STORED SOLAR ENERGY

(Mode 1)

SUN

HEAT STORAGE TANK

PUMP 2

HEATED SUPPLY AIR DUCT

AIR HANDLING UNIT

FAN

OUTSIDE VENTILATION AIR

COLD RETURN AIR DUCT

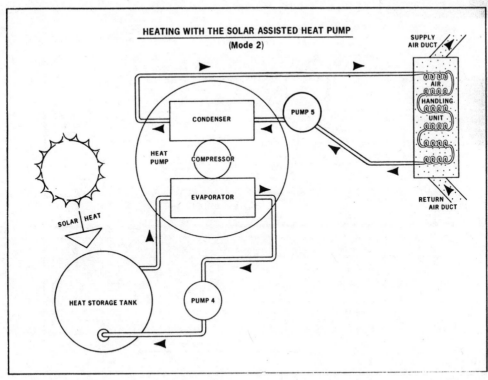

HEATING WITH THE SOLAR ASSISTED HEAT PUMP

(Mode 2)

CONDENSER

PUMP 5

HEAT PUMP

COMPRESSOR

EVAPORATOR

SOLAR HEAT

HEAT STORAGE TANK

PUMP 4

SUPPLY AIR DUCT

AIR HANDLING UNIT

RETURN AIR DUCT

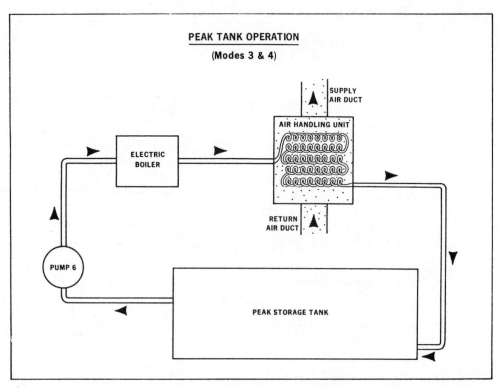

PEAK TANK OPERATION

(Modes 3 & 4)

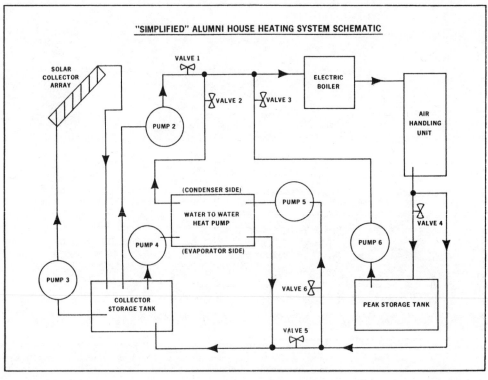

"SIMPLIFIED" ALUMNI HOUSE HEATING SYSTEM SCHEMATIC

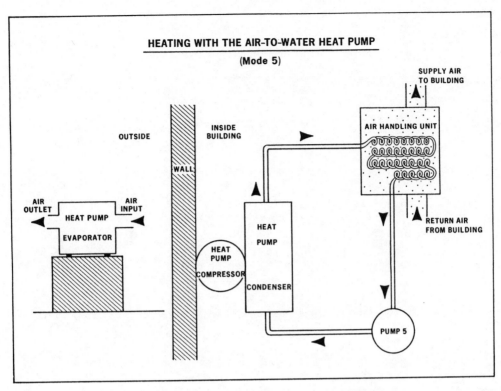

HEATING WITH THE AIR-TO-WATER HEAT PUMP

(Mode 5)

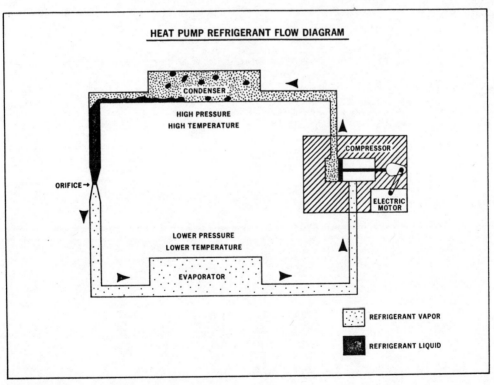

HEAT PUMP REFRIGERANT FLOW DIAGRAM

TITLE	ANGLESEY HOUSE
CODE	
KEYWORDS	HOUSE FLOOR SLAB WINDOW AREA STORAGE BAFFLES STRATIFICATION COPPER TUBES INSULATION
AUTHOR	S.V. SZOKOLAY
DATE & STATUS	THE ANGLESEY HOUSE WAS COMPLETED IN MID-1975.
LOCATION	ANGLESEY, NORTH WALES, GREAT BRITAIN
SCOPE	THE BUILDING IS A ONE-STORY, 2-BEDROOM, 82 SQ.M HOUSE, WITH A SMALL ATTIC BUT NO BASEMENT.
TECHNIQUE	THERE IS A LARGE WINDOW AREA ON THE SOUTH, AND SMALL WINDOW AREAS ON THE EAST, NORTH AND WEST. THE LIVING AREA WINDOWS ARE DOUBLE GLAZED; THE UTILITY AREA WINDOWS ARE SINGLE GLAZED. THE WALLS ARE MODERATELY INSULATED. THE ROOF IS WELL INSULATED, WITH 100 MM OF FIBERGLAS.
	THE COLLECTOR IS A 33 SQ.M (350 SQ. FT.), WATER TYPE COLLECTOR, MOUNTED ON THE ROOF SLOPING 28 DEGREES FROM THE HORIZONTAL. THERE ARE 16 PANELS. EACH IS 2.5 M X 0.82 M, EMPLOYS ALCOA ROLL-BOND ALUMINUM SHEET, AND IS GLAZED WITH ONE SHEET OF 4 MM, IRON-FREE GLASS. THE LIQUID CONSISTS OF WATER AND ANTI-FREEZE AND INHIBITOR. THE PUMP POWER IS 90 W.
	THE STORAGE SYSTEM CONSISTS OF 3 CUBIC METERS, 800 GALLONS, OF WATER IN A TALL, SLENDER, 3 M HIGH X 0.9 M X 0.9 M, STEEL TANK SITUATED IN THE CUPBOARD SPACE. THE INSULATION IS 100 MM OF FIBERGLAS. THE TANK CONTAINS BAFFLES TO ENHANCE STRATIFI-CATION. WHEN THE ROOMS NEED HEAT, HOT WATER FROM THE TANK IS CIRCULATED THROUGH THE COPPER TUBES EMBEDDED IN THE CONCRETE FLOOR, WHICH CONTRIBUTES TO THE STORAGE. CIRCULATION IS BY A SEPARATE, 90 W, PUMP.
MATERIAL	100 MM THICK CONCRETE FLOOR SLAB; DOUBLE GLAZED AND SINGLE GLAZED WINDOWS; 100 MM OF FIBERGLAS; ALCOA ROLL-BOND ALUMINUM SHEET; 4 MM, IRON-FREE GLASS; WATER, ANTIFREEZE, INHIBITOR; 800 GALLONS OF WATER; TALL, SLENDER STEEL TANK; 90 W PUMP
AVAILABILITY	ADDITIONAL INFORMATION MAY BE OBTAINED FROM STEVEN V. SZOKOLAY, DEPT. OF ARCHITECTURE, UNIVERSITY OF QUEENSLAND, ST. LUCIA, BRISBANE, QUEENSLAND 4067, AUSTRALIA.
REFERENCE	SZOKOLAY, STEVEN V. SOLAR ENERGY AND BUILDING HALSTEAD PRESS/ARCHITECTURAL PRESS, 1975, P 98

TITLE	ANNUAL CYCLE ENERGY SYSTEM HOME
CODE	ACES HOME
KEYWORDS	HOUSE HEAT PUMP HOT WATER TANK ICE COOL SOLAR
AUTHOR	ENERGY RESEARCH AND DEVELOPMENT ADMINISTRATION
DATE & STATUS	THIS BUILDING WAS COMPLETED IN MID-1976.
LOCATION	KNOXVILLE, TENNESSEE
SCOPE	THE ACES HOUSE IS ONE OF A PAIR OF TWO-STORY TEST HOUSES. THIS PARTICULAR HOUSE HAS A 1,500 SQ. FT. AREA. THE ADDED COST OF THE ACES SYSTEM FOR THIS 1,500 SQ. FT. HOME IS ESTIMATED AT $2,500, WITH ANNUAL ELECTRICITY SAVINGS OF $450.
TECHNIQUE	THE ACES HOME WILL USE AN INSULATED TANK OF WATER TO LOWER HEATING AND COOLING COSTS. IN WINTER, A HEAT PUMP WILL DRAW HEAT FROM THE WATER TO WARM THE HOME AND TO PROVIDE DOMESTIC HOT WATER. OVER A PERIOD OF MONTHS, HEAT REMOVAL WILL TURN THE WATER INTO ICE. IN SUMMER, WATER CHILLED BY THE ICE WILL COOL THE HOME WITHOUT THE OPERATION OF THE HEAT PUMP COMPRESSOR. THIS WILL MELT THE ICE OVER A PERIOD OF MONTHS, AND THUS STORE HEAT FOR USE IN WINTER.
MATERIAL	HEAT PUMP, INSULATED WATER TANK
AVAILABILITY	ADDITIONAL INFORMATION MAY BE OBTAINED FROM ERDA TECHNICAL INFORMATION CENTER, P.O.BOX 62, OAK RIDGE, TENNESSEE 37830.
REFERENCE	ENERGY RESEARCH AND DEVELOPMENT ADMINISTRATION SOLAR, GEOTHERMAL AND ADVANCED ENERGY SYSTEMS: SOLAR ENERGY AND CONSERVATION FOR HOME HEATING AND COOLING WASHINGTON, D.C., ENERGY RESEARCH AND DEVELOPMENT ADMINISTRATION PAMPHLET EDM-817 (5-76), 1976, 2 P

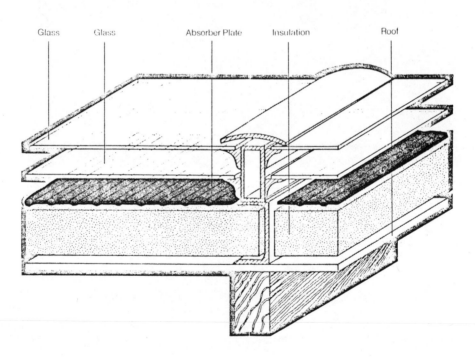

Glass   Glass   Absorber Plate   Insulation   Roof

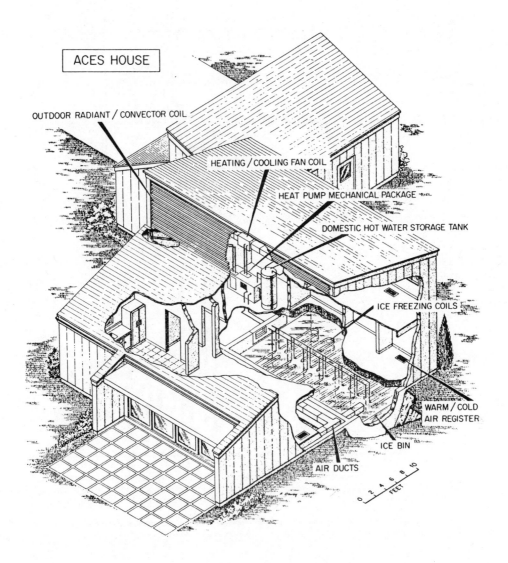

ACES HOUSE

OUTDOOR RADIANT / CONVECTOR COIL

HEATING / COOLING FAN COIL

HEAT PUMP MECHANICAL PACKAGE

DOMESTIC HOT WATER STORAGE TANK

ICE FREEZING COILS

WARM / COLD AIR REGISTER

ICE BIN

AIR DUCTS

0 2 4 6 8 10
FEET

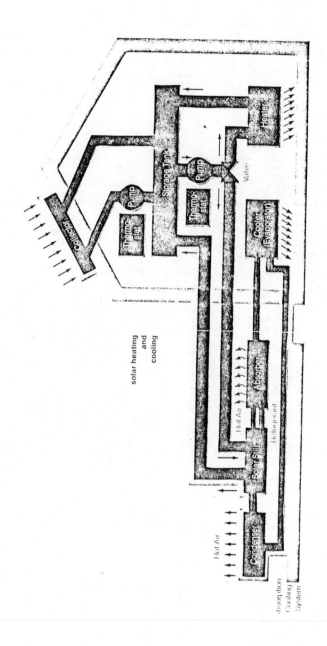

solar heating
and
cooling

TITLE	ARCOSANTI
CODE	
KEYWORDS	SOLAR ENERGIZED CITY GREENHOUSE CHIMNEY APSE EFFECTS ARCOLOGY
AUTHOR	PAOLO SOLERI
DATE & STATUS	A MODEL OF THE ARCOSANTI HAS BEEN COMPLETED AS OF 1975. A CERAMICS APSE, A SECOND VAULT, A CRAFTS VISITORS CENTER, AND RETAINING WALLS HAVE RECENTLY BEEN COMPLETED (1975). ALREADY IN USE ARE A FOUNDRY, STUDIOS, AND A VAULT FOR CONSTRUCTION WORK.
LOCATION	NEAR PHOENIX, ARIZONA
SCOPE	'ARCOLOGY' IS A HARMONIOUS BLEND OF ARCHITECTURE AND ECOLOGY. ARCOSANTI IS A SMALL ARCOLOGICAL CITY. NEARING COMPLETION IS A CRAFTS/VISITORS CENTER, AND SOON TO COMMENCE IS CONSTRUCTION ON A POOL AND SMALL LABORATORY BUILDING.
	WORKING DRAWINGS FOR A CULTURAL/LIVING CENTER CALLED THE TEILHARD DE CHARDIN CLOISTER HAVE BEEN STARTED. THIS PROJECT INCORPORATES ELEMENTS OF THE TWO SUNS ARCOLOGY - THE GREENHOUSE AND APSE EFFECTS.
TECHNIQUE	THE TWO SUNS ARCOLOGY IS BASED ON SEVERAL EFFECTS OF THE SUN. THE GREENHOUSE, CHIMNEY, AND APSE EFFECTS. AN ARCOLOGY (CITY) IS SITUATED DIRECTLY ABOVE A LARGE, TERRACED GREENHOUSE WHICH SENDS UP WARM, MOIST AIR THROUGH THE CHIMNEY EFFECT INTO THE CITY WHERE IT MAY BE USED OR STORED. MEANWHILE, APSE-SHAPED ELEMENTS OF THE CITY CATCH SUNLIGHT IN THE WINTER PROVIDING NATURAL HEAT; BUT IN THE SUMMER THEY SHIELD THE INHABITANTS FROM HOT RAYS, PROVIDING NATURAL COOLING.
MATERIAL	LANDSCAPING, RETAINING WALLS
AVAILABILITY	ADDITIONAL INFORMATION MAY BE OBTAINED FROM PAOLO SOLERI, TWO SUNS ARCOLOGY, CORDES JUNCTION, ARIZONA.
REFERENCE	PROGRESSIVE ARCHITECTURE SOLERI INTRODUCES TWO SUNS ARCOLOGY PROGRESSIVE ARCHITECTURE, NOVEMBER 1975, PP 17-18

TITLE	ASPEN AIRPORT TERMINAL
CODE	
KEYWORDS	AIRPORT TERMINAL EARTH BERMS BEADWALL COLLECTOR THERMAL RESISTANCE STORAGE NATURAL CONVECTION SKYLID
AUTHOR	ZOMEWORKS, INC., RON SHORE, AND COPLAN, FINHOLM, HAGMAN, AND YAW
DATE & STATUS	THE ASPEN AIRPORT TERMINAL WAS COMPLETED IN 1975.
LOCATION	ASPEN, COLORADO

SCOPE   THE BUILDING IS A ONE-STORY, 16,800 SQ. FT. COMPLEX, USED FOR AIRLINE OFFICES. IT HAS WAITING ROOMS, ETC., AND CONSISTS OF THREE RECTANGULAR CONNECTING PORTIONS IN A STAGGERED SOUTHEAST-TO-NORTHWEST ARRAY. EACH PORTION FACES 15 DEGREES EAST OF SOUTH. IN PLAN-VIEW, THE DIMENSIONS OF THE THREE PORTIONS ARE 70' X 70' (TWO PORTIONS), AND 100' X 70' (ONE PORTION). THE ROOFS ARE HORIZONTAL. THERE IS NO BASEMENT. THE COST OF THE BUILDING PROPER IS $600,000. THE BEADWALL AND SKYLIDS COST $30,000.

TECHNIQUE   THE FLOOR IS A 5" CONCRETE SLAB. THE EXTERIOR WALLS ARE MADE OF 8" CONCRETE BLOCKS WITH VOIDS FILLED WITH CONCRETE. THE OUTER FACE OF THE WALL IS INSULATED TO PROVIDE AN OVERALL THERMAL RESISTANCE OF R20. THE CEILINGS ALSO ARE R20. THE EASTERN WALLS ARE PROTECTED BY EARTH BERMS. THE SOUTHERN WALLS ARE OF BEADWALL. THE VIEW-WINDOWS ON THE SOUTHERN AND NORTHERN WALLS ARE DOUBLE GLAZED.

THE COLLECTOR AREA CONSISTS OF 750 SQ. FT. OF BEADWALL, AND 1,750 SQ. FT. OF SKYLIDS. THE THREE VERTICAL SOUTHERN WALLS, 70, 100, AND 70 FT. LONG RESPECTIVELY, INCLUDE 8, 12, AND 8 BEADWALL PANELS RESPECTIVELY. EACH PANEL, 7' X 4', WITH 27 SQ. FT. EFFECTIVE AREA, EMPLOYS TWO SHEETS OF 0.040" KALWALL SUN-LITE (POLYESTER REINFORCED FIBERGLAS) SPACED 2.7" APART. THIS SPACE CONTAINS, DURING SUNNY DAYS, AIR ONLY, AND TRANSMITS SOLAR RADIATION. IT HAS A THERMAL RESISTANCE OF R2 ONLY. AT NIGHT, IT CONTAINS ABOUT 180 MILLION 1/8" STYROFOAM BEADS, AND HAS A RESISTANCE OF R9. BEADS ARE INTRODUCED, OR REMOVED AND DE-LIVERED TO THE STORAGE TANK, IN ABOUT 10 MINUTES BY MEANS OF A SMALL BLOWER.

ON THE FLAT HORIZONTAL ROOF OF EACH BUILDING PORTION THERE ARE TWO LONG MONITORS, THE SOUTH FACES OF WHICH CONSIST LARGELY OF SKYLIGHTS. THE FIXED OUTER GLAZING CONSISTS OF TWO LAYERS OF FILON. BEHIND THE GLAZING, AND 18" FROM IT, ARE ZOMEWORKS, INC., SKYLIDS, I.E. SETS OF SIDE-BY-SIDE LOUVERS, EACH OF WHICH CAN BE TURNED ABOUT A HORIZONTAL AXIS SO AS TO BE OPEN, ADMITTING SOLAR RADIATION; OR CLOSED, BLOCKING RADIATION AND INTERPOSING FIBER-GLAS INSULATION WHICH, ON THE AVERAGE, IS 5" THICK. EACH LOUVER IS 16-1/2' LONG AND 32" WIDE; IN CROSS SECTION, EACH LOUVER IS OVAL, AND 8" THICK AT THE CENTER. EACH LOUVER IS SHEATHED WITH 0.05" OF ALUMINUM SHEET. THE TERMAL RESISTANCE OF THE ENTIRE ASSEMBLY (GLAZING AND SKYLID) IS R2 (OPEN) AND R5 (CLOSED). THE LOUVERS ARE GANGED TOGETHER MECHANICALLY AND ACTUATED BY A SOLAR-POWERED DEVICE (EMPLOYING FREON-FILLED CYLINDERS) WHICH IS ATTACHED TO THE MASTER LOUVER. THE MONITOR FACE IS 53 DEGREES FROM THE HORIZONTAL.

THE STORAGE SYSTEM OF THE BUILDING ITSELF, ESPECIALLY THE CONCRETE FLOOR-SLABS AND THE MASSIVE, EXTERNALLY INSULATED WALLS, STORES ENERGY. COOLING IN SUMMER IS ACCOMPLISHED BY NATURAL CONVECTION VIA OPENABLE WINDOWS AND OPENABLE VENTS IN THE ROOF-TOP MONITORS.

MATERIAL   5" CONCRETE SLAB; 8" CONCRETE BLOCKS; EARTH BERMS; BEADWALL; DOUBLE GLAZED WINDOWS; KALWALL SUN-LITE (POLYESTER REINFORCED FIBERGLAS); 180 MILLION 1/8" STYROFOAM BEADS; FILON; FIBERGLAS INSULATION; 0.05" ALUMINUM SHEET; FREON-FILLED CYLINDERS

AVAILABILITY   ADDITIONAL INFORMATION MAY BE OBTAINED FROM ZOMEWORKS, INC., P.O. BOX 712, ALBUQUERQUE, NEW MEXICO 87103.

REFERENCE   COMPRESSED AIR MAGAZINE
BEADWALLS
COMPRESSED AIR MAGAZINE, JUNE 1976, PP 10-11

In experimental shop setup, top photograph shows beads filling space between window panes. Picture immediately above shows the beads being drawn off.

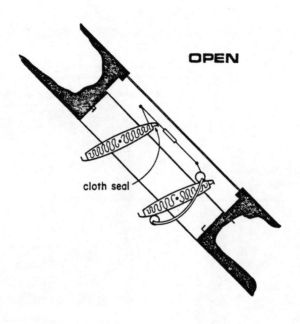

OPEN

cloth seal

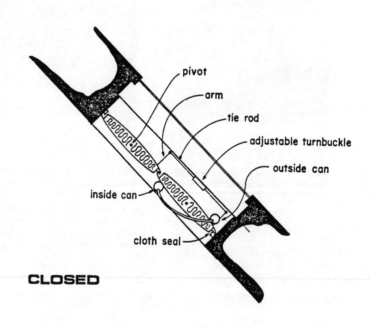

pivot

arm

tie rod

adjustable turnbuckle

outside can

inside can

cloth seal

CLOSED

83

TITLE	ASSOCIATION FOR APPLIED SOLAR ENERGY HOUSE
CODE	AFASE
KEYWORDS	COLLECTOR LOUVER HEATING STORAGE HEAT PUMP COOLING DRAPES SCREEN WALL THERMOSTAT
AUTHOR	PETER LEE
DATE & STATUS	THE AFASE HOUSE WAS COMPLETED IN THE SPRING OF 1958, BUT DUE TO A ZONING ORDINANCE, THE HOUSE WAS NEVER PUT TO ITS INTENDED PURPOSE.
LOCATION	SCOTTSDALE, ARIZONA
SCOPE	THE AFASE HOUSE HAS A 2,000 SQUARE FOOT FLOOR AREA, AND COST $30,000 TO BUILD IN 1958. THREE THOUSAND DOLLARS OF THE COST WAS FOR SOLAR APPARATUS.
TECHNIQUE	THE COLLECTORS TOOK THE FORM OF LOUVERS WITH COPPER TUBE HEAT ABSORBING SURFACES, MYLAR GLAZING AND ALUMINUM SHEET CASINGS. THE COLLECTORS WERE MOUNTED IN 15 ROWS OF FOUR EACH, SUPPORTED BY STEEL PIPES WHICH CARRIED THE WATER TO AND FROM THE COLLECTORS. SWIVEL JOINTS WERE USED TO CONNECT THESE PIPES TO THE HEADERS WHICH RAN ALONG THE EAST AND WEST MAIN STEEL BEAM OF THE STRUCTURE. THE PANELS COULD THUS BE ROTATED SO THAT THEY WOULD TAKE THE MOST ADVANTAGEOUS ANGLE EACH DAY, AND BE TURNED TO FACE DOWNWARD TO ELIMINATE DANGER OF FREEZING ON COLD WINTER NIGHTS AND TO PROVIDE SHADING IN SUMMER.
	SIXTY PANELS, EACH WITH A COLLECTOR SURFACE AREA OF 11.66 SQUARE FEET, PROVIDED 700 SQUARE FEET OF TOTAL SURFACE FOR HEATING THE HOUSE, AND 8 PANELS WITH A TOTAL COLLECTOR AREA OF 93 SQUARE FEET, WERE USED TO HEAT THE DOMESTIC HOT WATER. THE SWIMMING POOL HAD A 500 FOOT GRID OF REVERE COPPER TUBING EMBEDDED IN ITS BOTTOM GUNNITE. IT WAS INTENDED THAT IN MILD WEATHER THE COLLECTORS COULD PROVIDE HEAT BOTH FOR THE HOUSE AND FOR THE POOL. A 2000 GALLON STORAGE TANK WAS BURIED IN THE GROUND AT THE NORTH END OF THE HOUSE, ADJACENT TO THE INSTRUMENT ROOM WHICH PROVIDED SPACE FOR THE TWO CARRIER AIR-TO-AIR HEAT PUMPS. THESE WERE INTENDED TO PROVIDE AUXILIARY HEATING DURING THE WINTER, AND ALL OF THE COOLING DURING THE SUMMER. THE SWIMMING POOL WAS PIPED INTO THE CIRCUIT SO THAT ITS 21,000 GALLONS OF WATER COULD BE USED AS A HEAT SOURCE DURING PRO-TRACTED PERIODS OF BAD WINTER WEATHER, AND AS THE HEAT SINK DURING THE SUMMER.
	THERE ARE INTERCONNECTED STEEL "AIRFLOOR" PANELS UNDER THE CONCRETE FLOORS WHICH CARRY WARM AIR IN THE WINTER AND COOL AIR DURING THE SUMMER.
	THE SOUTH AND NORTH SURFACES OF THE LIVING ROOM WING, AND THE SOUTH SURFACE OF THE BEDROOM WING, WERE SINGLE-GLAZED WITH TINTED 1/4 INCH PLATE-GLASS. HEAVY DRAPES WERE USED AS INSULATION DURING THE NIGHT. A PERFORATED SCREEN WALL OF CONCRETE BLOCK WAS ERECTED ALONG THE EAST AND WEST SIDES OF THE ENTIRE STRUCTURE TO PROVIDE SHADE FROM MORNING AND AFTERNOON SUN IN SUMMER.
MATERIAL	COPPER TUBE-IN-STRIP HEAT ABSORBING SURFACES, MYLAR GLAZING, ALUMINUM SHEET CASINGS, LOUVERED SOLAR COLLECTORS, SWIVEL JOINTS, 2,000 GALLON STORAGE TANK, TWO CARRIER AIR-TO-AIR HEAT PUMPS, 21,000 GALLONS OF WATER IN THE SWIMMING POOL, SINGLE-GLAZED PPG SOLAR GREY 1/4 INCH PLATE GLASS, HEAVY DRAPES, PERFORATED SCREEN WALL OF CONCRETE BLOCK, INTERCONNECTED STEEL "AIRFLOOR" PANELS, RETURN GRILLES, HEAT PUMP, AEROFIN COIL, COMPRESSOR, ROOM THERMOSTAT
AVAILABILITY	ADDITIONAL INFORMATION ON THE AFASE SOLAR HOUSE MAY BE OBTAINED FROM THE REFERENCE BELOW.
REFERENCE	SOLAR ENERGY APPLICATIONS TEAM SOLAR ORIENTED ARCHITECTURE TEMPE, ARIZ., ARIZONA STATE UNIVERSITY, COLLEGE OF ARCHITECTURE JANUARY 16, 1975, PP 123-126

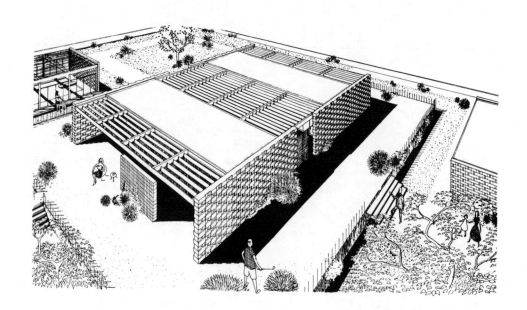

TITLE	ATLANTA OFFICE BUILDING
CODE	
KEYWORDS	OFFICE BUILDING MASS SHAPE SURFACE THERMAL BREAKS DARK VENETIAN BLINDS INSULATING MEAN RADIANT TEMPERATURE INSULATION CAPACITY
AUTHOR	SIZEMORE AND ASSOCIATES
DATE & STATUS	THE ATLANTA OFFICE BUILDING WAS MODIFIED IN 1976.
LOCATION	ATLANTA, GEORGIA
SCOPE	AN 18-STORY CURTAINWALL TOWER BUILT IN ATLANTA IN THE MIDDLE 1960S WAS PLAGUED WITH INSIDE TEMPERATURES IN EXCESS OF 82 DEGREES F. ALTHOUGH THE ALL-ELECTRIC BUILDING IS USED ONLY DURING NORMAL BUSINESS HOURS, IT IS NECESSARY TO RUN THE COOLING SYSTEM 24 HOURS A DAY DURING THE WARM SEASON. THE LIGHT MASS OF THE CURTAINWALL CAUSES THE BUILDING TO BE OVERLY COOL IN THE EARLY MORNING, BUT EXCESSIVELY HOT BY NOON DURING THE 8-MONTH WARM SEASON AND THROUGHOUT THE ENTIRE YEAR ON THE SOUTH ELEVATION. COMPLICATING THE PROBLEM, THE BUILDING HAS A CIRCULAR SHAPE. DURING SUNLIGHT HOURS APPROXIMATELY 50% OF THE SURFACE IS CONTINUALLY EXPOSED, AS COMPARED TO 25-38% EXPOSURE FOR A SQUARE BUILDING.
TECHNIQUE	THE KEY REASON FOR EXCESSIVE HEAT GAIN IS DUE TO THE CHARACTERISTICS OF THE CURTAINWALL. IT IS FABRICATED OF BRONZE ANODIZED EXTRUDED ALUMINUM FRAMING MEMBERS WITH INSULATED ALUMINUM SPANDREL PANELS IN THE SAME FINISH. THE WINDOWS HAVE 1-INCH HEAT ABSORBING DOUBLE GLAZING. OUTSIDE GLASS IS BRONZE-COLORED AND INSIDE GLASS IS CLEAR. THE CURTAINWALL FRAMING SYSTEM IS CONTINUOUS FROM THE EXTERIOR TO THE INTERIOR WITHOUT ANY THERMAL BREAKS--THE WAY CURTAINWALLS WERE POPULARLY BUILT DURING THE 1960S. (TODAY, TWO PIECE CONSTRUCTION WITH A POLYVINYL-CHLORIDE SEPARATOR STRIP CAN HELP REDUCE THE RATE OF HEAT FLOW.) WINDOWS ARE EQUIPPED WITH DARK COLORED VENETIAN BLINDS AND DRAPERIES.

SUBSTANTIAL HEAT WAS COLLECTED BY THE ALUMINUM SPANDRELS AND THEN TRANSFERRED TO THE SILLS AND MULLIONS.

BUILDINGS WITH HIGH THERMAL MASSES--FOR EXAMPLE MASONRY OR CONCRETE--CAN ABSORB AND STORE HEAT OVER LONGER PERIODS. THIS CAN OFTEN SERVE TO REDUCE PEAK AIR-CONDITIONING LOADS. LIGHTWEIGHT CURTAINWALL CONSTRUCTION, LACKING THERMAL MASS, TRANSFERS HEAT MORE RAPIDLY. WHEN IN USE ON SUNNY DAYS, THE DARK COLORED BLINDS ALSO HELP CONTRIBUTE TO HEAT GAIN. ACTING AS SOLAR COLLECTORS, THEY READILY ABSORB AND RADIATE HEAT TO THE INTERIOR.

TO REDUCE HEAT GAIN FROM GLASS AREAS BY HALF, THE ARCHITECTS RECOMMENDED USING REFLECTIVE FILM ON ALL THE WINDOWS EXCEPT THOSE ON THE NORTH SECTOR. IT WAS RECOMMENDED THAT INSULATING THE SPANDRELS, MULLIONS AND SILLS ON THE INSIDE WITH A FABRICATED COVER MADE OF A RIGID CLOSED-CELL INSULATING BOARD COVERED WITH VINYL FABRIC WOULD REDUCE THE HEAT GAIN FROM THE MULLIONS AND SPANDRELS. WITH THE INSTALLATION OF THE REFLECTIVE FILM AND INSULATION, THE OVERALL THERMAL TRANSFER VALUE (OTTV) WAS CALCULATED TO BE 29.55 BTU/SF/HR--NEARLY HALF THE 48.67 BTU/SF/HR OTTV FOR THE ORIGINAL CURTAINWALL DESIGN.

ANOTHER MEASURE WHICH HAD TO BE LOWERED BEFORE THE BUILDING COULD BE COMFORTABLY OCCUPIED WAS THE MEAN RADIANT TEMPERATURE - THE TEMPERATURE A PERSON FEELS, REGARDLESS OF THE AIR TEMPERATURE. AN ACCEPTABLE RANGE FOR MRT IN OFFICE BUILDINGS IS 65-80 DEGREES F. THIS STRUCTURE, HOWEVER, PRODUCED A MRT OF 88.7 DEGREES F., BASED ON AVERAGE JULY TEMPERATURES. UNDER THE SAME OUTSIDE TEMPERATURE CONDITIONS BUT WITH REFLECTIVE FILM AND INSULATION INSTALLED, THE ARCHITECTS CALCULATED THAT THE MRT WOULD BE REDUCED BY MORE THAN 10 DEGREES.

A TEST INSTALLATION OF THE FILM AND INSULATION WAS MADE IN ONE
TYPICAL OUTSIDE OFFICE. AN ANALYSIS OF A SERIES OF SURFACE
TEMPERATURES TAKEN OVER A THREE-HOUR PERIOD OF CLEAR-TO-
PARTIALLY-CLOUDY WEATHER WITH OUTSIDE AIR AT 79 DEGREES SHOWED
THAT THE UNINSULATED MULLIONS AND SILLS WERE 98-104 DEGREES,
WHILE THE INSULATED MULLIONS AND SILLS HELD STEADY AT 85 DEGREES.
THE TEMPERATURE OF THE GLASS WITH THE FILM APPLIED WAS THE
SAME TEMPERATURE AS THE UNTREATED GLASS--102 DEGREES. THE
MODIFIED SPANDREL WAS ABOUT TWO DEGREES COOLER THAN THE UNMODI-
FIED SPANDRELS. THE AIR TEMPERATURE IN THE TEST ROOM REMAINED
STEADY AT 82 DEGREES, WHILE THE ADJACENT ROOMS RANGED FROM
84-87 DEGREES DURING THE TEST PERIOD. THESE TEMPERATURES WERE
TAKEN WITH THE AIR-CONDITIONING OFF. THE LIGHTS WERE ON DURING
THE FIRST HALF OF THE TEST PERIOD AND OFF THE REMAINER OF THE
TIME. THE LIGHTS DID NOT SEEM TO AFFECT THE SURFACE TEMPERATURES.

OVER A 24-HOUR PERIOD WITH THE AIR CONDITIONING OFF, THE TEST
ROOM MAINTAINED A TEMPERATURE OF ABOUT TWO DEGREES LOWER THAN
THE ADJACENT ROOM DURING WARM DAYLIGHT HOURS. THE TEST ROOM,
HOWEVER, RAN ABOUT ONE DEGREE WARMER AT NIGHT, WHICH INDICATES
THAT THE INSULATION AND FILM INCREASED THE THERMAL CAPACITY
OF THE CURTAINWALL.

WHEN THE AIR CONDITIONING WAS TURNED ON, THE TEST ROOM MAIN-
TAINED A TEMPERATURE OF 76 DEGREES FOR MOST OF THE DAY, BUT
ROSE TO 80 DEGREES BY 5 P.M. MOREOVER, THE MRT IN THE TEST ROOM
WAS 78 DEGREES, ABOUT 4 DEGREES LESS THAN THE UNMODIFIED ROOM--
AND WELL WITHIN THE APPROVED COMFORT ZONE. THE TEMPERATURE IN
THE ROOM WITHOUT THE FILM OR INSULATION RANGED FROM 75-82
DEGREES BETWEEN 9 AND 2:30 P.M. AND THEN ROSE TO 84 DEGREES
BY 5 P.M.

THE RESULT OF THESE TESTS INDICATES THAT THE FILM AND INSULATION
ARE EFFECTIVE AT REDUCING BOTH SOLAR HEAT GAIN AND THE LOAD
ON THE AIR-CONDITIONING SYSTEM DURING THE DAY. BECAUSE OF THE
INCREASED THERMAL RESISTANCE OF THE CURTAINWALL, THE BUILDING'S
OPERATING ENGINEER PREDICTS THAT THE AIR-CONDITIONING SYSTEM
CAN BE SHUT DOWN AT THE END OF THE WORKING DAY AND BE TURNED
ON IN THE MORNING IN SUFFICIENT TIME TO ESTABLISH COMFORT
LEVELS BY THE START OF THE WORKING DAY. IN ADDITION, THE MODIFI-
CATIONS WILL ALSO PERMIT WEEKEND SHUTDOWNS BOTH IN THE WINTER
AND SUMMER--A SITUATION THAT WAS NOT POSSIBLE BEFORE.

MATERIAL   BRONZE ANODIZED EXTRUDED ALUMINUM FRAMING MEMBERS, INSULATED
                ALUMINUM SPANDREL PANELS, 1-INCH HEAT ABSORBING DOUBLE GLAZING,
                DARK COLORED BLINDS, REFLECTIVE FILM, RIGID CLOSED-CELL INSU-
                LATING BOARD COVERED WITH VINYL FABRIC

AVAILABILITY   ADDITIONAL INFORMATION MAY BE OBTAINED FROM M.M. SIZEMORE &
                ASSOCIATES, 860 PEACHTREE BATTLE CIRCLE N.W., ATLANTA,
                GEORGIA 30327.

REFERENCE   AMERICAN INSTITUTE OF ARCHITECTS
                CASE STUDY 11: ENERGY ANALYSIS AND MODIFICATIONS OF AN OFFICE
                BUILDING, ATLANTA
                WASHINGTON, D.C., IN AIA ENERGY NOTEBOOK, PUBLISHED BY THE
                AMERICAN INSTITUTE OF ARCHITECTS, 1975, PP CS-55 TO CS-62

TITLE	AVON RESIDENCE
CODE	
KEYWORDS	SOLAR PANEL HEAT COLLECTOR INSULATION GLAZING RESIDENCE DUMP
AUTHOR	BLUE SUN, LTD.; MINGES ASSOCIATES
LOCATION	209 NEW ROAD, AVON, CONNECTICUT
SCOPE	FOR THIS 2-STORY, SINGLE FAMILY RESIDENCE, 11 SOLAR PANELS AND 2 HEAT DUMP PANELS WERE USED. THE SYSTEM PROVIDES 50% OF THE ENERGY REQUIREMENTS OF THE HOUSE, WHICH CONTAINS 1,600 SQ. FT. OF PRIMARY LIVING SPACE, WITH AN ADDITIONAL 800 SQ. FT. AVAILABLE IN THE BASEMENT.
TECHNIQUE	THIS HOUSE INCLUDES SUCH ENERGY SAVING FEATURES AS TRIPLE GLAZED WINDOWS, NO NORTH FACING GLASS, NO FLOOR TO CEILING PARTITIONS TO PROMOTE THE GREATEST FLOW OF HEATED AIR, RIGID URETHANE INSULATION, AND A RECIRCULATOR IN THE ROOF PEAK TO CONTROL AND UTILIZE ACCUMULATED WARM AIR. THERE ARE 11 ROOF MOUNTED DAYSTAR 20 COLLECTORS USED FOR HEATING.
MATERIAL	SOLAR PANELS, HEAT DUMP PANELS, INSULATION, TRIPLE GLAZING
AVAILABILITY	ADDITIONAL INFORMATION MAY BE OBTAINED FROM DAYSTAR CORP., 90 CAMBRIDGE STREET, BURLINGTON, MASSACHUSETTS.
REFERENCE	DAYSTAR CORP. INSTALLATION FACT SHEET: RESIDENTIAL SPACE AND DOMESTIC HOT WATER HEATING BURLINGTON, MASSACHUSETTS, DAYSTAR CORP. PAMPHLET, IFS-3, 1976, 1 P

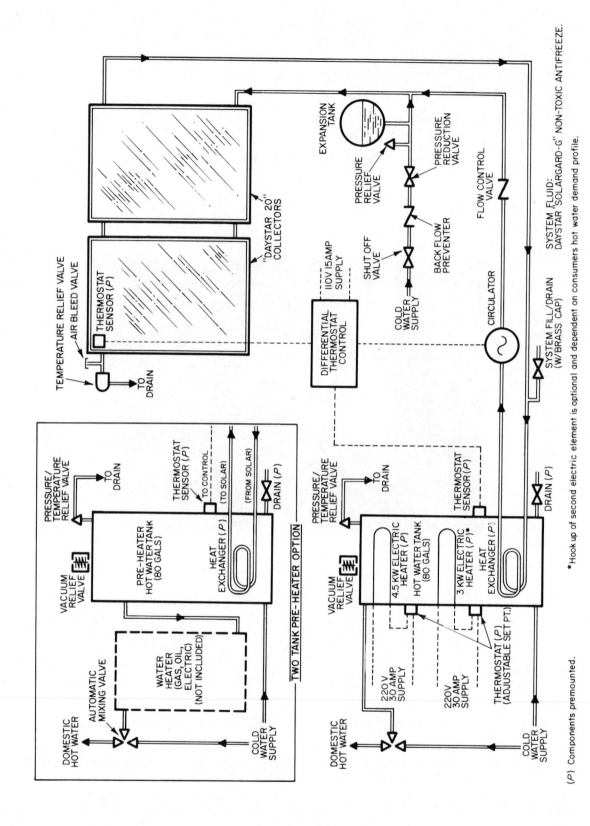

TEMPERATURE RELIEF VALVE

AIR BLEED VALVE

THERMOSTAT SENSOR (*P*)

TO DRAIN

"DAYSTAR 20" COLLECTORS

EXPANSION TANK

PRESSURE RELIEF VALVE

PRESSURE REDUCTION VALVE

SHUT OFF VALVE

BACK FLOW PREVENTER

COLD WATER SUPPLY

FLOW CONTROL VALVE

CIRCULATOR

110V 15AMP SUPPLY

DIFFERENTIAL THERMOSTAT CONTROL

SYSTEM FILL/DRAIN (W/BRASS CAP)

SYSTEM FLUID: DAYSTAR 'SOLARGARD-G" NON-TOXIC ANTIFREEZE.

PRESSURE/ TEMPERATURE RELIEF VALVE

TO DRAIN

THERMOSTAT SENSOR (*P*)

TO CONTROL

(TO SOLAR)

(FROM SOLAR)

DRAIN (*P*)

VACUUM RELIEF VALVE

PRE-HEATER HOT WATER TANK (80 GALS)

HEAT EXCHANGER (*P*)

AUTOMATIC MIXING VALVE

WATER HEATER (GAS, OIL, ELECTRIC) (NOT INCLUDED)

DOMESTIC HOT WATER

COLD WATER SUPPLY

TWO TANK PRE-HEATER OPTION

PRESSURE/ TEMPERATURE RELIEF VALVE

TO DRAIN

THERMOSTAT SENSOR (*P*)

DRAIN (*P*)

VACUUM RELIEF VALVE

4.5 KW ELECTRIC HEATER (*P*)

HOT WATER TANK (80 GALS)

3 KW ELECTRIC HEATER (*P*)*

HEAT EXCHANGER (*P*)

THERMOSTAT (*P*) (ADJUSTABLE SET PT.)

220V 30 AMP SUPPLY

220V 30 AMP SUPPLY

DOMESTIC HOT WATER

COLD WATER SUPPLY

*Hook up of second electric element is optional and dependent on consumers hot water demand profile.

(*P*) Components premounted.

91

TITLE	BAER HOUSE
CODE	
KEYWORDS	SKYLID WATER DRUM BARREL PASSIVE SOLAR REFLECTIVE PANEL HOUSE LOUVER WINDMILL
AUTHOR	STEVE BAER
DATE & STATUS	THE BAER HOUSE HAS BEEN OCCUPIED BY STEVE AND HOLLY BAER SINCE 1972.
LOCATION	THE HOME IS LOCATED AT THE PLATEAU BEHIND CORRALES, OVERLOOKING THE WHOLE VALLEY OF ALBUQEURQUE, NEW MEXICO
SCOPE	THE BAER HOUSE HAS BEEN REFERRED TO AS THE PROTOTYPE SOLAR HOUSE OF THE FUTURE. THE FACETED ALUMINUM PANELS OF THE "ZOME" GEOMETRY MAKE A PHOTOGENIC SUBJECT INSIDE AND OUT.
TECHNIQUE	THE BAER HOUSE USES LARGE GLASS WINDOWS TO ALLOW THE SUN TO WARM RACKS OF WATER DRUMS. AT NIGHT, THE GLASS WALLS ARE COVERED BY RAISING LARGE INSULATING SHUTTERS, WHICH, DURING THE DAY, LIE ON THE GROUND AS REFLECTORS. SUCH A PASSIVE SYSTEM USES ONLY A PULLEY TO RIG THE MOVABLE INSULATION WALLS. IT IS AN ARCHITECTURE THAT IS RUN LIKE A SAILING SHIP.
	SEVERAL OTHER DEVICES CONTRIBUTE TO THE THERMAL STABILITY: BOTH THE EXPOSED CONCRETE FLOOR AND ADOBE WALLS (INSIDE THE ALUMINUM PANELS) ADD TO THE THERMAL CAPACITY OF HEAT STORED IN THE WATER DRUMS.
	EACH "ZOME" HAS A VENT AT THE TOP THAT IS HAND-OPERATED BY A ROPE FOR VENTILATION, AND THE ROOF HAS SEVERAL "SKYLIDS". SKYLIDS ARE ANOTHER ZOMEWORKS PRODUCT AND CONSIST OF INSULATED LOUVERS UNDER A SKYLIGHT. WHEN THE TEMPERATURE BETWEEN THE LOUVER AND THE SKYLIGHT IS HIGHER THAN THE INTERIOR TEMPERATURE, THE LOUVERS AUTOMATICALLY STAY OPEN TO ALLOW THE HEAT INTO THE INTERIOR. OTHERWISE, THEY STAY CLOSED. THE BAER HOUSE ALSO USES SOLAR HOT WATER HEATERS AND A WINDMILL FOR PUMPING WATER.
	SOLAR HEAT IS STORED IN WATER DRUMS BY DAY, AND RETAINED AT NIGHT WITH LARGE MOVEABLE INSULATING SHUTTERS. THESE SHUTTERS ACT AS REFLECTORS DURING THE DAY.
MATERIAL	ADOBE WALLS, CONCRETE FLOORS, ALUMINUM REFLECTING PANELS, WATER DRUMS, SKYLIGHT, LOUVERS, WATER HEATER, WINDMILL
AVAILABILITY	FURTHER INQUIRIES MAY BE DIRECTED TO MR. STEVE BAER, ZOMEWORKS, P.O.BOX 712, ALBUQUERQUE, NEW MEXICO 87103.
REFERENCE	COOK, JEFFREY VARIED AND EARLY SOLAR ENERGY APPLICATIONS OF NORTHERN NEW MEXICO, THE AIA JOURNAL, AUGUST 1974, PP 37-42
	ARIZONA STATE UNIVERSITY, COLLEGE OF ARCHITECTURE BAER HOUSE WASHINGTON, D.C., SOLAR-ORIENTED ARCHITECTURE, PUBLISHED BY THE AIA RESEARCH CORPORATION, JANUARY 1975, P 41

TITLE	BARABOO HOUSE
CODE	
KEYWORDS	HOUSE SOLAR COLLECTOR THERMAL STORAGE
AUTHOR	BOWEN KANAZAWA PARTNERSHIP; KENDON CONSTRUCTION
DATE & STATUS	
LOCATION	BARABOO, WISCONSIN
SCOPE	THIS BUILDING IS A PRIVATE, 2-STORY, 2,350 SQ. FT. HOUSE, WITH A GARAGE.
TECHNIQUE	THE BUILDING'S EXTERIOR IS OF LOG AND STONE, WITH CEDAR ROOF SHAKES, AND 2" OF RIGID INSULATION. THERE IS 297 SQ. FT., 700 CFM, OF SUN STONE FREEZE PROOF TWO PASS HOT AIR COLLECTOR, MOUNTED ON THE ROOF.
	THE STORAGE SYSTEM CONSISTS OF 48,000 LB OF SUN STONE THERMAL STORAGE, 18 CUBIC YARDS OF STONES, LOCATED IN THE BASEMENT. THE BASEMENT SIZE OF STORAGE IS 10-1/2' X 14' X 7'. HEATED AIR FROM THE COLLECTORS IS FORCED THROUGH THE STORAGE, WHERE HEAT IS TRANSFERRED TO THE STORAGE MEDIA OR TO BE USED DIRECTLY FOR SPACE HEATING AS REQUIRED.
	THE STORAGE SIZE FOR THE DOMESTIC HOT WATER IS 82 GALLONS. HEATED AIR FROM THE COLLECTORS IS FORCED THROUGH A HEAT EX- CHANGER WHERE DOMESTIC WATER IS PREHEATED. THE DOMESTIC WATER IS CIRCULATED BETWEEN THE HEAT EXCHANGER AND THE STORAGE TANK.
	THE AUXILIARY ENERGY SYSTEM CONSISTS OF A WOOD BURNING/OIL FIRED FORCED AIR FURNACE.
MATERIAL	LOG, STONE, CEDAR ROOF SHAKES, 297 SQ. FT. OF SUN STONE FREEZE PROOF TWO PASS HOT AIR COLLECTOR, 48,000 LB OF SUN STONE THERMAL STORAGE, 82 GALLONS OF HOT WATER STORAGE, HEAT EXCHANGER, WOOD BURNING/OIL FIRED FORCED AIR FURNACE
AVAILABILITY	ADDITIONAL INFORMATION MAY BE OBTAINED FROM SUN STONE SOLAR ENERGY EQUIPMENT, A DIVISION OF SUN UNLIMITED RESEARCH CORP., P.O.BOX 941, SHEBOYGAN, WISCONSIN 53081.
REFERENCE	SUN STONE SOLAR ENERGY EQUIPMENT BARABOO HOUSE SHEBOYGAN, WISCONSIN, SUN STONE SOLAR ENERGY EQUIPMENT PAMPHLET, 1976, 1 P

TITLE	BARBASH HOUSE
CODE	
KEYWORDS	HOUSE INSULATED COLLECTOR TUBE STORAGE
AUTHOR	J.S. WHEDBEE
DATE & STATUS	THE BARBASH HOUSE WAS COMPLETED IN EARLY 1976.
LOCATION	QUOGUE, NEW YORK
SCOPE	THE BUILDING IS A 2-STORY, MODERN, 3,000 SQ. FT., 2-BEDROOM, WOODFRAME HOUSE. THE GROUND FLOOR INCLUDES A CARPORT, UTILITY ROOM, AND STUDIO. THE SECOND STORY, REACHED BY AN OPEN STAIRCASE AT THE SOUTH SIDE, CONTAINS A KITCHEN, LIVING ROOM, DINING ROOM, TWO BEDROOMS, A CENTRAL GREENHOUSE AREA, AND OFF OF THE LIVING ROOM AND MASTER BEDROOM AREAS ARE BALCONIES WITH SUNDECKS. A DUMB-WAITER CARRIES GROCERIES ETC. FROM THE GROUND FLOOR TO THE SECOND FLOOR. THERE IS NO BASEMENT OR ATTIC.
TECHNIQUE	MOST OF THE EXTERNAL SURFACES ARE FLAT, AND ARE 32-1/2 OR 57-1/2 DEGREES FROM THE HORIZONTAL. THE BUILDING WALLS AND CEILINGS ARE INSULATED WITH NOMINAL 6" OF FIBERGLAS. THE INSULATION UNDER THE FLOORS IS 9-1/2" THICK WITH AN EXTRA 1-1/2" LAYER OF HOMASOTE OVER THE OPEN PORT AREA.
	THE COLLECTOR CONSISTS OF 384 OWENS-ILLINOIS, INC., TUBE ASSEMBLIES OF TEMPERED GLASS. THE GROSS DIMENSIONS OF THE COLLECTOR ARE 32' X 16'; THE AREA IS 510 SQ. FT. THE NET AREA IS CONSIDERABLY LESS, BUT CANNOT BE DEFINED ACCURATELY. EACH TUBE ASSEMBLY CONSISTS OF AN OUTER TUBE, 2" IN DIAMETER, A 1.6" DIAMETER INTERMEDIATE TUBE, AND A 0.6" DIAMETER INNER TUBE. THE SPACE BETWEEN THE OUTER AND THE INTERMEDIATE TUBES VIRTUALLY ELIMINATES CONDUCTION AND CONVECTION TO THE OUTDOORS. BOTH SMALLER TUBES CONTAIN A COOLANT, THE INNER TUBE SERVING AS SUPPLY, AND THE INTERMEDIATE TUBE SERVING AS RETURN. BOTH OF THESE TUBES JOIN A MASSIVE, INSULATED, HORIZONTAL MANIFOLD, CONTAINING SUPPLY AND RETURN HEADERS. EACH MANIFOLD SERVES A LOWER ARRAY OF TUBE ASSEMBLIES AND ALSO AN UPPER ARRAY OF TUBE ASSEMBLIES. EACH ARRAY INCLUDES 48 TUBE ASSEMBLIES. THESE ARE 4" APART ON CENTERS. EACH TUBE IS ABOUT 4' LONG, AND THE END FARTHER FROM THE MANIFOLD IS FAIRLY FREE. THERE ARE NO COOLANT CONNECTIONS; NOTHING TO IMPEDE THERMAL EXPANSION. THE ARRAYS ARE PARALLEL TO THE SLOPING ROOF, WHICH IS 57-1/2 DEGREES FROM THE HORIZONTAL. EACH TUBE ASSEMBLY IS PARALLEL TO THIS PLANE AND PERPENDICULAR TO AN EAST-WEST LINE. ALL TUBE ASSEMBLIES ARE HYDRAULICALLY IN PARALLEL. THE OUTER SURFACE OF THE INTERMEDIATE TUBE IS COATED WITH A SELECTIVE BLACK COATING. BETWEEN THE ADJACENT TUBE ASSEMBLIES THERE IS AN OPEN SPACE, 2" WIDE. THE TUBES ARE THUS SPACED SOME DISTANCE APART TO REDUCE THE TENDENCY OF A GIVEN ASSEMBLY TO SHADE ITS NEIGHBOR AT TIMES SEVERAL HOURS BEFORE NOON OR AFTERNOON. NEAR NOON, HALF OF THE INCIDENT DIRECT RADIATION PASSES BETWEEN THE TUBES AND STRIKES A DIFFUSE WHITE SURFACE THERE, A MAJOR FRACTION OF THE DIFFUSELY REFLECTED RADIATION THEN STRIKES THE TUBE ASSEMBLIES. THE COOLANT IS WATER, WITH NO ANTIFREEZE OR INHIBITOR. ARRAYS AT DIFFERENT HEIGHTS ARE SERVED BY DIFFERENT CENTRIFUGAL PUMPS. THE PRESSURE IN EACH ARRAY IS ABOUT 15 TO 20 PSI. NEAR WHERE THE TUBES ENTER THE MAINFOLDS, THERE ARE SPRING-CONTROLLED PRESSURE-RELEASE PORTS THAT WOULD COME INTO PLAY IF THE PRESSURE WERE TO BECOME HIGHLY EXCESSIVE.
	THE STORAGE SYSTEM CONTAINS 1,000 GALLONS OF WATER IN A 6' DIAMETER SPHERICAL FIBERGLAS TANK.
MATERIAL	TEMPERED GLASS; OUTER TUBE, 2" IN DIAMETER; INTERMEDIATE TUBE, 1.6" IN DIAMETER; INNER TUBE, 0.6" IN DIAMETER; SELECTIVE BLACK COATING; DIFFUSE WHITE SURFACE; WATER; CENTRIFUGAL PUMPS; 1,000 GALLONS OF WATER; 6' DIAMETER SPHERICAL FIBERGLAS TANK
AVAILABILITY	ADDITIONAL INFORMATION MAY BE OBTAINED FROM P. BARBASH, DUNE ROAD, QUOGUE, NEW YORK 11959.
REFERENCE	SHURCLIFF, W.A. SOLAR HEATED BUILDINGS, A BRIEF SURVEY CAMBRIDGE, MASSACHUSETTS, W.A. SHURCLIFF, MARCH 1976, 212 P

TITLE	BARBER HOUSE
CODE	
KEYWORDS	SOLAR FLAT PLATE COLLECTOR WINDMILL LOUVERS AIR WATER NATURAL VENTILATION INSULATING SHUTTERS HOUSE
AUTHOR	PROF. EVERETT BARBER, IN COLLABORATION WITH CHARLES W. MOORE, ASSOCIATES
DATE & STATUS	BARBER'S HOUSE WAS COMPLETED IN JUNE 1975.
LOCATION	NEW HAVEN, CONNECTICUT
SCOPE	THE HOUSE IS AN ATTEMPT BY THE DESIGNERS TO MAKE THE BEST OF INSULATION AND NATURAL SOLAR HEATING IN COMBINATION WITH A NEW FLAT PLATE COLLECTOR SYSTEM DESIGNED BY BARBER. COST ESTIMATE OF A 1,300 SQUARE FOOT HOUSE IS $56,000. SOLAR ENERGY IS TO PROVIDE AN ESTIMATED 80 PERCENT OF ANNUAL HEATING REQUIREMENTS. THE HOUSE WILL HAVE LESS GLASS AREA THAN MANY CONTEMPORARY HOMES, AND ALL GLASS USED WILL BE DOUBLE PANE INSULATING GLASS, MOST OF IT LOCATED ON THE SOUTH EXPOSURE WHERE THE GREATEST AMOUNT OF SUNLIGHT CAN BE OBTAINED. THE HOUSE WILL BE METERED FOR A YEAR TO DETERMINE THE ENERGY NEEDS & CONSUMPTION.
TECHNIQUE	THE ROOF FACES 20 DEGREES WEST OF SOUTH AND IS TILTED AT A 57 DEGREE ANGLE FROM HORIZONTAL, AND HAS 340 SQUARE FEET OF USABLE COLLECTOR AREA. SOLAR HEAT WILL BE STORED IN A TANK, 5 SQUARE FEET IN DIAMETER AND 12 SQUARE FEET HIGH, LOCATED INSIDE THE HOUSE. A BELVEDERE ON TOP OF THE HOUSE WILL BE USED DURING HOT WEATHER TO VENT WARM AIR IN A CHIMNEY EFFECT. WARM AIR COMING THROUGH THE LOUVERS WILL MAKE WAY FOR COOLER AIR TO ENTER THROUGH OPEN WINDOWS AT THE GROUND FLOOR. AN OUTDOOR AIR INTAKE PROVIDES COMBUSTION AIR TO THE FIREPLACE, THUS CONSERVING HEATED INTERIOR AIR. THE CLOTHES DRYER IS SOLAR HEATED.
MATERIAL	FLAT PLATE COLLECTOR, CONCRETE BLOCK, SOLAR STORAGE TANK, DOUBLE GLASS WINDOWS, FAN COIL UNIT, BELVEDERE
AVAILABILITY	ADDITIONAL INFORMATION MAY BE OBTAINED FROM PROF. EVERETT BARBER, YALE UNIVERSITY, SCHOOL OF ARCHITECTURE, NEW HAVEN, CONNECTICUT 06520.
REFERENCE	ARCHITECTURAL RECORD ARCHITECT AND OWNER DEVELOP SOLAR ENERGY SYSTEM FOR NEW HOMES ARCHITECTURAL RECORD, FEBRUARY 1974, P 36
	CONNECTICUT ARCHITECT MAGAZINE ALTERNATE ENERGY SOURCES CONNECTICUT ARCHITECT MAGAZINE, APRIL-MAY 1974

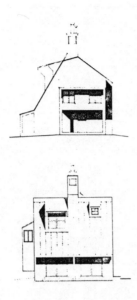

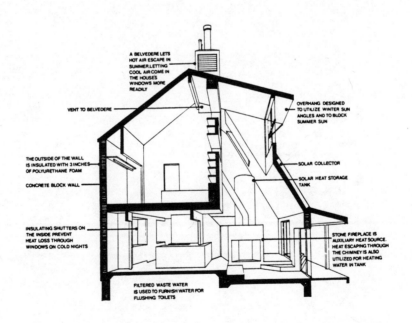

A BELVEDERE LETS
HOT AIR ESCAPE IN
SUMMER,LETTING
COOL AIR COME IN
THE HOUSES
WINDOWS MORE
READILY

OVERHANG DESIGNED
TO UTILIZE WINTER SUN
ANGLES AND TO BLOCK
SUMMER SUN

VENT TO BELVEDERE

SOLAR COLLECTOR

THE OUTSIDE OF THE WALL
IS INSULATED WITH 3 INCHES
OF POLYURETHANE FOAM

SOLAR HEAT STORAGE
TANK

CONCRETE BLOCK WALL

INSULATING SHUTTERS ON
THE INSIDE PREVENT
HEAT LOSS THROUGH
WINDOWS ON COLD NIGHTS

STONE FIREPLACE IS
AUXILIARY HEAT SOURCE.
HEAT ESCAPING THROUGH
THE CHIMNEY IS ALSO
UTILIZED FOR HEATING
WATER IN TANK

FILTERED WASTE WATER
IS USED TO FURNISH WATER FOR
FLUSHING TOILETS

98

TITLE	BARRACLOUGH HOUSE
CODE	
KEYWORDS	HOUSE WINDOW AREAS VENTILATION COLLECTOR INSULATION STORAGE
AUTHOR	K. PRICE OF BUFFALO SOLAR HEATING
DATE & STATUS	THE BARRACLOUGH HOUSE WAS COMPLETED IN DECEMBER 1975.
LOCATION	LEWISBURG, PENNSYLVANIA

SCOPE   THE BUILDING IS A TWO-STORY, 1,200 SQ. FT., WOOD-FRAME HOUSE THAT AIMS EXACTLY SOUTH. IT HAS A BASEMENT BUT NO ATTIC.

TECHNIQUE   THE FIBERGLAS INSULATION IS SAID TO BE STANDARD. THE DESIGN HEAT LOSS IS AT 0 DEGREES F. THE WINDOW AREAS ARE SMALL, AND ALL THE WINDOWS ARE THERMOPANE AND NON-OPENABLE. THE VENTILATION IS BY SEPARATE VENTS. THERE ARE SEVERAL LARGE TREES CLOSE TO THE SOUTH SIDE OF THE HOUSE. THEY BLOCK SOME SOLAR RADIATION IN WINTER.

THE COLLECTOR IS A 400 SQ. FT., AIR TYPE COLLECTOR, MOUNTED ON THE ROOF, SLOPING 45 DEGREES FROM THE HORIZONTAL. THE CORRUGATED ALUMINUM ABSORBER SHEET HAS A NON-SELECTIVE BLACK COATING OF LATEX PAINT. IT IS DOUBLE GLAZED WITH 0.025" OF KALWALL SUN-LITE ON THE OUTSIDE, AND 0.004" OF TEDLAR ON THE INSIDE. THERE IS A 1-1/2" AIRSPACE BETWEEN THE ALUMINUM SHEET AND THE GLAZING, AND BETWEEN THE ALUMINUM SHEET AND THE BACKING. THE BACKING IS 1/2" OF CELOTEX WITH FOIL ON THE UPPER FACE, AND 4" OF FIBERGLAS. THE DIRECTION OF THE AIRFLOW IS UPWARD. THE COLLECTOR WAS FABRICATED ON THE SITE, AND IS AN INTEGRAL PART OF THE ROOF. A 12" X 14" DUCT CARRIES HOT AIR FROM THE COLLECTOR TO THE WEST END OF THE BASE OF THE STORAGE SYSTEM. THE DUCT HAS 1" INSULATION. THE BLOWER IS A 1/10 HP, 400 CFM, DAYTON #4C445. IN SUMMER, THE COLLECTOR IS VENTED MANUALLY.

THE STORAGE SYSTEM HAS 28 TONS OF 1" DIAMETER STONES IN A RECTANGULAR, CONCRETE-BLOCK BIN, 17' X 8' X 5' HIGH, IN THE BASEMENT. THE BIN INSULATION ON THE SIDES IS 1" OF STYROFOAM ON THE OUTSIDE, AND 6" OF FIBERGLAS ON THE TOP. THE STONES REST ON A 14 GAUGE NETTING SUPPORTED BY SPACED BRICKS, ON THE EDGE, THAT PROVIDE A 4" HIGH PLENUM FOR HOT AIR FROM THE COLLECTOR, AND COOL AIR FROM THE ROOMS. THE RETURN AIR TO THE COLLECTOR EMERGES FROM THE OTHER END OF THE BASE OF THE BIN. THE HOT AIR TO BE SENT TO THE ROOMS EMERGES FROM THE TOP OF THE BIN AND FLOWS VIA A 10" X 12" DUCT, WHICH ALSO CONTAINS ELECTRIC RESISTANCE HEATERS. THE CIRCULATION TO THE ROOMS IS MAINTAINED, WHEN NEEDED, BY A MORE POWERFUL, 1/3 HP, 1090 CFM, DAYTON #4C058, BLOWER.

AUXILIARY HEAT IS PROVIDED BY ELECTRIC RESISTANCE HEATERS IN THE DUCT. A HEATILATOR FIREPLACE THAT HAS SPECIAL SCAVENGING COILS IS ALSO USED.

MATERIAL   THERMOPANE WINDOWS; CORRUGATED ALUMINUM ABSORBER SHEET; NON-SELECTIVE BLACK COATING OF LATEX PAINT; 0.025" KALWALL SUN-LITE; 0.004" TEDLAR; 1/2" CELOTEX WITH FOIL; 4" OF FIBERGLAS; 12" X 14" DUCT; 1/10 HP, 400 CFM, DAYTON #4C445 BLOWER; 28 TONS OF 1" DIAMETER STONES; RECTANGULAR, CONCRETE-BLOCK BIN; BRICKS; ELECTRIC RESISTANCE HEATERS; 1/3 HP, 1090 CFM, DAYTON #4C058 BLOWER

AVAILABILITY   ADDITIONAL INFORMATION MAY BE OBTAINED FROM K. PRICE, BUFFALO SOLAR HEATING, P.O.BOX 126, LEWISBURG, PENNSYLVANIA 17837.

REFERENCE   SHURCLIFF, W.A.
SOLAR HEATED BUILDINGS, A BRIEF SURVEY
CAMBRIDGE, MASSACHUSETTS, W.A. SHURCLIFF, MARCH 1976, 212 P

TITLE	BECKSTRAND HOUSE
CODE	
KEYWORDS	INSULATED HOUSE COLLECTOR FLUID STORAGE SOLAR WATER
AUTHOR	J. SCHULTZ
DATE & STATUS	THE BECKSTRAND HOUSE WAS COMPLETED IN JANUARY 1975.
LOCATION	DEL MAR, CALIFORNIA
SCOPE	THE BUILDING IS A TWO-STORY, 3-BEDROOM, 2,200 SQ. FT., WELL INSULATED HOUSE WITH A BASEMENT.

TECHNIQUE   THE COLLECTOR AREA IS 500 SQ. FT. THE COLLECTOR IS A WATER-TYPE COLLECTOR, SLOPING 23 DEGREES FROM THE HORIZONTAL. IT CONSISTS OF ONE LARGE, 36 FT. X 14 FT. PANEL WHICH EMPLOYS A NON-SELECTIVE BLACK ALUMINUM ABSORBER PLATE AND AN ARRAY OF 5/8" DIAMETER STAINLESS STEEL TUBES, 5" APART ON CENTERS, AFFIXED TO AN ALUMINUM PLATE BY MEANS OF THERMON CONDUCTIVE CEMENT. THE FLUID IN THE COLLECTOR IS WATER, WITH A ZINC CHROMATE INHIBITOR, INSTEAD OF ANTIFREEZE. THE SINGLE GLAZING IS REIN-FORCED FIBERGLAS WITH TEDLAR COATING. THE PANEL IS BUILT INTO THE ROOF AND HAS INSULATING (FIBERGLAS) BACKING. SENSORS TURN ON THE COLLECTION-SYSTEM PUMP (1/6 HP BELL & GOSSETT) WHENEVER THE COLLECTOR TEMPERATURE EXCEEDS THE STORAGE SYSTEM TEMPERATURE.

                       THE STORAGE SYSTEM CONTAINS 2,000 GALLONS OF WATER IN A HORI-ZONTAL, CYLINDRICAL STEEL TANK, WHICH IS 5 FT. IN DIAMETER AND 14 FT. LONG. IT IS MOUNTED IN THE BASEMENT. THE TANK INSULATION IS 12" MINERAL WOOL. WHENTHE ROOMS NEED HEAT, THE BLOWER BLOWS AIR THROUGH THE COIL FED WITH HOT WATER FROM THE MAIN STORAGE TANK OR FROM A SMALL TANK. THE SMALL TANK CONTAINS 75 GALLONS OF WATER. THE WATER IS HEATED BY GAS, AND SERVES AS A BACK-UP FOR THE HOUSE HEATING SYSTEM AND ALSO PROVIDES THE DOMESTIC HOT WATER.

MATERIAL   NON-SELECTIVE BLACK ALUMINUM ABSORBER PLATE; 5/8" DIAMETER STAINLESS STEEL TUBES; ALUMINUM PLATE; THERMON CONDUCTIVE CEMENT; REINFORCED FIBERGLAS WITH TEDLAR COATING; INSULATING (FIBERGLAS) BACKING; PUMP (1/6 HP BELL & GOSSETT); WATER; 12" MINERAL WOOL

AVAILABILITY   ADDITIONAL INFORMATION MAY BE OBTAINED FROM JACK SCHULTZ, SOLAR UTILITIES CO., 406 NO. CEDIOR STREET, SOLONA BEACH, CALIFORNIA 92075.

REFERENCE   SHURCLIFF, W.A.
SOLAR HEATED BUILDINGS, A BRIEF SURVEY
CAMBRIDGE, MASSACHUSETTS, W.A. SHURCLIFF, MARCH 1976, 212 P

TITLE	BENEDICTINE MONASTERY OFFICE BUILDING AND WAREHOUSE
CODE	
KEYWORDS	OFFICE BUILDING WAREHOUSE WINDOW AREA ORIENTATION EXTERNAL INSULATING SHUTTERS PASSIVE
AUTHOR	ZOMEWORKS CORP.
DATE & STATUS	THE BENEDICTINE MONASTERY OFFICE BUILDING AND WAREHOUSE WAS COMPLETED IN MID 1976.
LOCATION	PECOS, NEW MEXICO
SCOPE	THE BUILDING IS A ONE-STORY, 7,700 SQ. FT., 140' X 55', BUILDING, WITH A 3,000 SQ. FT. BASEMENT. OFFICES OCCUPY THE SOUTH PORTION OF THE BUILDING, AND THE NORTH PORTION IS A WAREHOUSE.
TECHNIQUE	THE FLOOR AREA IS MADE OF A 4" CONCRETE SLAB. A TYPICAL WALL IS MADE OF 8" CONCRETE BLOCKS WITH 2" OF BEADBOARD ON THE INSIDE. THE ROOF IS INSULATED WITH 6" OF FIBERGLAS. THE WINDOW AREAS ON THE EAST, NORTH, AND SOUTH ARE SMALL, 10 SQ. FT. IN ALL. THESE WINDOWS ARE SINGLE GLAZED. THE BUILDING FACES EXACTLY SOUTH.

THE SOLAR HEATING SYSTEM IS OF THE PASSIVE TYPE, AND EMPLOYS THREE LARGE AREAS OF WINDOWS.

THE LOWEST WINDOW AREA IS A 140' X 4' AREA OF VERTICAL THERMOPANE WINDOWS ALONG THE LOWER PORTION OF THE SOUTH SIDE OF THE BUILDING. THE EXTERNAL INSULATING SHUTTERS ARE MADE OF .050 ALUMINUM, AND ACT AS REFLECTORS DURING THE WINTER AND SHADES DURING THE SUMMER. THE REFLECTORS ARE HINGED AT THE BOTTOM, OPENING OUT ON THE GROUND. THEY ARE OPERATED MANUALLY, AND REMAIN CLOSED IN SUMMER TO EXCLUDE RADIATION AND REDUCE HEAT IN-FLOW. THEY MAY OR MAY NOT BE CLOSED ON COLD WINTER NIGHTS. ON WINTER DAYS, SOLAR RADIATION PASSING THROUGH THESE WINDOWS IMMEDIATELY STRIKES THE WATER-FILLED DRUMS. THERE ARE 138 DRUMS, EACH OF 55 GALLON CAPACITY, AND EACH MADE OF STEEL. CORROSION INHIBITOR IS USED. THE DRUMS ARE ARRANGED IN TWO ROWS, ONE ABOVE THE OTHER. THE LOWER ROW RESTS IN A 4' WIDE, 1-1/2' DEEP DEPRESSION. THE TOPS OF THE UPPER ROW DRUMS ARE 3' ABOVE FLOOR LEVEL. THE DRUMS ARE IN OPENABLE HOUSINGS THAT HAVE 4" TO 6" OF FIBERGLAS INSULATION.

THE INTERMEDIATE WINDOW AREA HAS A 140' X 4' AREA OF VERTICAL THERMOPANE WINDOWS SITUATED IMMEDIATELY ABOVE THE ABOVE-DISCUSSED LOWEST WINDOW AREA. BESIDES PROVIDING THE OCCUPANTS WITH A VIEW OF THE OUTDOORS, THE INTERMEDIATE WINDOW AREA ADMITS SOLAR RADIATION DEEP INTO THE ROOM, WHERE IT STRIKES AND WARMS THE CONCRETE-SLAB FLOOR AND THE EAST-WEST PARTITION WALL.

THE UPPERMOST WINDOW AREA HAS A 140' X 5' AREA OF VERTICAL THERMOPANE. THE RADIATION ENTERING HERE LIGHTS AND WARMS THE NORTH PORTION OF THE BUILDING.

MATERIAL	4" CONCRETE SLAB; 8" CONCRETE BLOCK; 2" BEADBOARD; 6" FIBERGLAS; SINGLE GLAZED WINDOWS; THERMOPANE; WATER-FILLED DRUMS; STEEL; CORROSION INHIBITOR; 4" TO 6" OF FIBERGLAS INSULATION
AVAILABILITY	ADDITIONAL INFORMATION MAY BE OBTAINED FROM ZOMEWORKS CORP., P.O.BOX 712, ALBUQUERQUE, NEW MEXICO 87103.
REFERENCE	SHURCLIFF, W.A. SOLAR HEATED BUILDINGS, A BRIEF SURVEY CAMBRIDGE, MASSACHUSETTS, W.A. SHURCLIFF, MARCH 1976, 212 P

TITLE	BIGHORN CANYON VISITORS CENTER
CODE	
KEYWORDS	VISITORS CENTER AIR TYPE COLLECTOR STORAGE
AUTHOR	WIRTH ASSOCIATES
DATE & STATUS	THE BIGHORN CANYON VISITORS CENTER WAS COMPLETED IN AUGUST 1976.
LOCATION	LOVELL, WYOMING. IT IS PART OF BIGHORN CANYON NATIONAL RECRE-ATIONAL AREA.
SCOPE	THE BUILDING IS A VISITORS CENTER AT THE ENTRANCE TO BIGHORN CANYON NATIONAL RECREATIONAL AREA. THE COST OF THE BUILDING AND SITE DEVELOPMENT IS ESTIMATED AT $1,300,000. THE FLOOR AREA IS 9,500 SQ. FT. THE HEAT LOAD IS 85,000 BTU PER DEGREE DAY.
TECHNIQUE	THE DESIGN OF THE COLLECTOR CALLS FOR 2,880 SQ. FT. OF AIR TYPE COLLECTOR ON A ROOF SLOPING 58 DEGREES FROM THE HORI-ZONTAL. THE COLLECTOR CONSISTS OF A LARGE NUMBER OF PANELS, EACH 24' X 2'. EACH IS SINGLE GLAZED, WITH A 1/2" THICK SPACE FOR AIR-FLOW. THE AIR IS DRIVEN BY A BLOWER, TURNED BY A TWO-SPEED 5 HP MOTOR. THE AIR-FLOW IS 5,000 CUBIC FT/MIN. AT A STATIC PRESSURE OF 2" OF WATER. THERE IS LESS POWER WHEN THE AIR IS CIRCULATED FROM THE COLLECTOR TO THE ROOMS, RATHER THAN TO THE STORAGE SYSTEM. THE COLLECTORS ARE BY R.M. PRODUCTS, DENVER, COLORADO. THE SITE DEVELOPMENT INCLUDES A REFLECTING POND ADJACENT TO THE COLLECTORS, WHICH SHOULD INCREASE THE COLLECTOR EFFICIENCY BY 15% ADDITIONAL REFLECTED SOLAR ENERGY. THE STORAGE SYSTEM CONSISTS OF ONE BIN OF STONES WITH A TOTAL MASS ON THE ORDER OF 75 TONS. TEMPERATURE STRATIFICATION OCCURS IN THE BIN OF STONES, PROVIDING CERTAIN ADVANTAGES. WHEN THE ROOMS NEED HEAT AND THE COLLECTOR IS NOT OPERATING, THE ROOM AIR IS CIRCULATED THROUGH THE BIN OF STONES IN THE OPPOSITE DIREC-TION. THE BIN IS LOCATED IN THE BASEMENT, THE BASEMENT WALLS FORMING PARTS OF THE ENCLOSURES. THE BIN IS LONG AND LOW WITH A HORIZONTAL AIR FLOW.
	AUXILIARY HEAT IS PROVIDED BY 3 GAS-FIRED, 200,000 BTU/HR, 3 HEATING ZONE FURNACES. IN SUMMER, THE BIN OF STONES IS COOLED AT NIGHT BY A STREAM OF COOL AIR FROM THE OUTDOORS. DURING THE DAY, COOL AIR FROM THE BIN IS CIRCULATED TO THE ROOMS.
MATERIAL	SINGLE GLAZING WITH TEMPERED GLASS; TWO-SPEED 5 HP MOTOR; 75 TONS OF STONES
AVAILABILITY	ADDITIONAL INFORMATION MAY BE OBTAINED FROM WIRTH ASSOCIATES, 1739 GRAND AVENUE, BILLINGS, MONTANA 59102.
REFERENCE	SHURCLIFF, W.A. SOLAR HEATED BUILDINGS, A BRIEF SURVEY CAMBRIDGE, MASSACHUSETTS, W.A. SHURCLIFF, MARCH 1976, 212 P

NORTHWEST VIEW

SOUTHWEST VIEW

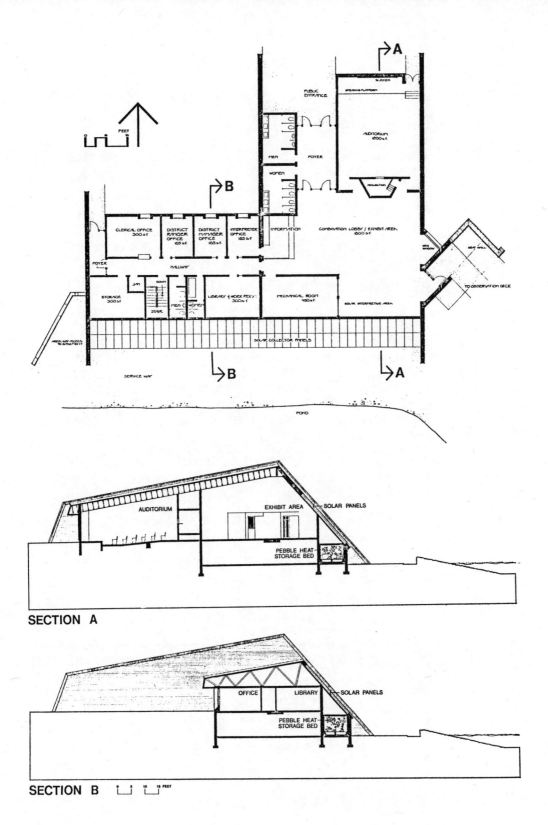

SECTION A

SECTION B

TITLE	BIO SHELTER
CODE	
KEYWORDS	AQUACULTURE GREENHOUSE METHANE GENERATOR SOLAR STILL COLLECTOR MEMBRANE LIGHTING CLIMATIC ENVELOPE THERMAL REGULATION COMPUTER EVAPORATIVE
AUTHOR	SEAN WELLESLEY-MILLER AND DAY CHAHROUDI
DATE & STATUS	BIO SHELTER IS A CONCEPT FOR A TOTAL LIVING ENVIRONMENT. IT CAN BE DESIGNED ON ANY SCALE. THE LARGEST PROJECT YET DESIGNED USING THE PASSIVE DIRECT INSOLATION SYSTEM IS THE AMHERST COMMUNITY CENTER. AS OF 1974 THIS PROJECT IS DESIGNED, BUT NOT BUILT.
LOCATION	AMHERST, MASSACHUSETTS
SCOPE	THE BIO SHELTER IS CAPABLE OF PROVIDING SHELTER FROM THE WEATHER, SOME FOOD, FRESH WATER, LIQUID AND SOLID WASTE DISPOSAL, SPACE HEATING. AND COOLING, POWER FOR COOKING AND REFRIGERATION, ELECTRICITY FOR COMMUNICATIONS, LIGHTING AND HOUSEHOLD APPLIANCES. IN ADDITION, IT PREVENTS HEAT FROM ESCAPING WHEN THE INTERIOR MICRO-CLIMATE IS TOO COOL, AND IT REFLECTS SUNLIGHT AND DUMPS HEAT OUT INTO THE NIGHT SKY WHEN ITS INTERIOR IS TOO WARM.
	THE BIO SHELTER CAN TAKE THE FORM OF SINGLE, CONVENTIONAL APARTMENT COMPLEXES, SINGLE-FAMILY OR LARGE-SPAN CLIMATIC ENVELOPES.
TECHNIQUE	SUNLIGHT ENTERS THE SYSTEM THROUGH THE GREENHOUSE BY PASSING THROUGH A SOLAR MEMBRANE. THIS CAN BE EITHER TRANSPARENT INSULATION OR MOVEABLE INSULATION THAT COVERS THE TRANSPARENT PORTIONS OF THE GREENHOUSE. THE MEMBRANE LETS IN ENERGY IN THE FORM OF LIGHT AND PREVENTS ITS ESCAPE IN THE FORM OF HEAT. BUT BEFORE IT DEGRADES TO HEAT, SOME OF THE SUNLIGHT ENTERING THE MODEL GREENHOUSE IS CONVERTED TO FOOD AND FRESH WATER.
	AFTER PASSING THROUGH THE MEMBRANE, SUNLIGHT IS ABSORBED BY THE PLANTS IN THE GREENHOUSE. THE PLANTS HEAT UP THE AIR SURROUNDING THEM, AND THIS HOT AIR IS BLOWN THROUGH THE HEAT STORAGE BATTERY BY A FAN. THE AREA OF THE HEAT STORAGE UNIT MUST BE LARGER THAN THE FLOOR AREA IN ORDER TO GET HEAT INTO AND OUT OF IT EASILY. IT IS THUS BEST INCORPORATED INTO THE BUILDING IN THE FORM OF WALLS, CEILINGS, OR UNDER THE FLOOR. HEAT IS TRANSFERRED FROM THE STORAGE MATERIAL TO THE LIVING SPACE BY CIRCULATING ROOM AIR THROUGH STORAGE.
	THE GREENHOUSE HELPS TO COOL THE LIVING SPACE DURING THE SUMMER BY ADMITTING ONLY THE MINIMUM AMOUNT OF LIGHT NEEDED BY THE PLANTS AND BY VENTILATING WITH OUTSIDE AIR THE HEAT STORAGE UNIT, LIVING SPACE, AND GREENHOUSE, WHENEVER OUTDOOR TEMPERATURES ARE BELOW INDOOR TEMPERATURES. WHEN IT'S COOLER INSIDE, HEAT STORAGE AND THE LIVING AREA ARE KEPT AS THERMALLY ISOLATED AS POSSIBLE FROM THE ENVIRONMENT.
	THE GREENHOUSE IS USED TO PROVIDE PRODUCE FOR THE BIO SHELTER. THE PARTS OF THE PLANTS NOT USED BY THE OCCUPANTS CAN BE FED TO CHICKENS IN A SEPARATE BUILDING OR TO FISH TO PRODUCE PROTEIN. THE AQUA-CULTURE PONDS ARE IN THE GREENHOUSE. FISH AND CHICKENS ARE FAVORED BECAUSE OF THEIR EXTREMELY HIGH PLANT-TO-PROTEIN CONVERSION RATIO. THE ORGANIC WASTE FROM THE PEOPLE, FISH AND CHICKENS ARE DISPOSED OF AND STERILIZED BY DECOMPOSITION IN A METHANE GENERATOR. THIS CAN BE SUPPLEMENTED WITH ORGANIC MATTER SUCH AS DEAD LEAVES FROM OUTSIDE THE BUILDING. THE STERILE BYPRODUCTS OF DECOMPOSITION ARE USED AS FERTILIZER FOR THE PLANTS IN THE GREENHOUSE. THE METHANE PRODUCED IS USED FOR COOKING, AS IS WASTE PAPER.

THE GREENHOUSE ALSO ACTS AS A SOLAR STILL. IT IS DIVIDED INTO
THREE SECTIONS, ONLY ONE OF WHICH OPERATES AS A STILL AT A
TIME. SECONDARY WASTE WATER IS POURED INTO THE SOIL, FROM
WHICH IT EVAPORATES. THE HUMIDITY REACHES SATURATION, AND
MOISTURE CONDENSES WHEREVER HEAT LEAVES THIS SECTION OF THE
GREENHOUSE. THE CONDENSATION DRIPS DOWN INTO TROUGHS WHERE
IT IS COLLECTED AND STORED FOR REUSE. LOSSES ARE REPLACED WITH
RAIN WATER COLLECTED FROM THE ROOF.

A HIGH TEMPERATURE SOLAR COLLECTOR IS MOUNTED BEHIND PART OF
THE SOLAR MEMBRANE. USED AS A COVER WINDOW ON A COLLECTOR,
THE TRANSPARENT INSULATION PERMITS HIGH EFFICIENCIES AT HIGH
TEMPERATURES WITHOUT MIRRORS. A BRINE SOLUTION IS HEATED IN
THE COLLECTOR AND THE HOT BRINE USED TO DRIVE THE WATER HEATER,
THE REFRIGERATOR, AND IN HOT HUMID AREAS, THE SPACE COOLING
OR DEHUMIDIFYING SYSTEMS.

WIND GENERATED ELECTRICITY IS USED FOR COMMUNICATIONS, LIGHTING
AND APPLIANCES. THE DIRECT CURRENT ENERGY IS STORED IN BATTERIES
AND CONNECTED TO ALTERNATING CURRENT WITH AN INVERTER.

INSTRUMENTATION WILL BE PART OF THE SENSORY-CONTROL LOOPS
THAT ALLOW THE BUILDING TO FUNCTION VIABLY AS AN ORGANISM
REGULATING INTERNAL ENERGY FLOWS AND EXCHANGES OF ENERGY AND
MATTER WITH THE EXTERNAL ENVIRONMENT, AND WITH EMPHASIS ON LOCAL
FEEDBACK.

MATERIAL   A MULTI-LAYER SOLAR MEMBRANE THAT TRANSMITS 76 PERCENT OF
               SHORT-WAVE SOLAR RADIATION, YET IS VIRTUALLY OPAQUE TO
               LONG-WAVE THERMAL RADIATION, A THERMOSTATIC CLOUD GEL

AVAILABILITY   ADDITIONAL INFORMATION ON THE BIO SHELTER MAY BE OBTAINED
               FROM SUNTEK, INC., 500 TAMAL PLAZA, TAMAL VISTA BLVD.,
               CORTE MADERA, CALIFORNIA 94925.

REFERENCE   WELLESLEY-MILLER, SEAN; CHAHROUDI, DAY; VILLECO, MARGUERITE
               BIO SHELTER
               ARCHITECTURE PLUS, NOVEMBER/DECEMBER 1974, PP 90-95

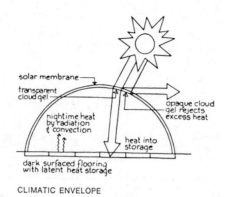

solar membrane

transparent
cloud gel

nightime heat
by radiation
& convection

opaque cloud
gel rejects
excess heat

heat into
storage

dark surfaced flooring
with latent heat storage

CLIMATIC ENVELOPE

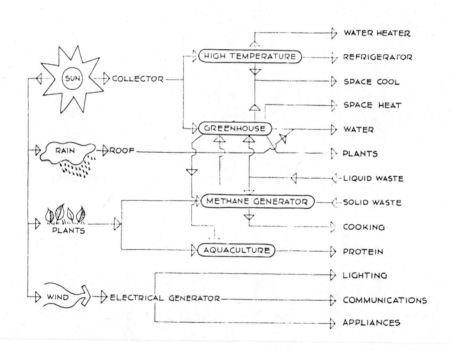

SUN → COLLECTOR

HIGH TEMPERATURE

WATER HEATER
REFRIGERATOR
SPACE COOL
SPACE HEAT

GREENHOUSE

WATER
PLANTS

RAIN → ROOF

LIQUID WASTE

METHANE GENERATOR

SOLID WASTE
COOKING

PLANTS

AQUACULTURE

PROTEIN

WIND → ELECTRICAL GENERATOR

LIGHTING
COMMUNICATIONS
APPLIANCES

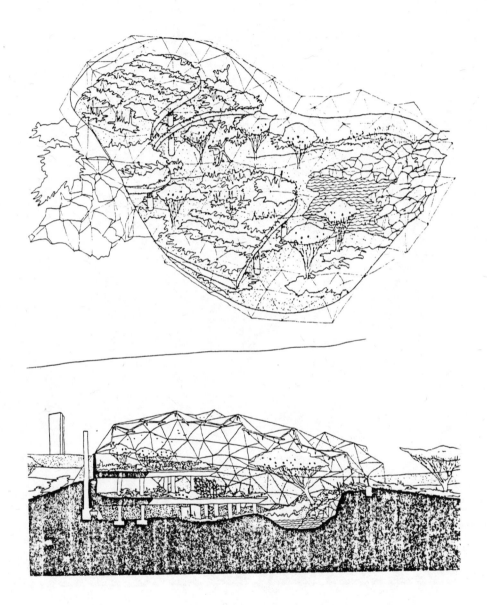

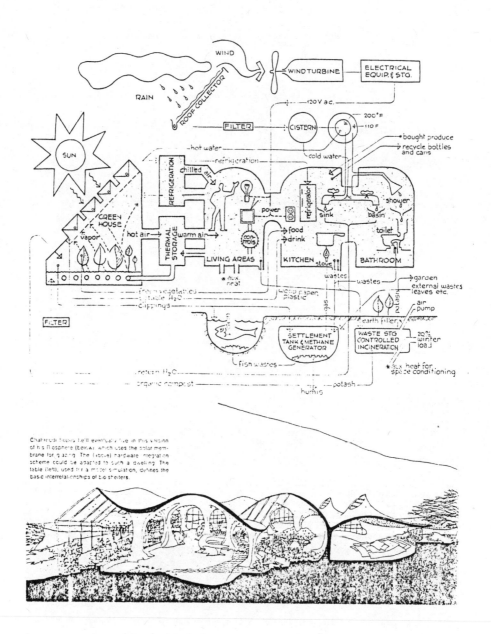

WIND

RAIN

ROOF COLLECTOR

WIND TURBINE

ELECTRICAL EQUIP. & STG.

120 V a.c.

FILTER

CISTERN

200°F

110 F

SUN

bought produce

recycle bottles and cans

hot water

refrigeration

cold water

chilled air

REFRIGERATION

refrigerator

power

sink

basin

shower

GREEN HOUSE

vapor

hot air

warm air

THERMAL STORAGE

con- trols

food

drink

LIVING AREAS

KITCHEN

stove

toilet

BATHROOM

Aux heat

wastes

wastes

garden

external wastes leaves etc.

fresh vegetables

potable H₂O

clippings

old paper plastic

gas

potash

air pump

earth filter

FILTER

SETTLEMENT TANK & METHANE GENERATOR

WASTE STG CONTROLLED INCINERATION

20% winter load

Aux heat for space conditioning

fish wastes

return H₂O

organic compost

potash

humus

Chahroudi hopes he'll eventually live in this version of his Biosphere (below) which uses the solar mem- brane for glazing. The (above) hardware integration scheme could be adapted to such a dwelling. The table (left), used for a model simulation, defines the basic interrelationships of bio-shelters.

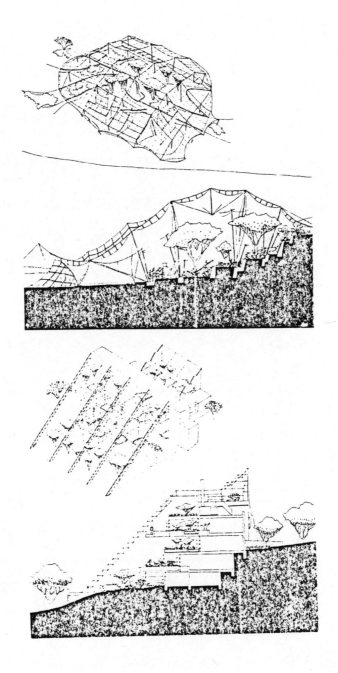

TITLE	BISHOPRICK SOLAR HEATED RESIDENCE
CODE	
KEYWORDS	HOUSE AIR TYPE COLLECTOR STORAGE
AUTHOR	WILLIAM BISHOPRICK
DATE & STATUS	THE BISHOPRICK SOLAR HEATED RESIDENCE WAS COMPLETED IN APRIL 1976. THE SYSTEM WILL BE MONITORED IN 1977-78 BY PORTLAND GENERAL ELECTRIC CO.
LOCATION	OREGON
SCOPE	THE BUILDING IS A PRIVATE HOUSE.
TECHNIQUE	THE COLLECTOR IS A 460 SQ. FT., AIR-TYPE COLLECTOR, MOUNTED ON THE WALL, SLOPING 60 DEGREES FROM THE HORIZONTAL. THE PANELS EMPLOY GALVANIZED STEEL AND ARE DOUBLE GLAZED.
	THE STORAGE SYSTEM CONSISTS OF 1,200 CUBIC FEET OF 1-1/2" TO 3" DIAMETER STONES IN A RECTANGULAR, WATERPROOFED, INSULATED, 18' X 12' X 6' HIGH BIN BENEATH THE HOUSE. IN SUMMER, THE STONES ARE COOLED AT NIGHT BY CIRCULATION OF COOL NIGHT AIR. DURING A HOT DAY, ROOM AIR IS CIRCULATED THROUGH THE BIN OF STONES.
MATERIAL	GALVANIZED STEEL; DOUBLE GLAZED; 1,200 CUBIC FEET OF 1-1/2" TO 3" DIAMETER STONES; RECTANGULAR, WATERPROOFED, INSULATED BIN
AVAILABILITY	ADDITIONAL INFORMATION MAY BE OBTAINED FROM WILLIAM BISHOPRICK, C/O PAYNE, SETTECASE AND SMITH & PARTNERS, 725 COMMERCIAL STREET, SE, SALEM, OREGON 97301.
REFERENCE	BISHOPRICK, WILLIAM BISHOPRICK SOLAR HEATED RESIDENCE SALEM, OREGON, WILLIAM BISHOPRICK LEAFLET, 1976, 4 P

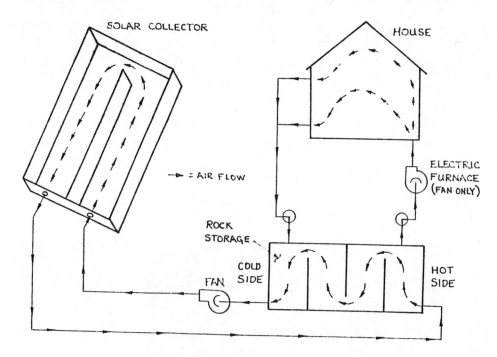

HEATING MODE (WHEN STORAGE TEMP. IS GREATER THAN 80°)

SOLAR COLLECTOR

HOUSE

→ = AIR FLOW

ELECTRIC FURNACE (FAN ONLY)

ROCK STORAGE

COLD SIDE

HOT SIDE

FAN

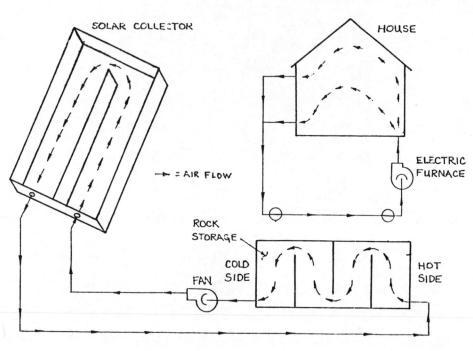

AUX HEATING MODE (WHEN STORAGE TEMP. IS LESS THAN 80°)

SOLAR COLLECTOR

HOUSE

→ = AIR FLOW

ELECTRIC FURNACE

ROCK STORAGE

COLD SIDE

HOT SIDE

FAN

TITLE   BLAKEWOOD ELEMENTARY SCHOOL

CODE

KEYWORDS   ROTARY COMPRESSOR ELEMENTARY SCHOOL DOUBLE BUNDLED CONDENSER
            HEAT PUMP RECOVERY STORAGE SOLAR COLLECTOR PANELS

AUTHOR   WALTER R. RATAI

DATE & STATUS   THE BLAKEWOOD ELEMENTARY SCHOOL IS BUILT AND OPERATING
                AS OF 1972.

LOCATION   MILWAUKEE, WISCONSIN

SCOPE   FIRST COST FOR THE HEAT RECOVERY SYSTEM, INCLUDING EQUIPMENT
            AND INSTALLATION, WAS $310,000. A CONVENTIONAL HVC SYSTEM
            WITH GAS HEATING AND ELECTRIC AIR CONDITIONING WAS ESTIMATED AT
            $248,000. HOWEVER, IN THE FIRST TWO YEARS OF OPERATION, THE
            BOARD OF EDUCATION HAS FOUND IT'S SAVING NEARLY $10,000 A YEAR
            WITH THE HEAT RECOVERY SYSTEM. IF THE PRICE OF SOLAR COLLECTORS
            WOULD COME DOWN, THEN THEY COULD BE INSTALLED ON BLAKEWOOD
            ELEMENTARY SCHOOL WITH ONLY MINOR ADAPTATIONS, BECAUSE THE
            STORAGE CAPACITY ALREADY EXISTS.

TECHNIQUE   KEY TO THE HEAT RECOVERY SYSTEM IS THE ROTARY SCREW, DOUBLE-
            BUNDLED COMPRESSOR/CONDENSER UNIT INSTALLED WITH CLOSED CIRCUIT
            WATER LOOPS. THIS MACHINE IS INCORPORATED INTO THE CENTRAL-
            TYPE, WATER-TO-WATER HEAT PUMP, HEAT RECOVERY, AND STORAGE
            SYSTEM THAT PROVIDES HVC FOR THE 90,000 SQUARE FOOT SCHOOL.

            HOT WATER FROM THE HEATING CONDENSER AND CHILLED WATER FROM
            AN EVAPORATOR SUPPLY AIR HEATING AND COOLING COILS IN A
            BUILT-UP, MULTI-ZONE UNIT FOR ACADEMIC AREAS AND A SINGLE-ZONE,
            DRAW-THROUGH UNIT FOR THE GYMNASIUM.

            HEAT RECOVERED FROM LIGHTS, PEOPLE, AND EQUIPMENT, IS RECLAIMED
            VIA HEAT RECLAIM AIR HANDLING UNITS, AND STORED IN A 15,000
            GALLON WATER TANK FOR USE AS SUPPLEMENTAL HEAT WHENEVER THE
            BUILDING HEAT LOSS EXCEEDS INTERNAL HEAT GAINS. ELECTRIC
            HEATING COILS IN THE TANK PROVIDE DIRECT ELECTRIC HEATING
            WHEN NECESSARY, AND IN CASE THE HEAT PUMP FAILS, IN-LINE
            ELECTRIC HEATERS ALSO WERE INSTALLED.

            TEMPERATURE CONTROLS IN INDIVIDUAL ROOMS ALLOW OCCUPANTS IN
            SEPARATE ZONES TO HAVE HEATING AND COOLING SIMULTANEOUSLY.

            IN DAILY OPERATION, THE BALANCED HEAT PUMP RECOVERY STORAGE
            SYSTEM REMAINS ON THE HEATING CYCLE DAY AND NIGHT AS LONG AS
            THE OUTDOOR TEMPERATURE IS BELOW 70 DEGREES. DURING THE HEATING
            SEASON, A PRE-SET MASTER CONTROLLER REGULATES THE HOT WATER
            SUPPLY WITHIN A 90 TO 120 DEGREE RANGE.

            THE BUILDING'S AIR DISTRIBUTION IS VIA A MEDIUM-VELOCITY,
            DOUBLE-DUCT SYSTEM INSTALLED ABOVE CORRIDOR CEILINGS. TWO
            SEPARATE AIR SUPPLIES - ONE MODULATED DOWNWARD FROM 95 TO 75
            DEGREES AS OUTDOOR TEMPERATURE RISES, AND THE OTHER COOLED TO
            54 DEGREES - ARE CARRIED INDEPENDENTLY FROM THE CENTRAL UNIT
            TO A NETWORK OF CONSTANT-VOLUME MIXING BOXES.

            DURING SUMMERTIME AIR CONDITIONING, THE COOLING TOWER CIRCUIT
            ON THE CONDENSER IS OPEN. THE HOT WATER CIRCUIT IS CLOSED.
            THE HEAT PUMP EVAPORATOR SUPPLIES CHILLED WATER TO AIR COOLING
            COILS IN THE TWO CENTRAL AIR HANDLING UNITS. INTERIOR BUILDING
            HEAT IS REJECTED THROUGH THE ROOF-MOUNTED COOLING TOWER.

            IN WINTER, THE COOLING TOWER CIRCUIT ON THE CONDENSERS CLOSES.
            THIS PERMITS ALL THE HEAT ENERGY, INCLUDING THAT GENERATED BY
            MOTORS, COMPRESSORS AND CONDENSERS, TO BE RECLAIMED THROUGH THE
            SYSTEM'S HEATING WATER SIDE. IN SUB-ZERO WEATHER CONDITIONS,
            WHEN THERE IS NOT SUFFICIENT HEAT AVAILABLE FROM INSIDE
            SOURCES, ADDITIONAL HEAT IS SUPPLIED BY THE HEATING ELEMENTS
            IN THE WATER STORAGE TANKS.

114

MATERIAL   ROTARY COMPRESSOR, HEAT PUMP, AIR HANDLING UNITS, 15,000 GALLON
                WATER TANK, ELECTRIC HEATING COILS, ELECTRIC HEATERS, VARIABLE
                PITCH FAN, DOUBLE BUNDLED CONDENSER, ADAPTABLE SYSTEM SOLAR
                COLLECTOR PANELS

AVAILABILITY   ADDITIONAL INFORMATION ON THE HEAT RECOVERY SYSTEM MAY BE
                OBTAINED FROM WALTER R. RATAI, INC., CONSULTING ENGINEERS,
                6659 NORTH SIDNEY PLACE, MILWAUKEE, WISCONSIN 53209.

REFERENCE   BUILDING DESIGN AND CONSTRUCTION
                ROTARY COMPRESSOR ANSWER TO HEAT RECOVERY AND COSTS, A
                BUILDING DESIGN AND CONSTRUCTION, JUNE 1975, PP 52-53

FLOW DIAGRAM OF HEAT RECOVERY SYSTEM

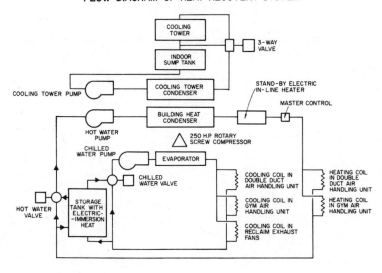

116

TITLE	BLISS HOUSE
CODE	
KEYWORDS	BUNGALOW SOLAR COLLECTOR RADIATING
AUTHOR	R.W. BLISS AND M.K. DONOVAN
DATE & STATUS	THE BLISS HOUSE WAS COMPLETED IN 1954.
LOCATION	AMADO, ARIZONA
SCOPE	THE EXISTING ONE-STORY BUNGALOW (700 SQ. FT. FLOOR AREA) WAS CONVERTED TO 100% SOLAR HEATING. THE SYSTEM, ALTHOUGH A VISUAL "MONSTROSITY" AND ECONOMICALLY UNATTRACTIVE (BECAUSE OF HIGH FIRST COST, WHICH INCLUDED $4,000 FOR MATERIALS AND LABOR), SUCCESSFULLY DEMONSTRATED THE PRINCIPLES INVOLVED AND MADE THIS THE FIRST 100%-SOLAR-HEATED HOUSE IN USA.
TECHNIQUE	THE COLLECTOR IS A 315 SQ. FT., AIR-TYPE COLLECTOR. IT IS MOUNTED ADJACENT TO THE HOUSE. THE TILT IS AROUND 50 DEGREES FROM HORIZONTAL. SINGLE GLAZING WAS USED. RADIATION IS ABSORBED IN DEPTH IN A SET OF FOUR LAYERS OF BLACK COTTON CLOTH WITH 1/4" SPACES BETWEEN THE LAYERS. THE STORAGE SYSTEM CONSISTS OF 300 CUBIC FEET (65 TONS) OF 4" DIAMETER FIELD ROCK IN AN INSULATED UNDERGROUND BIN NEAR THE HOUSE. COOLING IN SUMMER IS ACCOMPLISHED BY THE SAME ROCK BIN, BLOWERS, DUCTS USED, BUT A COLLECTOR IS NOT USED. AT NIGHT, COOL AIR IS DRAWN DOWNWARD THROUGH A BLACK CLOTH (RADIATING HEAT TO THE CLEAR SKY) COVERING A SEPARATE HORIZONTAL BED OR PLENUM AND IS THEREBY COOLED AN ADDITIONAL 2 DEGREES F. THIS AIR IS DELIVERED TO THE ROCK BIN. DURING A HOT DAY, COOL AIR FROM THE ROCK BIN IS CIRCULATED TO THE ROOMS. THE COOLING SYSTEM SOMETIMES FAILED TO KEEP ROOMS AS COOL AS DESIRED.
MATERIAL	SINGLE GLAZING, BLACK COTTON CLOTH, 65 TONS OF 4" DIA. FIELD ROCK, INSULATED UNDERGROUND BIN, BLOWERS, DUCTS
AVAILABILITY	ADDITIONAL INFORMATION MAY BE OBTAINED FROM U.S. FOREST SERVICE AT DESERT GRASSLAND STATION, AMADO, ARIZONA.
REFERENCE	SHURCLIFF, W.A. SOLAR HEATED BUILDINGS, A BRIEF SURVEY CAMBRIDGE, MASSACHUSETTS, W.A. SHURCLIFF, MARCH 1976, 212 P

TITLE   BLOOMER RESIDENCE

CODE

KEYWORDS   WINDMILL AESTHETIC USED MACHINE PARTS

AUTHOR   YALE ARCHITECTURAL STUDENTS: D. GARDNER, R. GODSHALL, D. QUINTO,
                AND C. PUCCI

DATE & STATUS   THE WINDMILL WAS ERECTED IN 1974.

LOCATION   CONNECTICUT

SCOPE   THE WINDMILL WAS BUILT AT A MATERIALS COST OF APPROXIMATELY
                $1,500. IT WAS BUILT TO PROVIDE ALTERNATING CURRENT FOR DOMESTIC
                HOT WATER HEATING. THE DESIGN WAS DEVELOPED WITH A HIGH REGARD
                FOR THE AESTHETIC POSSIBILITIES OF SUCH A STRUCTURE. A PLAY-
                HOUSE AND A GAZEBO ARE INCORPORATED INTO THE WINDMILL BASE.

TECHNIQUE   THE EQUIPMENT WAS FABRICATED FROM USED MACHINE PARTS RECOVERED
                FROM JUNKYARDS AND, IN THE CASE OF THE WINDMILL GEAR, FROM A
                NEARBY SITE, WHERE WINDMILLS WERE COMMONLY USED UP TO THE 1940S.

MATERIAL   WINDMILL

AVAILABILITY   ADDITIONAL INFORMATION MAY BE OBTAINED FROM E.M. BARBER, JR.,
                GUILFORD, CONNECTICUT 06437.

REFERENCE   WATSON, DONALD / BARBER JR., EVERETT
                ENERGY CONSERVATION IN ARCHITECTURE, PART 2: ALTERNATIVE
                ENERGY SOURCES
                CONNECTICUT ARCHITECT, MAY-JUNE, 1974, P 4

TITLE	BLUE CROSS AND BLUE SHIELD BUILDING
CODE	
KEYWORDS	RHOMBOID BUILDING MIRRORED REFLECTIVE GLASS
AUTHOR	ODELL ASSOCIATES
DATE & STATUS	THE BLUE CROSS AND BLUE SHIELD BUILDING WAS COMPLETED IN 1974.
LOCATION	DURHAM, NORTH CAROLINA

SCOPE
THE BLUE CROSS AND BLUE SHIELD BUILDING HAS A FLOOR AREA
OF 225,000 SQUARE FEET AND ACCOMODATES 1000 EMPLOYEES. THE
SITE IS A 39-ACRE PARCEL OF COUNTRYSIDE. A THIRD OF THE HEAD-
QUARTERS VOLUME CONTAINS BACKUP SERVICES, MECHANICAL EQUIPMENT
AND STORAGE. THIS IS ACCOMODATED BELOW GRADE. THE OFFICES
FOR OPERATIONS VISUALLY FLOAT, AT THE HIGHLY VISIBLE CREST OF A
HILL, IN A THREE-DIMENSIONAL RHOMBOID WHICH HAS ITS LONG SIDES
EXPOSED TO MAJOR HIGHWAYS ON THE NORTH AND SOUTH.

THE HEADQUARTERS HAS THE INTERNAL FUNCTIONING OF A VERTICAL
BUILDING. MECHANICAL AND ELECTRICAL RISERS AND ELEVATORS CAN GO
STRAIGHT UP TO THE TOP, AND THE AREAS OF THE VARIOUS FLOORS
REMAIN CONSTANT. THE CLIENTS REQUIRED A LARGE PERCENTAGE OF
OPEN-PLAN WORK AREAS. THERE ARE FEW ENCLOSED OFFICE SPACES.
LARGE AREAS ARE FREE OF COLUMNS AND ACCOMMODATE FLEXIBLE
PLANNING. CONSTRUCTION COST WAS 9.3 MILLION DOLLARS.

TECHNIQUE
THE SLOPE OF THE EXTERIOR WALLS IS MORE THAN AN EYE CATCHING
DESIGN. IT ROUGHLY PARALLELS THE SUN'S RAYS, AND REDUCES THE
DIRECT ENERGY GAIN BY HALF OF THAT IN A VERTICAL, GLASS-ENCLOSED
BUILDING (DESPITE THE GREATER AREA OF EXPOSURE). COUPLED WITH
THE REFLECTIVE VALUE OF THE CHROME-PLATED GLASS, THE CONDITIONS
HERE REDUCE THE REQUIRED AIR COOLANT COMPENSATION FOR SOLAR
HEAT GAIN BY 90 PER CENT.

MATERIAL
REFLECTIVE CHROME-PLATED GLASS, TRIANGULAR RIGID FRAME, STRUC-
TURAL MEMBERS COVERED WITH WHITE VINYL TUBES

AVAILABILITY
ADDITIONAL INFORMATION ABOUT THE BLUE CROSS AND BLUE SHIELD
BUILDING MAY BE OBTAINED BY WRITING TO ODELL ASSOCIATES INC.,
222 SOUTH CHURCH ST., CHARLOTTE, N.C. 28202.

REFERENCE
ARCHITECTURAL RECORD
BLUE CROSS AND BLUE SHIELD
ARCHITECTURAL RECORD, MAY 1974, PP 134-135

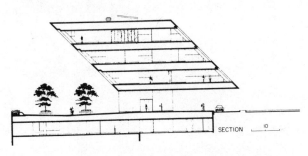

SECTION |____ 10 ___|

TITLE	BOLEYN HOUSE
CODE	
KEYWORDS	HOUSE INSULATION COLLECTOR STORAGE THERMAL SEGREGATION
AUTHOR	D. BOLEYN
DATE & STATUS	THE BOLEYN HOUSE WAS COMPLETED IN DECEMBER 1975.
LOCATION	GLADSTONE, OREGON
SCOPE	THE BUILDING IS A 2-STORY, WOOD-FRAME HOUSE WITH AN UNFINISHED BASEMENT, BUT NO ATTIC. THE FLOOR AREA, NOT INCLUDING THE BASEMENT, IS 1,800 SQ. FT. THERE ARE 3 BEDROOMS AND 2 BATHS.
TECHNIQUE	THERE IS EXCELLENT INSULATION; THE WALLS ARE R-11, AND THE ROOF OR CEILING IS R-19. THE WINDOW AREA IS SMALL, AT 250 SQ. FT. DOUBLE GLAZING IS USED. THE HOUSE AIMS 6 DEGREES WEST OF SOUTH.

THE COLLECTOR IS A 430 SQ. FT. NET, WATER-TYPE COLLECTOR, MOUNTED ON THE ROOF, SLOPING 60 DEGREES FROM THE HORIZONTAL. IT INCLUDES 22 PANELS BY REVERE COPPER & BRASS CO., 3' X 6-1/2', ARRANGED IN TWO ROWS OF 11 EACH. THE BLACK COATING ON THE COPPER SHEET IS NON-SELECTIVE. THE RECTANGULAR COPPER TUBES ARE ATTACHED TO THE SHEET BY MEANS OF CLIPS AND ADHESIVES. EACH PANEL IS DOUBLE GLAZED WITH TEMPERED GLASS. THE BACKING IS 2" OF FIBERGLAS. THE LIQUID IS WATER AND 30% ETHYLENE GLYCOL. THE FLOWRATE IS ABOUT 17 GPM. THE FLOW OCCURS WHENEVER THE RHO-SIGMA DIFFERENTIAL THERMOSTAT FINDS THE COLLECTOR TEMPERATURE TO BE HIGHER THAN THE COLDEST PART OF THE STORAGE SYSTEM, ABOUT 20 DEGREES F.

THE STORAGE SYSTEM CONTAINS 3,750 GALLONS OF WATER IN THREE 1,250 GALLON, INSULATED, SIDE-BY-SIDE TANKS IN THE BASEMENT ROOM, WHICH IS INSULATED WITH 6" OF FIBERGLAS. EACH TANK IS MADE OF FIBERGLAS AND POLYESTER BY HOFFMAN FIBERGLAS CO., AND EACH TANK IS A VERTICAL CYLINDER, 6' IN DIAMETER AND 6-1/2' HIGH. THE THREE TANKS ARE CONNECTED IN SERIES IN SUCH A MANNER AS TO PERMIT A LARGE DEGREE OF THERMAL SEGREGATION. THE ROOMS ARE HEATED BY A FAN-COIL SYSTEM IN THE MAIN DUCT. THE FAN-COIL SYSTEM HAS THREE COILS: A, B AND C. A IS FED DIRECTLY FROM THE LIQUID CIRCULATING FROM THE COLLECTOR; B IS FED FROM THE MAIN STORAGE SYSTEM; AND C IS FED FROM THE AUXILIARY-HEATING-SYSTEM TANK OR, IN THE COLDEST WEATHER, FROM THE ELECTRIC HEATER ITSELF. DURING A SUNNY DAY, A OFTEN SUFFICES; IF NOT, B IS USED; AND IF IT DOES NOT SUFFICE, C IS USED. A SECOND CENTRIFUGAL PUMP, 10 GPM, SERVES THE LOOP INCLUDING THE HEAT-EXCHANGER AND THE MAIN STORAGE SYSTEM, AND A THIRD PUMP, 9 GPM, SERVES THE MAIN STORAGE SYSTEM AND COIL B.

AUXILIARY HEAT IS PROVIDED BY A 24 KW ELECTRIC HEATER, OPERATED OFF-PEAK. IT USUALLY KEEPS A SPECIAL, 1,250 GALLON AUXILIARY-HEATING-SYSTEM TANK AT A HIGH TEMPERATURE. THE WATER IS DRIVEN BY A 15 GPM CENTRIFUGAL PUMP.

A DOMESTIC HOT WATER PREHEAT SYSTEM, EXTRACTING HEAT VIA A HEAT EXCHANGER FROM THE MAIN STORAGE TANKS, IS INCLUDED. THE DOMESTIC HOT WATER PREHEAT SYSTEM INCLUDES A 40 GALLON STOREX TANK WITH HEAT EXCHANGER, A DELTA T CONTROLLER, AND A 1/20 HP GRUNDFOS CIRCULATING PUMP.

MATERIAL	DOUBLE GLAZED WINDOWS; NON-SELECTIVE BLACK COATING; COPPER SHEET; COPPER TUBES; TEMPERED GLASS; 2" FIBERGLAS; WATER; 30% ETHYLENE GLYCOL; RHO-SIGMA DIFFERENTIAL THERMOSTAT; THREE 1,250 GALLON INSULATED TANKS; 4" FIBERGLAS; FIBERGLAS AND POLYESTER; FAN-COIL SYSTEM; CENTRIFUGAL PUMP; 24 KW ELECTRIC HEATER; 1,250 GALLON AUXILIARY-HEATING-SYSTEM TANK; 15 GPM CENTRIFUGAL PUMP; 40 GALLON STOREX TANK; HEAT EXCHANGER; DELTA T CONTROLLER; 1/20 HP GRUNDFOS CIRULATING PUMP
AVAILABILITY	ADDITIONAL INFORMATION MAY BE OBTAINED FROM D. BOLEYN, 17610 SPRINGHILL PL., GLADSTONE, OREGON 97027.
REFERENCE	SHURCLIFF, W.A. SOLAR HEATED BUILDINGS, A BRIEF SURVEY CAMBRIDGE, MASSACHUSETTS, W.A. SHURCLIFF, MARCH 1976, 212 P

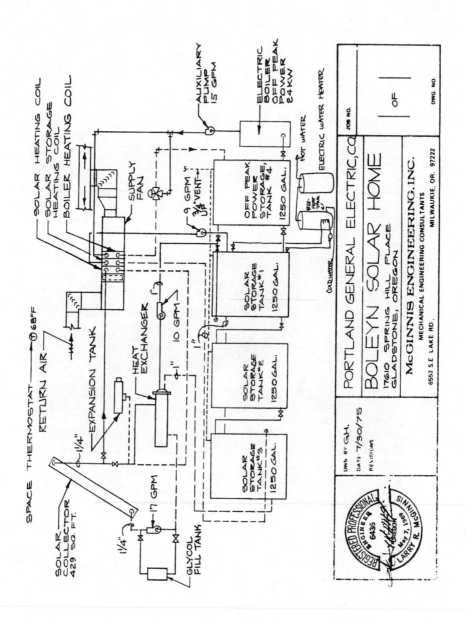

SPACE THERMOSTAT — 68°F

RETURN AIR

SOLAR HEATING COIL
SOLAR STORAGE HEATING COIL
BOILER HEATING COIL

SUPPLY FAN

AUXILIARY PUMP 15 GPM

ELECTRIC BOILER OFF PEAK POWER 24 KW

ELECTRIC WATER HEATER

HOT WATER

9 GPM

¾" VENT

OFF PEAK POWER STORAGE TANK #4 1250 GAL.

PRE-HEAT TANK

COLD WATER

EXPANSION TANK

HEAT EXCHANGER

1" 10 GPM

SOLAR STORAGE TANK #1 1250 GAL.

1"

SOLAR STORAGE TANK #2 1250 GAL.

1¼"

SOLAR STORAGE TANK #3 1250 GAL.

SOLAR COLLECTOR 429 SQ. FT.

17 GPM

1¼"

GLYCOL FILL TANK

PORTLAND GENERAL ELECTRIC, CO.

BOLEYN SOLAR HOME
1710 SPRING HILL PLACE
GLADSTONE, OREGON

McGINNIS ENGINEERING, INC.
MECHANICAL ENGINEERING CONSULTANTS
6552 S.E. LAKE RD.   MILWAUKIE, OR. 97222

JOB NO.

OF

DWG. NO.

DWN. BY G.H.
DATE: 7/30/75
REVISIONS

REGISTERED PROFESSIONAL ENGINEER 6435 OREGON MAY 7, 1967 LARRY R. McGINNIS SINNIGIS

TITLE	BOSTON SOLAR HOME
CODE	
KEYWORDS	COMPACT SOLAR COLLECTOR EAST WEST AXIS ORIENTATION THERMAL MASS NATURAL VENTILATION DOUBLE GLAZED SHUTTER BEADWALL GREENHOUSE HOUSE TEMPERATE ROOF TILT OPEN STAIRWAY
AUTHOR	LEANDRE POISSON
DATE & STATUS	THE BOSTON SOLAR HOUSE WAS DESIGNED IN 1975 TO BE THE PROTOTYPE OF A HOUSE THAT CAN BE BUILT IN THE BOSTON AREA. IT HAS NOT BEEN BUILT.
LOCATION	THE BOSTON SOLAR HOUSE IS DESIGNED TO RUN AT MAXIMUM EFFICIENCY IN THE BOSTON AREA, OR IN ANY OTHER AREA WITH A TEMPERATE CLIMATE AND SIMILAR WEATHER CONDITIONS.
SCOPE	IT WAS FELT THAT THE DAYLIGHT LIVING AND WORKING SPACES SHOULD BE CLUSTERED ON ONE FLOOR. THE UPPER FLOOR (LEVEL 2) WAS CHOSEN BECAUSE IT HAD THE BEST ACCESS TO THE STREET LEVEL. IT ALSO PROVIDED ACCESS TO THE MOST AMOUNT OF SUNLIGHT (AFTER THE COLLECTOR LEVEL) DURING THE WINTER MONTHS. THE BEDROOMS ARE LOCATED ON LEVELS ABOVE AND BELOW THIS AREA FOR REASONS OF PRIVACY. THE LIVINGROOM IS ON THE LOWER LEVEL BECAUSE IT IS OUT OF THE TRAFFIC AND FUNCTIONS AREA.

THE HEAT LOSS FOR THE HOUSE IS 43,990,000 BTU. THE SYSTEM PROVIDES 60% OF THE SEASONAL LOAD. THE BALANCE OF THE HEAT LOAD IS PROVIDED BY A BACKUP SYSTEM, WHICH IS AN OIL FIRED HOT AIR SYSTEM.

THE BOSTON SOLAR HOUSE HAS A GROSS FLOOR AREA OF 1638 SQUARE FEET. OF THAT AREA 906 SQUARE FEET ARE HEATED. THE GENERAL CON-FIGURATION OF THE HOUSE IS SQUARE AND RECTOLINEAR ON A VERTICAL CLUSTER WITH SOME PLANTED PLANES. IT IS ORIENTED ON A DIRECT NORTH, EAST, SOUTH, WEST AXIS. THE PITCHED ROOF AREA CONSISTS OF 660 SQUARE FEET, AND THE FLAT (UNHEATED) ROOF IS 150 SQUARE FEET IN AREA. THE WINDOW AREA IS 256 SQUARE FEET, AND IS MOSTLY FIXED GLASS. THE WALL AREA OF THE HEATED SPACE IS 1601 SQUARE FEET, AND THE INSULATED FLOOR OVER THE BASEMENT AREA IS 384 SQUARE FEET.

TECHNIQUE	THE DESIGN OF THE BOSTON SOLAR HOUSE WAS DEVELOPED FROM THE EXISTING CLIMATIC CONDITIONS AND THE CONFIGURATION OF THE SITE. THE HEATING SEASON IN BOSTON IS 5,630 DEGREE DAYS. IN ORDER TO INSURE AN ECONOMICAL SOLAR SYSTEM, A HEAT LOSS RATE OF NOT MORE THAN 9.5 BTU/SQUARE FOOT/DEGREE DAY WAS SET AS AN UPPER LIMIT THE DESIGN WAS REQUIRED TO MEET.

THE SITE WAS A 75 FOOT BY 150 FOOT PLOT SLOPED DIRECTLY NORTH. ON THE SOUTH, EAST, AND WEST WERE EXISTING STRUCTURES WHICH PROVIDED SERIOUS SHADING EFFECTS, THE WORST OF WHICH OCCURRED ON DECEMBER 22. THE TOTAL EFFECTS OF THESE ENCUMBRANCES WAS ANALOGOUS TO HAVING A FORTY-FOOT WALL RUNNING FROM EAST TO WEST ON THE FRONT OF THE PROPERTY. BECAUSE THE SUN'S ANGLE WAS SLIGHTLY GREATER THAN THE SLOPE'S, IT PROVED ADVANTAGEOUS TO MOVE THE HOUSE DOWN THE HILL IN ORDER TO MAXIMIZE THE AMOUNT OF SUN REACHING THE HOUSE'S SOUTH SURFACE. IT ALSO PROVED USEFUL IN MINIMIZING THE HEIGHT THAT THE HOUSE HAD TO ATTAIN IN ORDER TO COLLECT THE SUNSHINE. THESE CONSTRAINTS WOULD SEEM TO ENCOURAGE A ROW HOUSE CLUSTERING IN AN URBAN SITUATION. BECAUSE OF THE HEIGHT RE-QUIREMENTS, IT SEEMS THAT GOING TO MASONRY CONSTRUCTION WOULD HELP STRUCTURALLY AND WOULD ALSO HELP AMPLIFY THE THERMAL MASS OF THE CLUSTERED UNITS.

THE SOLAR HEATING SYSTEM USES AIR AS A TRANSFER MEDIUM. IT WAS FELT THAT FOR THIS CLIMATE IT HAD MANY ADVANTAGES OVER A WATER SYSTEM. THE PROBLEMS OF FREEZING ARE THEN TOTALLY ELIMINATED, AND THE RESOURCE REQUIREMENTS, BOTH TECHNICAL AND MATERIAL, ARE MINIMIZED. THE COLLECTOR IS 345 SQUARE FEET, AND IS TILTED TO AN ANGLE OF 60'. THIS ANGLE IS SUCH THAT THE SUN'S RAYS STRIKE THE COLLECTOR AT A 90 DEGREE ANGLE ON THE 21ST OF JANUARY. IT FACES DUE SOUTH, AND OPERATES AT A TEMPERATURE OF 90 DEGREES. THE HEAT STORAGE IS LOCATED IN THE BASEMENT AND CONSISTS OF 350 CUBIC FEET

OF FIST-SIZED ROCKS. THE HEATED AIR ENTERS THE TOP OF THE STORAGE
CONTAINER, TRANSFERS ITS HEAT TO THE STONES, AND EXITS AT THE
BOTTOM OF THE STORAGE CONTAINER.

THE DOMESTIC HOT WATER TANK IS ENCASED IN THE STONES INSIDE THE
STORAGE CONTAINER. DEPENDING ON THE TEMPERATURE OF THE STORAGE,
THE WATER IS HEATED OR ONLY PREHEATED PRIOR TO BEING FULLY
HEATED BY THE FURNACE.

COOLING IS ACHIEVED IN TWO DIFFERENT WAYS WHICH FUNCTION EITHER
SEPARATELY OR TOGETHER. THE SIMPLEST METHOD IS ACHIEVED BY
NATURAL VENTILATION. THE HOUSE, BECAUSE OF ITS HEIGHT AND SIZE,
FUNCTIONS SLIGHTLY AS A CHIMNEY. SPIRAL STAIRCASES AND AN OPEN
STAIRWAY WERE DELIBERATELY UTILIZED IN ORDER TO PERMIT THE EASY
FLOW OF AIR FROM ONE LEVEL TO ANOTHER. OPENING THE WINDOW ON THE
TOP LEVEL AND ANY OTHER DOOR OR WINDOW ON A LOWER LEVEL WILL
TRIGGER THIS NATURAL SYSTEM. THE SPIRAL STAIRCASE THAT GOES TO
THE BASEMENT IS CLOSED BY A HATCH. WINDS FLOW BY THE BUILDING
ASSISTING THIS NATURAL AIR MOVEMENT THROUGHOUT THE HOUSE.

THE OTHER WAY OF COOLING THE HOUSE CAN BE EMPLOYED IN THE
SUMMER-TIME. THE COLLECTOR WILL HAVE TO BE VENTED DURING
THE SUMMER BECAUSE OF THE POSSIBLE HIGH TEMPERATURE BUILD-
UP. BY OPENING UP VENTS IN THE RETURN PIPE, AIR CAN BE
PULLED OUT OF THE LIVING SPACES BY THE HEATED AIR LEAVING THE
COLLECTOR. THIS MOVING AIR WILL HAVE A COOLING EFFECT ON THE
INTERIOR SPACES AND WILL ALSO COOL DOWN THE COLLECTOR.

ALL WINDOWS ARE DOUBLE GLAZED AND ARE SHUTTERED OR FOAM-FILLED
WITH THE "BEADWALL" SYSTEM. IT IS ASSUMED THAT THESE GLAZED AREAS
WOULD BE SHUTTERED ONE-THIRD OF THE TIME, GIVING AN AVERAGE U-
VALUE OF 0.35. THE BEADWALL SYSTEM USES STYRENE BEAD INSULATION
THAT IS BLOWN INTO THE 6 INCH CAVITY BETWEEN THE TWO PANES OF
GLASS OF THE 5 FOOT BY 5 FOOT WINDOWS. THE BEADS ARE ALSO BLOWN
INTO THE 3 INCH CAVITY OF THE GREENHOUSE FIBERGLAS GLAZING.
BEADS CAN BE PUT IN OR TAKEN OUT AT WILL OR THE SYSTEM CAN BE
AUTOMATED TO EMPTY AT SUNRISE AND FILL AT SUNSET. THE BEADS ARE
STORED IN CYLINDRICAL TANKS HANGING OUTSIDE THE BUILDING.

MATERIAL   SOLAR COLLECTORS, ROCKS, WARM AIR, OIL FURNACE, SMALL FANS,
                   DOUBLE GLAZING, SPIRAL STAIRCASE, GLASS, INSULATED FLOOR,
                   STYRENE BEAD INSULATION, FIBERGLAS GLAZING, CYLINDRICAL TANKS

AVAILABILITY   ADDITIONAL INFORMATION ON THE BOSTON SOLAR HOUSE CAN BE FOUND
                   IN "SOLAR ENERGY HOME DESIGN IN FOUR CLIMATES", BY TOTAL
                   ENVIRONMENTAL ACTION, BOX 47, HARRISVILLE, NEW HAMPSHIRE 03450.

REFERENCE   TOTAL ENVIRONMENTAL ACTION
                   BOSTON SOLAR HOME, THE
                   HARRISVILLE, NEW HAMPSHIRE, IN SOLAR ENERGY HOME DESIGN IN
                   FOUR CLIMATES, PUBLISHED BY TOTAL ENVIRONMENTAL ACTION,
                   MAY 1975, PP 67-83

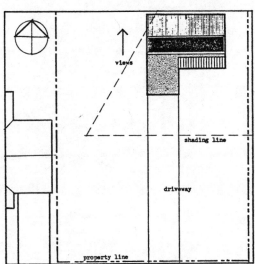

SITE PLAN OF BOSTON SOLAR HOME

SECTION

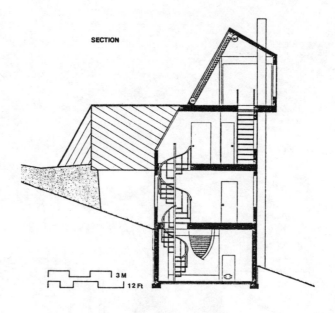

```
            ⎍‾⎍_  3 M
            ⎍__⎍  12 Ft
```

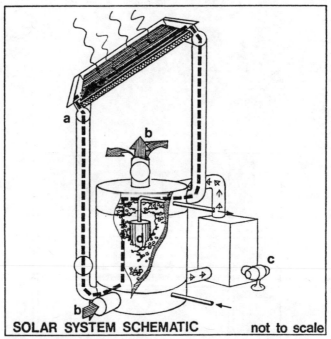

SOLAR SYSTEM SCHEMATIC   **not to scale**

KEY

a: <u>SOLAR COLLECTION</u>: Air enters the bottom of the collector and exits the top. The heated
   air is drawn down through the rock storage, heating it, and then blown back up through
   the collector.

b: <u>INTERIOR SPACE HEATING</u>: Air is drawn from the living spaces and enters the bottom of
   the storage chamber. It is heated up while passing through the rock storage and is
   circulated to the living spaces.

c: <u>BACK-UP HEATING</u>: Oil fired burner heats air that comes from the living spaces via the
   bottom plenum of the storage battery. The heated air is circulated to the living spaces
   via the top plenum of the storage battery.

d: <u>DOMESTIC HOT WATER</u>: A tank in the storage which acts as a heater or pre-heater depending
   on the temperature of the storage.

129

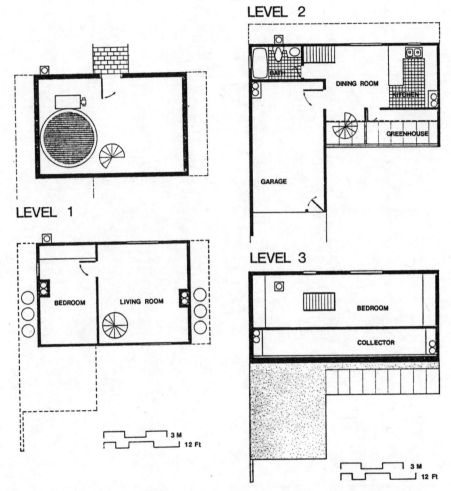

LEVEL 2

BATH

DINING ROOM

KITCHEN

GREENHOUSE

GARAGE

LEVEL 3

BEDROOM

COLLECTOR

LEVEL 1

BEDROOM

LIVING ROOM

3 M
12 Ft

3 M
12 Ft

BASEMENT PLAN AND LEVEL ONE PLAN OF BOSTON SOLAR HOME

LEVEL TWO PLAN AMD LEVEL THREE PLAN OF BOSTON SOLAR HOME

130

TITLE	BOULDER HOUSE
CODE	
KEYWORDS	BLACKENED METAL TROUGH SOLAR COLLECTOR AIR INSULATION GRAVEL OVERLAPPED PLATE HOUSE
AUTHOR	GEORGE LOF
DATE & STATUS	THE COLLECTOR WAS DESIGNED IN 1950, AND ADDED TO AN ALREADY EXISTING BUNGALOW.
LOCATION	BOULDER, COLORADO
SCOPE	THE COLLECTOR WAS APPLIED TO AN ALREADY EXISTING FIVE-ROOM, 1,000 SQUARE FOOT BUNGALOW IN BOULDER, COLORADO. THE PRIMARY OBJECTIVE IN THE DESIGN WAS THE MAINTENANCE OF SIMPLICITY AND ECONOMY IN CONSTRUCTION AND THE DEVELOPMENT OF A COLLECTOR SUITABLE FOR LARGE-SCALE FACTORY PRODUCTION.
	EFFICIENCY OF HEAT COLLECTION RANGES FROM 30-65 PERCENT. AS AIR VELOCITY INCREASES, EFFICIENCY RISES BUT EXIT AIR TEMPERATURE DECREASES. FIFTY PERCENT EFFICIENCY IS OBTAINED AT AN AIR-FLOW RATE OF 1.6 CUBIC FEET PER SQUARE FOOT OF COLLECTOR SURFACE. WITH SURFACE TREATED LOW REFLECTIVE GLASS AND TWO COVER PLATES, THIS EFFICIENCY IS INCREASED TO 59 PERCENT.
TECHNIQUE	THE SOLAR COLLECTOR UNIT CONSISTS OF A SHEET METAL TROUGH APPROXIMATELY 3 INCHES DEEP, 2 FEET WIDE, AND 4 FEET LONG, CONTAINING A SERIES OF SINGLE-STRENGTH GLASS PLATES ARRANGED IN A STAIR-STEP FASHION AND SEPARATED BY 1/4 INCH SPACES. EACH PANE OF GLASS IS 24 INCHES WIDE, 18 INCHES LONG, AND BLACKENED WITH BLACK PAINT OR A BLACK GLASS COATING IN AN AREA 6 INCHES BY 24 INCHES. THE GLASS IS ARRANGED SO THAT EACH BLACK SURFACE IS BENEATH TWO CLEAR SURFACES. ONE OR MORE SINGLE-STRENGTH COVER GLASSES, 2 FEET BY 4 FEET IN SIZE, ARE SUPPORTED ON THE TOP EDGES OF THE TROUGH AND FORM A NEARLY AIR-TIGHT ENCLOSURE CONTAINING THE OVERLAPPED PLATES. BY MEANS OF THIS ARRANGEMENT, SOLAR ENERGY IS TRANSMITTED THROUGH THE TRANSPARENT SURFACES AND ABSORBED IN THE BLACK AREAS. THE "GREENHOUSE EFFECT" CAUSES THE BLACK SURFACES TO REACH A RELATIVELY HIGH TEMPERATURE. AIR TO BE HEATED ENTERS THE LOWER END OF THE TROUGH AT A LOW VELOCITY, AND EXITS AT THE UPPER AT TEMPERATURES APPROACHING THAT OF THE BLACK AREAS. THE BEST PERFORMANCE RESULTS WHEN THE AIR ENCOUNTERS FOUR SETS OF GLASS PLATES BETWEEN ENTERING AND LEAVING THE TROUGH.
	A COLLECTOR, 463 SQUARE FEET, IS MOUNTED ON THE ROOF (FACING SOUTH AT A 27 DEGREE ANGLE WITH THE HORIZONTAL) AND SEPARATED FROM THE SHINGLES BY A ONE-HALF INCH LAYER OF CELOTEX INSULATION. THE 180 CUBIC FEET BASEMENT STORAGE BED CONSISTS OF 8.3 TONS OF 3/4 INCH GRAVEL. WARMED AIR FROM THE COLLECTOR IS GATHERED AT THE ROOF RIDGE AND IS TRANSPORTED TO THE STORAGE. IT PASSES THROUGH THE BED TO RETURN TO THE LOWER END OF THE COLLECTOR, BECOMING COOLER AS IT TRANSFERS ITS HEAT TO THE GRAVEL.
MATERIAL	GRAVEL BED, SHEET METAL TROUGH, SINGLE STRENGTH PLATE GLASS, BLACK PAINT, SOLAR COLLECTOR
AVAILABILITY	FURTHER INFORMATION MAY BE OBTAINED FROM BRUCE ANDERSON, TOTAL ENVIRONMENTAL ACTION, BOX 47, HARRISVILLE, NEW HAMPSHIRE 03450.
REFERENCE	ANDERSON, BRUCE BOULDER HOUSE HARRISVILLE, N.H., IN SOLAR ENERGY AND SHELTER DESIGN, JANUARY 1973, PP 131-132

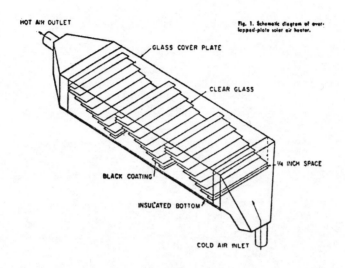

HOT AIR OUTLET

Fig. 1. Schematic diagram of over-lapped-plate solar air heater.

GLASS COVER PLATE

CLEAR GLASS

¼ INCH SPACE

BLACK COATING

INSULATED BOTTOM

COLD AIR INLET

TITLE   BOUWCENTRUM HOUSE

CODE

KEYWORDS   HOUSE WINDOW INSULATION COLLECTOR STORAGE

AUTHOR   J.M. VAN HEEL

DATE & STATUS THE BOUWCENTRUM HOUSE WAS COMPLETED IN MID-1975.

LOCATION   OSS, NETHERLANDS

SCOPE   THE BUILDING IS A 2-1/2-STORY, 200 SQ.M HOUSE. IT IS ONE OF
               FOUR CONTIGUOUS HOUSES CONSTITUTING A ROW OF EXPERIMENTAL,
               LOW-COST, SOLAR HEATED HOUSES IN WHICH TWO DIFFERENT KINDS OF
               SOLAR HEATING ARE USED. IT HAS AN ATTIC, WHICH INCLUDES A
               COLLECTOR-AND-STORAGE SYSTEM, AND MANY DUCTS. THERE IS ALSO
               ONE BEDROOM, BUT NO BASEMENT. THE ENTRANCE HALL AND TOILET
               ROOM PROJECT TO THE SOUTH.

TECHNIQUE   THE HOUSE FACES DUE SOUTH. IT HAS BRICK WALLS, WHICH ARE
               VIRTUALLY AIR-TIGHT, AND CONCRETE FLOORS. THE WINDOW AREA IS
               MODERATELY LARGE. MOST OF THE WINDOWS ARE NON-OPENABLE, AND
               DOUBLE GLAZED. THE OPENABLE WINDOWS ARE SINGLE GLAZED. UNUSUALLY
               THICK INSULATION IS USED.

               THERE IS A COMBINATION COLLECTOR-AND-STORAGE SYSTEM, WITH AN
               AREA OF 25 SQ.M. THE DIMENSIONS ARE 5 M X 5 M. THE SYSTEM IS
               MOUNTED ON THE SOUTH ROOF, SLOPING 60 DEGREES FROM THE HORI-
               ZONTAL. THE HEART OF THE SYSTEM IS A 30 CM THICK CONCRETE SLAB
               THAT HAS A NON-SELECTIVE BLACK COATING ON THE UPPER FACE, AND
               INCLUDES A PLANAR ARRAY OF PARALLEL, 7 CM IN DIAMETER AIR-
               CHANNELS, 25 CM APART ON CENTERS, THAT RUN UP AND DOWN THE
               SLAB AND ARE SLIGHTLY NEARER THE UPPER FACE THAN THE LOWER FACE.
               THE HOT AIR IN THE CHANNELS FLOWS UPWARD INTO A HEADER DUCT
               NEAR THE TOP OF THE CENTER OF THE HOUSE, THEN FLOWS, VIA THE
               AUXILIARY HEATING SYSTEM WHICH INCLUDES A 1/4 HP BLOWER, TO
               OUTLETS BENEATH THE VARIOUS WINDOWS. THE RETURN AIR FLOWS, VIA
               THE ATTIC, TO THE HEADER ALONG THE LOWER EDGE OF THE SLAB. THE
               GLAZING ABOVE THE SLAB IS DOUBLE, AND CONSISTS OF GLASS SHEETS,
               4 MM THICK. THERE IS A 2-1/2 CM DEAD-AIR SPACE BETWEEN THE
               TWO GLASS SHEETS, AND A 10 CM DEAD-AIR SPACE BETWEEN THE
               INNER SHEET AND THE CONCRETE SLAB. THE SLAB IS BACKED BY 10 CM
               OF INSULATION. AT NIGHT, A 1 CM INSULATING SHEET IS INSTALLED
               IN THE 10 CM DEAD-AIR SPACE ABOVE THE CONCRETE SLAB. NEAR THE
               BASE OF THE SLAB THERE IS AN 80 CM WIDE SERVICE WALKWAY AND
               A 60 CM HIGH WALL.

MATERIAL   BRICK WALLS; CONCRETE FLOORS; 30 CM THICK CONCRETE SLAB; NON-
               SELECTIVE BLACK COATING; PLANAR ARRAY OF PARALLEL, 7 CM DIAMETER
               AIR-CHANNELS; 1/4 HP BLOWER; 4 MM THICK GLASS SHEETS; 1 CM
               INSULATING SHEET

AVAILABILITY   ADDITIONAL INFORMATION MAY BE OBTAINED FROM J.M. VAN HEEL,
               BOUWCENTRUM, WEENA 700, POSTBUS 299, ROTTERDAM BUILDING
               CENTER, ROTTERDAM, NETHERLANDS.

REFERENCE   SHURCLIFF, W.A.
               SOLAR HEATED BUILDINGS, A BRIEF SURVEY
               CAMBRIDGE, MASSACHUSETTS, W.A. SHURCLIFF, MARCH 1976, 212 P

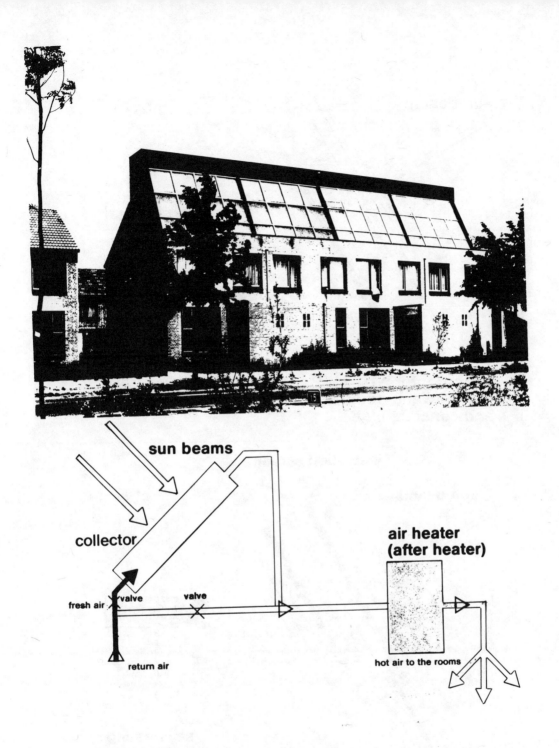

sun beams

collector

air heater
(after heater)

fresh air   valve   valve

return air

hot air to the rooms

principle solar energy installation
in houses in oss

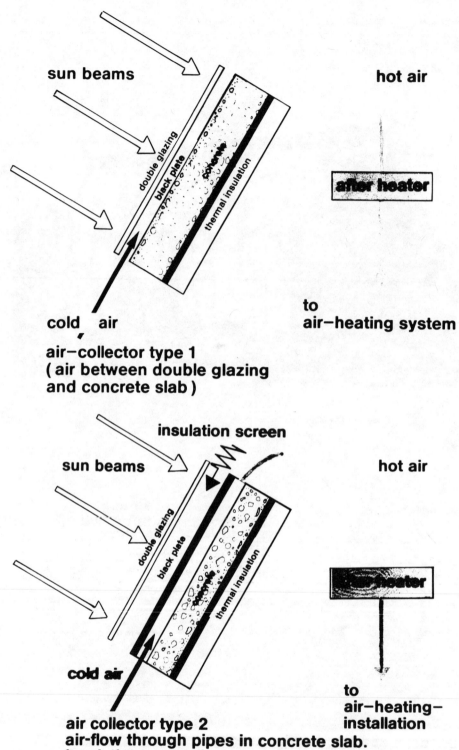

sun beams

hot air

double glazing

black plate

concrete

thermal insulation

after heater

cold air

to
air-heating system

air-collector type 1
(air between double glazing
and concrete slab)

insulation screen

sun beams

hot air

double glazing

black plate

concrete

thermal insulation

after heater

cold air

to
air-heating-
installation

air collector type 2
air-flow through pipes in concrete slab.
insulation screen between double glazing
and concrete slab

TITLE	BREUCH HOUSE
CODE	
KEYWORDS	HOME RETROFIT SOLAR WINDOWS SLIDING DOORS STORAGE SYSTEM
AUTHOR	R.A. BREUCH
DATE & STATUS	THE BREUCH HOUSE WAS RETROFITTED WITH SOLAR COLLECTORS IN 1975. THE ENERGY USED TO HEAT THE HOUSE FOR THE 1975-76 WINTER SEASON WAS 35% OF THE PREVIOUS 5 YEAR AVERAGE ENERGY USED IN THE WINTER SEASONS BEFORE THE INSTALLATION OF A SOLAR HEATER, I.E. SOLAR HEATING CARRIED 65% OF THE THERMAL SPACE HEATING LOAD LAST WINTER.
LOCATION	LOS ALTOS, CALIFORNIA
SCOPE	THE BUILDING IS A ONE-STORY HOME WITH RETROFIT SOLAR HEATING. IT INCLUDES A MAIN STRUCTURE AND TWO ELLS. THE BUILDING RESTS ON A 40-TON CONCRETE SLAB. THE HEATED LIVING AREA OCCUPIES 1,600 SQ. FT. IT INCLUDES 3 BEDROOMS AND 2 BATHS. THE VERTICAL SOUTHERN WALL OF THE MAIN STRUCTURE HAS 130 SQ. FT. OF SINGLE-GLAZED WINDOWS AND SLIDING GLASS DOORS. THERE IS SOME ATTIC SPACE AND A BUILT-IN GARAGE, BUT NO BASEMENT. COOLING IN SUMMER IS NOT PROVIDED.
TECHNIQUE	CEILING INSULATION CONSISTS OF 6 TO 8" OF FIBERGLAS. THE WALLS (OF THICK STUCCO) ARE NOT INSULATED. THE COLLECTOR OCCUPIES 300 SQ. FT. GROSS (15 FT. X 20 FT.). IT IS OF THE AIR TYPE, MOUNTED ON THE ROOF AND SLOPES 10 DEGREES FROM THE HORIZONTAL. THE 12 BLACK ABSORBING PANELS EXTEND EAST-WEST AND SLOPE 35 DEGREES FROM THE HORIZONTAL. THE BLACK COATING ON THE 28-GAGE SHEET METAL IS A CARBON-BLACK, ACRYLIC PAINT, AND IS NON-SELECTIVE. THE SINGLE GLAZING IS OF 5-OZ.-GRADE FIBERGLAS-REINFORCED, ACRYLIC-MODIFIED, POLYESTER, THE UPPER SURFACE OF WHICH IS TEDLAR-COATED. THE BACK OF THE PANEL IS INSULATED, AND A STRUCTURE FOR SUPPRESSING CONVECTION IS SITUATED BETWEEN THE BLACK ABSORBING SURFACE AND THE GLAZING. THE TRIANGULAR-CROSS-SECTION SPACES BETWEEN THE BLACK ABSORBING SURFACES AND THE ROOF SERVE AS DUCTS AND COMPRISE AN EAST-WEST, WEST-EAST SERPEN-TINE PASSAGE FOR AIRFLOW WITH TURN-AROUNDS AT PANEL ENDS. THE AIR IS DRIVEN BY A 4-SPEED BLOWER WITH A MAXIMUM POWER OF 1/3 HP. USUALLY, THE BLOWER IS RUN AT THE LOWEST SPEED, GIVING ABOUT 500 CFM FLOW. THE HOT AIR FROM THE COLLECTOR PASSES THROUGH A PAIR OF AIR-TO-WATER HEAT EXCHANGERS IMMEDIATELY ADJACENT TO THE COLLECTOR AND THE BLOWER, AND THEN TRAVELS BACK TO THE COLLECTOR. THE WATER, HEATED BY THIS PROCESS, TRAVELS INTO THE MAIN STORAGE SYSTEM, AND FROM HERE IT EVENTUALLY TRAVELS BACK TO THE PAIR OF HEAT EXCHANGERS NEAR THE COLLECTOR.
	THE STORAGE SYSTEM CONSISTS OF A 40 TON CONCRETE SLAB. THE ROOMS ARE HEATED BY THE CONCRETE FLOOR SLAB CONTAINING 3/4" DIAMETER COPPER PIPES, 12" APART ON CENTERS, IN WHICH HOT WATER FROM THE HEAT EXCHANGER FLOWS. THE SLAB IS AN IMPORTANT PART OF THE ENERGY STORAGE SYSTEM.
MATERIAL	SINGLE-GLAZED WINDOWS; SLIDING GLASS DOORS; STUCCO; 12 BLACK ABSORBING PANELS; 28-GAGE SHEET METAL; CARBON-BLACK, ACRYLIC PAINT; 5-OZ.-GRADE FIBERGLAS-REINFORCED, ACRYLIC-MODIFIED, POLYESTER; TEDLAR; A PAIR OF AIR-TO-WATER HEAT EXCHANGERS; 40 TON CONCRETE SLAB; 3/4" DIAMETER COPPER PIPES
AVAILABILITY	ADDITIONAL INFORMATION MAY BE OBTAINED FROM R.A. BREUCH, 590 SPARGUR DRIVE, LOS ALTOS, CALIFORNIA 94022.
REFERENCE	SHURCLIFF, W.A. SOLAR HEATED BUILDINGS, A BRIEF SURVEY CAMBRIDGE, MASSACHUSETTS, W.A. SHURCLIFF, MARCH 1976, 212 P

TITLE	BRICK ASSOCIATION OF NORTH CAROLINA HOUSE
CODE	
KEYWORDS	SINGLE FAMILY HOUSE SOLAR COLLECTORS WIND GENERATORS EARTH BERMS WINDMILL STORE REFLECT ILLUMINATION RECTANGULAR FORMS HEXOID ORIENTATION OCTOID SHADING THERMAL CONDUCTION INSULATION VENTILATION
AUTHOR	ARNOLD J. AHO
DATE & STATUS	THE BRICK ASSOCIATION HOUSE HAS BEEN DESIGNED AS A STUDY HOUSE AND WILL BE BUILT AT A FUTURE DATE.
LOCATION	PIEDMONT, NORTH CAROLINA

SCOPE

THE BRICK ASSOCIATION HOUSE IS A PROTOTYPE MIDDLE INCOME SINGLE FAMILY HOUSE. MORE SPECIFICALLY, IT IS A 3 BEDROOM STUDY HOUSE OF ABOUT 1,600 SQ. FT., ADAPTABLE TO VARIOUS SITES WITHIN THE PIEDMONT REGION OF NORTH CAROLINA. THE HOUSE HAD TO BE CURRENTLY FEASIBLE, USING CURRENTLY AVAILABLE ENVIRONMENTAL CONTROL SYSTEMS; AND IT HAD TO BE CAPABLE OF ACCEPTING SUCH FUTURE ENERGY SOURCES AS SOLAR COLLECTORS OR WIND-DRIVEN GENERATORS. IN EITHER SITUATION, THE HOUSE HAD TO CONSERVE ENERGY. RESEARCH INTO THESE NATURAL ENERGIES LED THE ARCHITECT TO CHOOSE A NEW TYPE OF BRICK RIBBED WALL FOR THE HOUSE, AND TO PARTIALLY BURY THE HOUSE AND SURROUND IT WITH EARTH BERMS. THE INTERIOR STRUCTURE CONSISTS OF PIERS AND WOOD BEAMS, WITH A PREFABRICATED ROOF TRUSS SYSTEM. THE MECHANICAL SYSTEMS WOULD BE HOUSED IN A TOWER, WHICH COULD ALSO PROVIDE STORAGE FOR SOLAR HEATED WATER OR A SUPPORT FOR A POSSIBLE WINDMILL; STRUCTURAL PROVISIONS WERE ALSO MADE FOR SOLAR COLLECTORS.

WHEN THE STUDY HOUSE WAS COMPARED TO THE TYPICAL WOOD FRAME HOUSE IN THE NORTH CAROLINA PIEDMONT, IT WAS DISCOVERED THAT FOR A TYPICAL 24 HOUR PERIOD IN WINTER (DEC. 21) TOTAL HEAT FLOW IN BTUS WAS SIGNIFICANTLY DECREASED: 38,881 BTU'S FOR OUR STUDY HOUSE AGAINST 189,873 BTU'S FOR THE TYPICAL WOOD FRAME HOUSE, OR AN 80% DECREASE. FOR THE SUMMER, THE STUDY SHOWED A 20% DECREASE: 246,733 BTUS FOR THE STUDY HOUSE, 306,648 BTUS FOR THE TYPICAL FRAME HOUSE. THESE REDUCTIONS ARE PARTICULARLY SIGNIFICANT, SINCE THE AVERAGE FAMILY IN THIS PART OF NORTH CAROLINA EXPENDS 47% OF ITS TOTAL ENERGY BUDGET IN HEATING, AND 14% ON COOLING. IN ADDITION, A COMPARISON OF THE PEAK COOLING PERIOD (JULY 21, 4 PM) SHOWED A 35% REDUCTION IN PEAK DEMAND; BTUS PER HOUR FOR THE STUDY HOUSE WERE 18,438 COMPARED TO 28,479 FOR THE FRAME HOUSE.

TECHNIQUE

A CONCEPT THAT HAD A STRONG INFLUENCE ON THE DESIGN WAS THE WALL AS USABLE SPACE. THIS ALLOWED FOR THE MODERATION OF HEAT FLOW, THE STORAGE OF HEAT AND THE REFLECTION OF HEAT. THIS CONCEPT ALSO LED TO THE CONTROL OF DAYLIGHT FOR ILLUMINATION WHILE SHADING THE INTERIOR FROM DIRECT SOLAR RADIATION, AS WELL AS TO THE DEVELOPMENT OF THE STRUCTURAL SYSTEM. THE "SPACE-WALL" ALSO PROVIDED ROOM FOR ALL THE PROGRAMMED ELEMENTS THAT DO NOT NEED AN ENERGY INTENSIVE, ENERGY-COSTLY TEMPERED HUMAN ENVIRONMENT, SUCH AS DUCT AND PIPE SPACES, CLOSED STORAGE, CLOSETS, AND MECHANICAL EQUIPMENT.

RESEARCH INTO THE FORM OF THE BUILDING INDICATED THAT OPTIMUM RECTANGULAR FORMS WERE PROPORTIONED BETWEEN 1: 1.1 AND 1: 1.6, ELONGATED DIRECTLY ALONG THE EAST-WEST AXIS.

FURTHER INVESTIGATIONS SHOWED THAT CERTAIN HEXOID AND OCTOID SHAPES WERE 5-10% BETTER THAN THE OPTIMUM RECTANGLE. THE HEXOID SHAPES WERE SELECTED AS A DIRECTION FOR THE DESIGN OF THE FLOOR PLANS BASED ON THEIR HEAT FLOW PERFORMANCE AND ALSO BECAUSE THEY COULD BE GENERATED FROM A 90 DEGREE GRID, GIVING USABLE INTERIOR SPACES WITHOUT PIE-SHAPED ROOMS.

ADDITIONAL RESEARCH INDICATED THAT SHADING WINDOWS DURING THE SUMMER COULD PROVIDE A 37% REDUCTION IN HEAT GAIN PER SQUARE FOOT; WINTER SUNLIGHT, ON THE OTHER HAND, COULD PROVIDE VALUABLE NATURAL HEAT -- 2.29 TIMES THE AMOUNT OF HEAT BLOCKED BY THE SUMMER SHADE PROTECTION. THE 2 FT. 8 IN. THICK WALL SYSTEM ALLOWED THE PROVISION OF APPROPRIATE SHADING.

THE MATERIALS WERE SELECTED NOT ONLY FOR THEIR INSULATION VALUE
BUT ALSO FOR THEIR HEAT STORAGE AND RADIATION CHARACTERISTICS.
THIS LED TO THE SELECTION OF A DOUBLE-WYTHE BRICK WALL 2 FT.
8 IN. THICK WITH 2-1/2 IN. OF RIGID FOAM (STYRENE OR URETHANE)
INSULATION APPLIED TO THE EXTERIOR WYTHE, INSIDE THE 2 FT.
CAVITY, GIVING THE WALL A U-VALUE OF 0.07. THE PEAK HEAT GAIN
FOR THIS WALL CONSTRUCTION IS 42% OF THAT FOR AN EQUALLY INSU-
LATED 2 X 4 WOOD FRAME CONSTRUCTION.

SOIL IS AN EFFECTIVE INSULATOR. AT A DEPTH OF APPROXIMATELY
24 INCHES, ALL DAILY TEMPERATURE FLUCTUATIONS DISAPPEAR, WHICH
ALLOWS THE DESIGNER TO USE A 24-HOUR AVERAGE TEMPERATURE. IN
PIEDMONT NORTH CAROLINA, THIS MEANS DESIGNING TO A SUMMER
AVERAGE OF 78 DEGREES F RATHER THAN 88 DEGREES, AND IN WINTER,
TO AN AVERAGE OF 41 DEGREES INSTEAD OF 31 DEGREES. FURTHER DOWN,
BELOW 24 INCHES DEEP, THE SEASONAL SOIL TEMPERATURE VARIATIONS
AVERAGE OUT AT THE RATE OF 1 DEGREE F FOR EVERY 4 INCH DEPTH.
ABOUT 10 FEET DEEP, THE TEMPERATURE STAYS AT A CONSTANT 59 DE-
GREES ALL YEAR.

IN THIS STUDY, THE HOUSE IS SET 3 FEET INTO THE GROUND, USING
THE SOIL FROM THE EXCAVATION TO BUILD UP EARTH BERMS AROUND
THE BUILDING.

PREVAILING WINDS IN THE PIEDMONT AREA ARE FROM THE SOUTHWEST. SO
WHILE THE WALLS OF THE HOUSE ARE ALIGNED WITH THE EAST-WEST AXIS
(THE OPTIMUM SOLAR ORIENTATION), THE ROOF IS ORIENTED PERPENDI-
CULARLY TO THE SOUTHWEST AXIS. THIS ALLOWS THE LOCATION OF
CLERESTORY WINDOWS ON THE NORTHEAST SIDE OF THE HOUSE TO PUT THE
VENTILATION OUTLETS ON THE LOW PRESSURE SIDE. BECAUSE THEY ARE
LOCATED ON THE ROOF, THE CLERESTORY WINDOWS ALSO WORK AS STACKS
TO DRAW OFF HOT AIR WITHIN THE HOUSE. BECAUSE THE INLETS ARE
SMALLER THAN THE OUTLETS, THE AIR VELOCITY IS BOOSTED, AND AIR
FLOW IS DIRECTED TO THE LIVING AREAS OF THE HOUSE.

MATERIAL   SOLAR COLLECTORS, WIND-DRIVEN GENERATORS, BRICK RIBBED WALL,
                EARTH BERMS, PIERS, WOOD BEAMS, PREFABRICATED ROOF TRUSS SYSTEM,
                SOLAR HEATED WATER, WINDMILL, SPACE-WALL, DUCT, PIPE, DOUBLE-
                WYTHE BRICK, RIGID FOAM (STYRENE OR URETHANE) INSULATION,
                CLERESTORY WINDOWS, HOT AIR

AVAILABILITY   FURTHER INFORMATION MAY BE OBTAINED FROM ARNOLD J. AHO, AIA,
                P.O.BOX 5291, MISSISSIPPI STATE, MISSISSIPPI 39762.

REFERENCE   AHO, ARNOLD J.
                DESIGNING WITH NATURAL ENERGIES
                WASHINGTON, D.C., IN THE ENERGY NOTEBOOK, PUBLISHED BY THE
                AMERICAN INSTITUTE OF ARCHITECTS, 1975, PP SA-91 TO SA-103

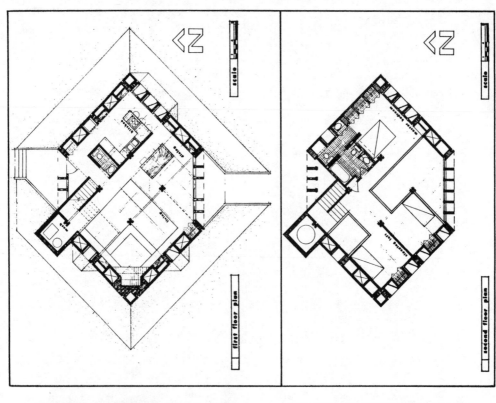

first floor plan

second floor plan

scale

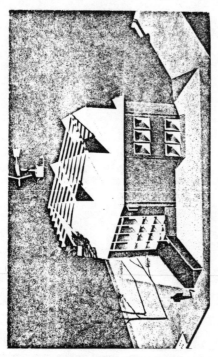

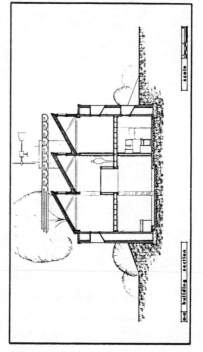

building section

scale

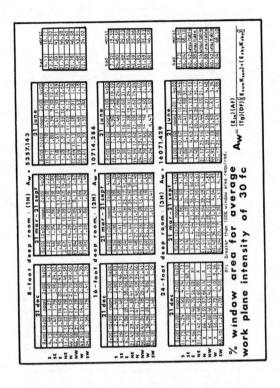

8-foot deep room : (1H) Aw = 5357.143

16-foot deep room : (2H) Aw = 10714.286

24-foot deep room : (3H) Aw = 16071.429

X%: Greater than 100% window area required.

% window area for average work plane intensity of 30 fc

$$A_w = \frac{(E_{th})(A_f)}{(T_g)(DF)[(E_{sun}K_{sun})+(E_{sky}K_{sky})]}$$

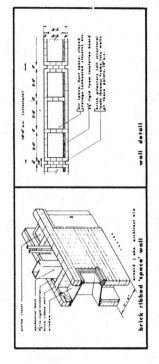

wall detail

brick ribbed 'space' wall

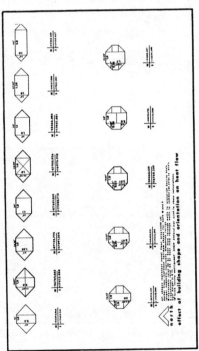

north

effect of building shape and orientation on heat flow

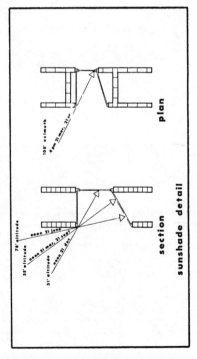

plan

section

sunshade detail

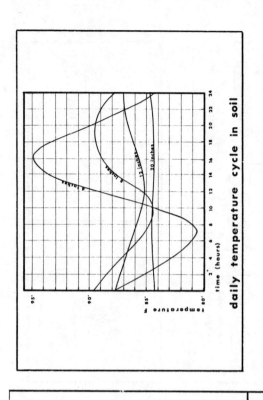

daily temperature cycle in soil

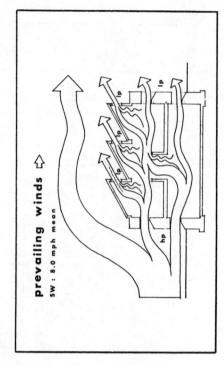

prevailing winds ⇧
SW : 8.0 mph mean

21 july   36° N lat

single pane glass   U=1.06   SC=0.88

heat loss/gain per square foot

°F|H	S	SE	E	NE	N	NW	W	SW
1	-4.24	-4.24	-4.24	-4.24	-4.24	-4.24	-4.24	-4.24
2	-5.30	-5.30	-5.30	-5.30	-5.30	-5.30	-5.30	-5.30
3	-6.63	-6.63	-6.63	-6.63	-6.63	-6.63	-6.63	-6.63
4	-7.42	-7.42	-7.42	-7.42	-7.42	-7.42	-7.42	-7.42
5	-6.36	-6.31	-5.36	-5.36	-6.36	-6.36	-6.36	-6.36
6	+3.70	+55.20	+119.20	+109.70	+29.70	+3.70	+3.70	+3.70
7	+15.26	+116.26	+196.76	+160.76	+29.26	+14.26	+14.26	+14.26
8	+26.38	+150.38	+213.88	+153.38	+27.38	+23.88	+21.88	+23.88
9	+46.62	+162.12	+195.62	+119.12	+35.12	+33.12	+33.12	+33.12
10	+71.80	+151.80	+152.30	+72.30	+41.30	+40.30	+40.80	+41.30
11	+99.54	+151.86	+92.54	+45.54	+48.54	+45.54	+46.42	+48.54
12	+95.46	+52.66	+51.04	+49.16	+49.66	+52.66	+49.66	+49.66
13	+80.28	+49.28	+48.78	+48.78	+80.78	+160.78	+160.28	+173.78
14	+58.28	+44.78	+44.78	+46.78	+130.78	+208.28	+208.72	+165.22
15	+41.22	+38.72	+38.72	+42.22	+168.22	+176.10	+132.98	+41.22
16	+17.48	+17.48	+17.48	+43.48	+123.48	+123.48	+63.98	+17.48
17	+10.10	+10.10	+10.10	+29.10	+44.10	+10.10	+7.36	+10.10
18	+6.36	+6.36	+6.36	+6.36	+6.36	+6.36	+6.36	+6.36
19	+1.06	+1.06	+1.06	+1.06	+1.06	+1.06	+1.06	+1.06
20	0	0	0	0	0	0	0	0
21	-2.12	-2.12	-2.12	-2.12	-2.12	-2.12	-2.12	-2.12
22	-3.18	-3.18	-3.18	-3.18	-3.18	-3.18	-3.18	-3.18
sum	+694.15	+1043.15	+917.65	+509.15	+918.15	+223.65	+1043.15	

(shaded)

S	SE	SW
-4.24	-4.24	-5.30
-5.30	-5.30	-6.63
-6.63	-6.63	-6.63
-7.42	-7.42	+3.70
-6.36	-6.36	+14.26
+3.70	+3.70	+21.88
+14.26	+15.26	+33.12
+23.88	+26.38	+40.80
+35.12	+35.12	+45.42
+41.30	+41.30	+46.06
+45.42	+45.42	+49.66
+48.54	+48.54	+46.78
+49.66	+49.66	+41.22
+46.78	+44.78	+17.48
+38.72	+38.72	+17.48
+29.10	+29.10	+10.10
+6.36	+6.36	+6.36
+1.06	+1.06	+1.06
0	0	0
-2.12	-2.12	-2.12
-3.18	-3.18	-3.18
+420.15	+420.65	+420.65

hourly thermal impact through glass in btus/sf

21 july   36° N lat

thermal insulated glass (¼"-¼"-¼")   U=0.61   SC=0.88

heat loss gain per square foot

°F|H	S	SE	E	NE	N	NW	W	SW
1	-2.44	-2.44	-2.44	-2.44	-2.44	-2.44	-2.44	-2.44
2	-3.05	-3.05	-3.05	-3.05	-3.05	-3.05	-3.05	-3.05
3	-3.66	-3.66	-3.66	-3.66	-3.66	-3.66	-3.66	-3.66
4	-4.27	-4.27	-4.27	-4.27	-4.27	-4.27	-4.27	-4.27
5	-3.66	-2.78	-3.66	-3.66	-3.66	-3.66	-3.66	-3.66
6	+14.72	+50.19	+106.51	+98.15	+27.75	+4.87	+4.87	+4.87
7	+21.86	+103.60	+174.44	+142.76	+27.04	+13.84	+13.84	+13.84
8	+40.38	+132.38	+188.86	+135.62	+24.74	+21.66	+21.66	+21.66
9	+58.57	+142.02	+172.38	+104.18	+30.26	+28.50	+28.50	+28.50
10	+64.69	+131.97	+132.41	+62.01	+34.73	+33.85	+33.85	+34.29
11	+66.45	+139.17	+78.73	+38.73	+42.01	+37.71	+37.71	+37.71
12	+77.09	+35.21	+39.71	+39.81	+39.61	+42.79	+39.81	+39.81
13	+32.40	+30.20	+30.20	+42.79	+66.89	+137.29	+179.09	+148.73
14	+80.67	+22.38	+22.38	+35.58	+110.89	+197.40	+197.40	+141.52
15	+21.26	+22.38	+22.38	+35.68	+151.30	+182.98	+114.44	+58.12
16	+12.80	+12.80	+12.80	+36.08	+114.16	+126.08	+114.44	+23.26
17	+2.44	+2.44	+2.44	+10.44	+44.54	+2.44	+2.44	+12.80
18	+0.61	+0.61	+0.61	+0.61	+0.61	+0.61	+0.61	+2.44
19	0	0	0	0	0	0	0	+0.61
20	-1.22	-1.22	-1.22	-1.22	-1.22	-1.22	-1.22	0
21	-1.83	-1.83	-1.83	-1.83	-1.83	-1.83	-1.83	-1.22
22	+557.49	+1053.65	+789.37	+429.89	+789.37	+1058.66	+899.81	-1.83
sum	+557.49	+1053.65	+789.37	+429.89	+789.37	+1058.66	+899.81	

(shaded)

S	SE	SW
-2.44	-2.44	-3.05
-3.05	-3.05	-3.66
-3.66	-3.66	-3.66
-3.66	-3.66	+4.87
+4.87	+4.72	+13.84
+13.84	+23.86	+21.66
+21.66	+30.26	+28.50
+30.26	+32.71	+33.85
+34.73	+34.73	+37.71
+37.71	+37.71	+39.61
+39.61	+39.81	+39.81
+36.97	+35.21	+33.85
+30.20	+30.20	+12.80
+22.38	+22.38	+12.80
+12.80	+12.80	+2.44
+2.44	+2.44	+0.61
+0.61	+0.61	0
0	0	-1.22
-1.22	-1.22	-1.83
-1.83	-1.83	+552.01
+351.57	+352.01	+552.01

hourly thermal impact through glass in btus/sf

TITLE   BRIDGERS & PAXTON OFFICE BUILDING

CODE

KEYWORDS   BUILDING CONTROL ROOM SOLAR WATER COLLECTORS HEAT STORAGE PUMP
                 REFRIGERATION EVAPORATIVE COOLER

AUTHOR   BRIDGERS & PAXTON AND STANLEY & WRIGHT

DATE & STATUS   THE BRIDGERS & PAXTON OFFICE BUILDING WAS COMPLETED IN 1956.

LOCATION   ALBUQUERQUE, NEW MEXICO

SCOPE   THE BUILDING RECEIVED OVER 90% OF ITS HEAT FROM THE SUN, AND THE
                 SYSTEM RAN SUCCESSFULLY FOR A NUMBER OF YEARS. BUT WHEN THE
                 COLLECTORS DEVELOPED LEAKS, AND WHEN THE COST OF NATURAL GAS WAS
                 LOW, THE SYSTEM WAS CONVERTED TO CONVENTIONAL FUEL. NOW THE
                 SOLAR COLLECTORS HAVE BEEN REPAIRED AND THE SYSTEM RESTORED TO
                 ITS ORIGINAL STATE. THE BUILDING HOUSES THE MECHANICAL
                 ENGINEERING OFFICE OF BRIDGERS & PAXTON.

TECHNIQUE   ASIDE FROM THE SOLAR COLLECTORS, THE BUILDING USES ITEMS OF
                 STANDARD MANUFACTURE. THE SYSTEM INCLUDES THREE DAY HEAT STORAGE,
                 A HEAT PUMP FOR REFRIGERATION AND TO BOOST HEATING EFFICIENCIES
                 AND AN EVAPORATIVE WATER COOLER. THE INTERIOR CAN BE COOLED
                 WHILE THE SUN'S HEAT IS BEING STORED FOR LATER USE.

MATERIAL   SOLAR COLLECTORS, HEAT PUMP, EVAPORATIVE WATER COOLER

AVAILABILITY   ADDITIONAL INFORMATION ON THE BRIDGERS & PAXTON OFFICE BUILDING
                 MAY BE OBTAINED FROM BRIDGERS & PAXTON, 213 TRUMAN N.E.
                 ALBUQUERQUE, NEW MEXICO.

REFERENCE   COOK, JEFFREY
                 VARIED AND EARLY SOLAR ENERGY APPLICATIONS OF NORTHERN NEW
                 MEXICO, THE
                 AIA JOURNAL, AUGUST 1974, P 38

TITLE	BROOKHOLLOW CORPORATION BUILDING
CODE	
KEYWORDS	OFFICE SOLAR COLLECTOR LOCATE SPANDREL SOUTHERN ELEVATION GLASS VERTICAL
AUTHOR	BRANDY AND FREEMAN, ENGINEERS / HARWOOD K. SMITH & PARTNERS, INC.
DATE & STATUS	THIS BUILDING WAS DESIGNED IN EARLY 1976.
LOCATION	DALLAS, TEXAS
SCOPE	THIS 15-STORY, 275,000 SQ. FT. OFFICE BUILDING WILL UTILIZE SOLAR PANELS AS A MEANS OF HEAT COLLECTION. THE SYSTEM WILL NOT HAVE THE CAPACITY FOR POWERING THE AIR CONDITIONING SYSTEM. THE COST OF THE COLLECTOR SYSTEM AND ADDITIONAL PIPING WOULD BE OFFSET BY REDUCING THE BOILER EQUIPMENT NEEDED. THIS SYSTEM IS EXPECTED TO CUT DOWN ON THE FANTASTIC FIRST COST OF OTHER APPROACHES TO SOLAR ENERGY. THIS CONCEPT SIMPLIFIES THE DESIGN OF A BUILDING USING SOLAR ENERGY, BECAUSE IT TAKES THE COLLECTORS OFF THE ROOF AND LOCATES THEM BEHIND A SURFACE THAT IS ALREADY THERE.
TECHNIQUE	IN THIS SCHEME, SOLAR COLLECTORS ARE MOUNTED ON SOME OF THE SPANDRELS AND NON VISION GLASS AREAS ON THE SOUTHERN ELEVATION OF THE BUILDING. GLASS COVERED, PAINTED COPPER SHEETS WITH WATER FLOWING THROUGH THEM, AND INSULATED BOXES BEHIND THEM, REPLACE THE DARK, NON-VISION GLASS BACKED BY INSULATION NORMALLY USED IN THE SPANDREL AREAS. THIS TECHNIQUE COULD BE ADAPTED TO NORTHERN CLIMATES, WHERE MORE CAPACITY IS NEEDED BY MOUNTING COLLECTORS ON COLUMNS.
MATERIAL	SOLAR COLLECTOR PANELS
AVAILABILITY	ADDITIONAL INFORMATION MAY BE OBTAINED FROM HARWOOD K. SMITH AND PARTNERS, INC., 2900 SOUTHLAND CENTER, DALLAS, TEXAS 75201.
REFERENCE	ENGINEERING NEWS RECORD LOW COST SOLAR COLLECTORS MOUNTED ON BUILDING FACADE ENGINEERING NEWS RECORD, FEBRUARY 19, 1976, P 9

TITLE	BROSS UTILITIES SERVICE CORPORATION BUILDING
CODE	
KEYWORDS	SOLAR COLLECTOR HEATING COOLING WATER EXCHANGER ABSORPTION UNIT STORAGE BACKUP
AUTHOR	BROSS UTILITIES SERVICE CORP., SOLAR DIVISION
DATE & STATUS	THIS HEATING AND COOLING SYSTEM IS PRESENTLY BEING TESTED. AFTER ALL EVALUATIONS HAVE BEEN COMPLETED, THE SYSTEM WILL BE BUILT AND MARKETED FOR COMMERCIAL AND RESIDENTIAL APPLICATION.
LOCATION	BLOOMFIELD, CONNECTICUT
SCOPE	THE HEATING AND COOLING SYSTEM IS DESIGNED TO SERVE 12,800 SQ. FT. OF OFFICE SPACE. PROJECTED FUEL SAVINGS ARE ESTIMATED AT BETWEEN 50% AND 70%.
TECHNIQUE	THIS BUILDING UTILIZES 108 FLAT PLATE COLLECTORS ARRANGED IN 12 ROWS OF 9 EACH, PLACED AT A 52 DEGREE SOUTH FACING ANGLE ON THE ROOF OF THE BUILDING. FOUR DIFFERENT TYPES OF GLASS ARE BEING USED FOR TEST PURPOSES. DURING THE HEATING PERIOD, THE SYSTEM IS CAPABLE OF GENERATING 300,000 BTUS PER HOUR, AND DURING THE COOLING PERIOD 420,000 TO 600,000 BTUS CAN BE GENE-RATED. THE SYSTEM HAS A 3 TO 4 DAY THERMAL ENERGY STORAGE CAPACITY WITH A CONVENTIONAL BACKUP. THE HEAT TRANSFER SOLUTION IS PROPYLENE GLYCOL. A 5,000 GALLON WATER STORAGE TANK IS USED IS PROPYLENE GLYCOL AND WATER WITH INHIBITOR. A 5,000 GALLON WATER STORAGE TANK IS USED ALONG WITH TWO TWO-PASS HEAT EX-CHANGERS. THE COOLING UNIT USES TWO WATER-FIRED ABSORPTION UNITS, CAPABLE OF CHILLING THE WATER IN THE SYSTEM TO 40 DE-GREES F.
MATERIAL	FLAT PLATE COLLECTOR, REDWOOD CONTAINER BOX, THERMAL STORAGE TANK, FILON PLASTIC, G.E. LAXON, WATER-WHITE CRYSTAL GLASS, WINDOW GLASS, PROPYLENE GLYCOL WITH WATER AND INHIBITOR, TWO PASS HEAT EXCHANGERS, WATER-FIRED ABSORPTION UNITS WATER-FIRED ABSORPTION UNITS
AVAILABILITY	ADDITIONAL INFORMATION MAY BE OBTAINED FROM BROSS UTILITIES, SOLAR DIVISION, 42 EAST DUDLEY TOWN ROAD, BLOOMFIELD, CONNECTI-CUT 06002.
REFERENCE	BROSS UTILITIES SERVICE CORP. NEW CONCEPT IN ENERGY CONSERVATION: SOLAR ENERGY INSTALLATION FOR HEATING AND COOLING AT BROSS UTILITIES SERVICE CORP. OFFICE BUILDING, BLOOMFIELD, CONNECTICUT, A BLOOMFIELD, CONNECTICUT, BROSS UTILITIES SERVICE CORP. PAMPHLET, 1976, 6 P

VIEW OF THE BUILDING ROOF WITH THE
SOLAR ABSORBER ARRAY

PARTIAL VIEW OF THE MECHANICAL ROOM

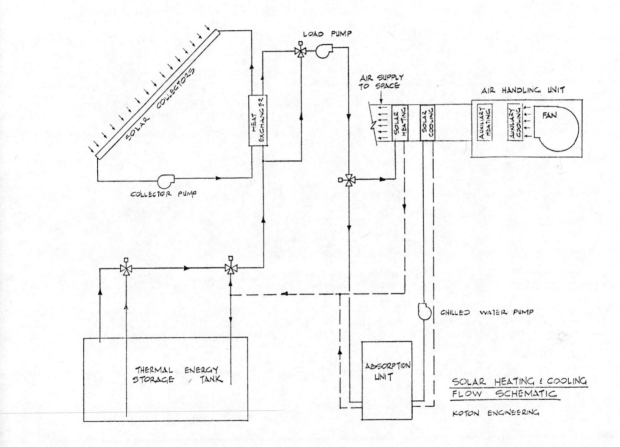

LOAD PUMP

AIR SUPPLY
TO SPACE

AIR HANDLING UNIT

SOLAR COLLECTORS

HEAT EXCHANGER

COLLECTOR PUMP

SOLAR HEATING

SOLAR COOLING

AUXILIARY HEATING

AUXILIARY COOLING

FAN

CHILLED WATER PUMP

THERMAL ENERGY STORAGE TANK

ABSORPTION UNIT

SOLAR HEATING & COOLING
FLOW SCHEMATIC

KOTON ENGINEERING

148

TITLE   BUSHNELL HOUSE

CODE

KEYWORDS   HOUSE INSULATED RETROFIT SOLAR DETACHED FREESTANDING COLLECTOR

AUTHOR   R.H. BUSHNELL

DATE & STATUS   THE COLLECTOR STARTED OPERATING IN DECEMBER 1974. THE STORAGE
                 SYSTEM WAS INSTALLED IN MID-1975.

LOCATION   BOULDER, COLORADO

SCOPE   THE BUILDING IS A ONE-STORY HOUSE WITH A PARTIAL BASEMENT. THE
                 TOTAL HEATED AREA IS 2,000 SQ. FT.

TECHNIQUE   WALLS ARE MODERATELY WELL INSULATED. THE CEILING IS INSULATED
                 WITH 6 TO 15" OF FIBERGLAS. THE BUILDING IS RETROFIT-SOLAR-HEATED
                 BY MEANS OF A DETACHED, FREESTANDING COLLECTOR ABUTTING THE
                 SOUTH END OF THE HOUSE.

                 THE AIR-TYPE COLLECTOR OCCUPIES 345 SQ. FT. IT IS ON A CONCRETE
                 FOUNDATION ADJACENT TO THE SOUTH END OF THE HOUSE. IT SLOPES
                 62-1/2 DEGREES FROM THE HORIZONTAL. THE AZIMUTH AIMS 4 DEGREES
                 WEST OF SOUTH. THE COLLECTOR FACE IS 12' X 28'. THE ABSORBING
                 SURFACE IS A BLACK COTTON-AND-POLYESTER CLOTH, 4 TO 18", BEHIND
                 THE GLAZING. THE AIR IS DRIVEN BY A 3/4 HP BLOWER, AND TRAVELS
                 UPWARD BETWEEN THE BLACK CLOTH AND THE GLAZING. THE GLAZING IS
                 DOUBLE. THE INNER LAYER IS OF SINGLE STRENGTH GLASS; THE OUTER
                 LAYER IS OF 4' X 12' VERTICAL SHEETS OF FILON SUPREME PLASTIC
                 (TEDLAR-COATED, POLYESTER-REINFORCED, FIBERGLAS). THE SPACE
                 BETWEEN THE INNER AND OUTER GLAZING LAYERS IS 1". THE COLLECTOR
                 BACK IS INSULATED WITH 6" OF FIBERGLAS.

                 THE STORAGE SYSTEM CONTAINS ABOUT 5 TONS OF RUBBLE, BRICKS,
                 CONCRETE BLOCKS, AND NON-CIRCULATING WATER IN STEEL DRUMS. THIS
                 IS EQUIVALENT TO 6 TONS OF STONES. IT IS SITUATED IN A BIN IN THE
                 HOUSE BASEMENT AREA CLOSE TO THE WEST END OF THE COLLECTOR.

                 FOR SUMMER COOLING, THE STORAGE SYSTEM IS COOLED AT NIGHT BY
                 FORCED CIRCULATION OF COOL OUTDOOR AIR. DURING A HOT DAY, THE
                 ROOM AIR IS CIRCULATED THROUGH THE BIN.

MATERIAL   6 TO 15" OF FIBERGLAS; CONCRETE; BLACK COTTON-AND-POLYESTER
                 CLOTH; 3/4 HP BLOWER; SINGLE STRENGTH GLASS; FILON SUPREME
                 PLASTIC (TEDLAR-COATED, POLYESTER-REINFORCED, FIBERGLAS);
                 RUBBLE; BRICKS; CONCRETE BLOCKS; WATER

AVAILABILITY   ADDITIONAL INFORMATION MAY BE OBTAINED FROM R.H. BUSHNELL,
                 502 ORD DRIVE, BOULDER, COLORADO 80303.

REFERENCE   SHURCLIFF, W.A.
                 SOLAR HEATED BUILDINGS, A BRIEF SURVEY
                 CAMBRIDGE, MASSACHUSETTS, W.A. SHURCLIFF, MARCH 1976, 212 P

149

TITLE	C & I BANK
CODE	
KEYWORDS	PLAZA GREENHOUSE
AUTHOR	GASSNER - NATHAN - BROWNE
DATE & STATUS	THE C & I BANK WAS COMPLETE AS OF 1972.
LOCATION	MEMPHIS, TENNESSEE
SCOPE	THE 44,000 SQUARE FOOT TRIANGULAR STRUCTURE IS BRIGHTENED BY A LARGE, RICHLY PLANTED ENTRANCE PLAZA ON ITS SOUTH SIDE. THE PLAZA IS COVERED IN GRAY TINTED SOLAR GLASS, SUPPORTED BY A SYSTEM OF SLOPING, STEEL PIPE TRUSSES. THE VOLUME ENCLOSED IS STRONGLY SCULPTURED, GENEROUSLY TUFTED WITH GREENERY, AND PATTERNED WITH A SHIFTING FILAGREE OF SHADOWS CAST BY THE OVER-HEAD TRUSS SYSTEM. THE MAIN BANKING FLOOR IS ADJACENT TO THE GARDEN. THE TELLERS' AREA IS LOCATED SO THAT IT CAN ALSO SERVE DRIVE-IN CUSTOMERS AT THE REAR. ABOVE THE BANKING FLOOR ARE THREE LEVELS OF AUXILIARY SPACE THAT OVERLOOK THE ENCLOSED GARDEN. THE FIFTH FLOOR IS FOR EXECUTIVE OFFICE SPACE. THE REST OF THE SPACE IS FOR MECHANICAL EQUIPMENT.
TECHNIQUE	THE TINTED GLASS ON THE SOUTH FACADE OF THE GREENHOUSE MINI-MIZES THE INTENSITY OF THE SUMMER SUN. IN THE WINTER, THE GREENHOUSE ACTS AS A BUFFER BETWEEN THE EXTERIOR AND THE BUILDING, AND REDUCES HEAT LOSS TO THE EXTERIOR AND PENETRATION OF COLD TO THE INTERIOR.
MATERIAL	GREY TINTED SOLAR GLASS, STEEL PIPE TRUSSES, GREENERY, EXHAUST FAN, POURED IN PLACE CONCRETE
AVAILABILITY	ADDITIONAL INFORMATION MAY BE OBTAINED BY WRITING GASSNER - NATHAN - BROWNE, 265 COURT AVE., MEMPHIS, TENNESSEE 38103.
REFERENCE	ARCHITECTURAL RECORD C & I BANK: A WEDGE FOR MEMPHIS ARCHITECTURAL RECORD, MAY 1972, PP 109-111

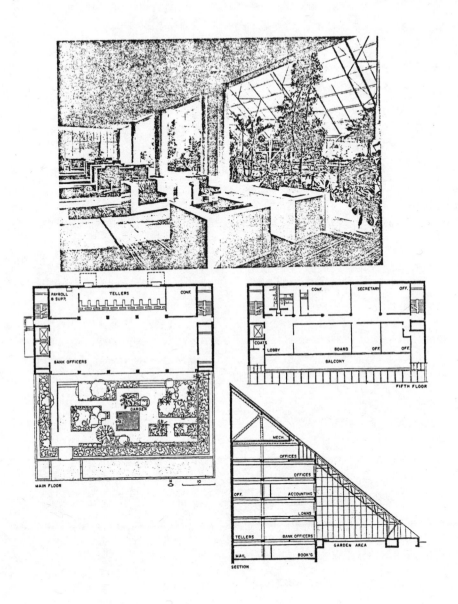

PAYROLL & SUPP.   TELLERS   CONF.

BANK OFFICERS

GARDEN

MAIN FLOOR   N   10

CONF.   SECRETARY   OFF.

COATS   LOBBY   BOARD   OFF.   OFF.

BALCONY

FIFTH FLOOR

MECH.

OFFICES

OFFICES

OFF.   ACCOUNTING

LOANS

TELLERS   BANK OFFICERS   GARDEN AREA

MAIL   BOOK'G

SECTION

TITLE	CAINE ECO HOUSE
CODE	
KEYWORDS	ANAEROBIC DIGESTER METHANE ALGAE FUEL HOUSE GRRENHOUSE TIMBER PLASTIC SHEET SOLAR WATER HEATING SYSTEM CONSERVATORY HYDROPONIC HORTICULTURE RADIATORS RAINWATER
AUTHOR	DESIGNED BY GRAHAME CAINE OF STREET FARMHOUSE, ARCHITECTS
DATE & STATUS	THE CAINE ECO HOUSE WAS BUILT DURING 1972-73.
LOCATION	ELTHAM, SOUTH LONDON, ENGLAND
SCOPE	THE CAINE ECO HOUSE IS PLANNED AS A SELF-CONTAINED WORKING SYSTEM THAT INCORPORATES PLANT AND ANIMAL LIFE IN HARMONIOUS INTERDEPENDENCE. THE 37 X 40 FOOT TIMBER AND PLASTIC DWELLING WILL INCLUDE A GARDEN THAT IS TO SUPPLY MOST OF THE HOUSEHOLD'S FOOD.

THE HOUSE IS DESIGNED TO USE ANAEROBIC DIGESTERS TO CONVERT HUMAN WASTE MATERIAL INTO METHANE. IT ALSO HAS A SOLAR WATER HEATING SYSTEM AND A CONSERVATORY WITH HYDROPONIC HORTICULTURE.

THE PURPOSE IS TO DEMONSTRATE A VISIBLE WORKING EXAMPLE OF A HOUSE THAT CAN BE BUILT CHEAPLY, BY ANYONE, WITHOUT SPECIAL SKILLS. IT COSTS UNDER $2,000 TO BUILD.

TECHNIQUE	THE WATER SOURCE FOR THE GREENHOUSE PLANTS IS THE RAIN WHICH FALLS ON THE 600 SQUARE FOOT ROOF. IN LONDON, THE RAIN WILL PROVIDE ABOUT 20 GALLONS OF WATER A DAY. SINCE THE AVERAGE AMOUNT OF WATER REQUIRED PER PERSON IS AROUND 40 GALLONS, PIPES CONNECTING TO THE CITY WATER SUPPLY ARE BEING INSTALLED AS BACKUP. BEFORE THE RAINWATER CAN BE USED, IT IS PASSED THROUGH A SAND FILTER THAT WILL REMOVE DUST AND OTHER IMPURITIES. IN THE FUTURE, MORE SOPHISTICATED DEVICES MAY BE USED TO REMOVE SOME OF THE HIGH LEAD CONTENT OF THE PRECIPITATION FROM A DIRTY URBAN SKY.

ALL THE WASTE WATER FROM THE KITCHEN AND BATHROOM, AS WELL AS ALL ORGANIC SOLID WASTES ARE PASSED ON TO AN INGENIOUS SYSTEM THAT CONVERTS THE MATERIAL TO METHANE GAS FOR COOKING FUEL AND LIQUID NUTRIENTS FOR THE GREENHOUSE. THE WASTES FIRST ENTER A TWO-COMPARTMENT DIGESTER. THE DOUBLE CONTAINER RESEMBLES AN ORDINARY SEPTIC TANK, WITH TWO IMPORTANT DIFFERENCES. FIRST, GRAHAME'S VERSION IS AIRTIGHT TO PROVIDE A HAPPY HOME FOR THE BACTERIA THAT CONVERT PART OF THE RAW SEWAGE AND GARBAGE INTO METHANE. ALSO, SINCE THE OPTIMUM TEMPERATURE FOR THE GAS-PRODUCING BUGS IS BETWEEN 70 AND 90 DEGREES FAHRENHEIT, THE RECEPTACLE IS INSULATED AND EQUIPPED WITH A SOLAR HEATING PANEL.

THE MATERIAL FROM THE LIQUID COMPARTMENT IN THE DIGESTER NEXT PASSES TO ANOTHER TANK ... THIS TIME AIRY AND SUNLIT TO ACCOMO-DATE THE TINY, FAST-MULTIPLYING ALGAE PLANTS THAT DEVOUR ORGANIC MATTER STILL IN THE WASTE. THE ALGAE ALSO ADD OXYGEN THAT HELPS BACTERIA DIGEST THE SEWAGE AND NITROGEN THAT ENRICHES THE END PRODUCT (WHICH EVENTUALLY BECOMES PLANT FOOD). ALSO, ANY DANGEROUS MICRO-ORGANISMS IN THE HUMAN EXCRETA PORTION OF THE SEWAGE SHOULD PERISH AT THIS STAGE, KILLED BY THE OXYGEN AND THE ULTRAVIOLET RADIATION OF THE SUN. SINCE THE ALGAE NEED WARMTH AS WELL AS LIGHT AND AIR, GRAHAME WILL KEEP THEM COMFORTABLE IN THE DULL, COLD LONDON WINTER BY EXTENDING PART OF THEIR TANK INTO HIS HOUSE IN THE FORM OF A LOOP. THIS INDOOR SECTION OF THE ALGAE POOL IS TO BE CONTINUOUSLY ILLUMINATED WITH ELECTRIC LIGHT (ONLY 500 FOOT CANDLES ARE NEEDED, SO ONE 40 WATT FLUORESCENT TANK SHOULD SUFFICE).

WHEN THE PLANTS HAVE DONE THEIR WORK, THE LIQUID PASSES ON – ALGAE AND ALL – TO A FINAL TANK, ANOTHER DIGESTER IN WHICH THE CONTENTS ARE KEPT AT 110 DEGREES F. BY MEANS OF A SOLAR PANEL. THIS TEMPERATURE KILLS THE ALGAE (WHICH ARE SOON REPLACED BY THE FAST GROWING COLONY IN THE OPEN POOL), AND DECOMPOSES THEM TO FURNISH MORE METHANE.

THE GAS FROM THE FIRST AND SECOND DIGESTERS COLLECTS IN THE TOP
OF THE TANKS AND IS PIPED TO GRAHAME'S KITCHEN STOVE, WHICH WILL
ALSO HAVE CYLINDERS OF STORE BOUGHT FUEL STANDING BY.
THE HEAT OF THE SUN IS TRAPPED BY AN ARRAY OF BLACK-PAINTED
HOT-WATER RADIATORS ON THE SOUTH WALL OF THE DWELLING. THE
HEATING SYSTEM IS ALSO BEING HOOKED INTO THE PUBLIC ELECTRICITY
SUPPLY AS A BACKUP SOURCE OF HEAT.

GRAHAME IS CURRENTLY USING PUBLIC-UTILITY ELECTRICITY AT THE
OUTSET TO LIGHT HIS ECO-HOUSE AND TO HELP HIS RADIATOR ARRAY ON
COLD DAYS. EVENTUALLY, HOWEVER, HE HOPES TO INSTALL A WIND
GENERATOR AND POWER-STORAGE SYSTEM TO MAKE THE HOUSE INDEPENDENT
OF THE ELECTRIC GRID AND FOSSIL FUELS.

MATERIAL   RAINWATER, SAND FILTER, STORAGE TANK, ACTIVATED CHARCOAL
                PURIFIER, ORGANIC SOLID WASTES, METHANE GAS, LIQUID NUTRIENTS FOR
                THE GREENHOUSE, TWO-COMPARTMENT DIGESTER, CYLINDERS OF STORE
                BOUGHT METHANE, 500 SQUARE FOOT GREENHOUSE, VEGETABLES, BLACK-
                PAINTED HOT-WATER RADIATORS

AVAILABILITY   REFER TO THE FOLLOWING REFERENCE FOR ADDITIONAL INFORMATION.

REFERENCE   HUGHES, F. P.
                ECO-HOUSE, THE
                MOTHER EARTH NEWS, NO. 20, PP 62-65

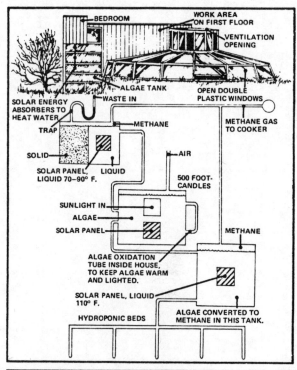

BEDROOM

WORK AREA ON FIRST FLOOR

VENTILATION OPENING

ALGAE TANK

OPEN DOUBLE PLASTIC WINDOWS

SOLAR ENERGY ABSORBERS TO HEAT WATER

WASTE IN

TRAP

METHANE

METHANE GAS TO COOKER

SOLID

AIR

SOLAR PANEL, LIQUID 70–90° F.

LIQUID

500 FOOT-CANDLES

SUNLIGHT IN

ALGAE

SOLAR PANEL

METHANE

ALGAE OXIDATION TUBE INSIDE HOUSE, TO KEEP ALGAE WARM AND LIGHTED.

SOLAR PANEL, LIQUID 110° F.

ALGAE CONVERTED TO METHANE IN THIS TANK.

HYDROPONIC BEDS

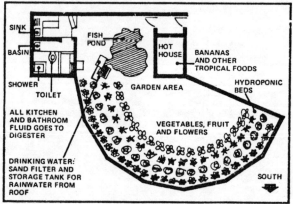

SINK

BASIN

FISH POND

HOT HOUSE

BANANAS AND OTHER TROPICAL FOODS

SHOWER

TOILET

GARDEN AREA

HYDROPONIC BEDS

ALL KITCHEN AND BATHROOM FLUID GOES TO DIGESTER

VEGETABLES, FRUIT AND FLOWERS

DRINKING WATER: SAND FILTER AND STORAGE TANK FOR RAINWATER FROM ROOF

SOUTH

TITLE	CAMBRIDGE SCHOOL SOLAR BUILDING
CODE	
KEYWORDS	SCHOOL WINDOWS NATURAL LIGHTING SOLAR HEATING PASSIVE SKYLIGHT STORAGE MASSIVE WALLS
AUTHOR	N.B. SAUNDERS; AND DAVIES, WOLF & BIBBINS
DATE & STATUS	THE CAMBRIDGE SCHOOL SOLAR BUILDING WAS COMPLETED IN APRIL 1976.
LOCATION	WESTON, MASSACHUSETTS
SCOPE	THE BUILDING IS A PARTLY NEW CONSTRUCTION (THE WEST END), AND PARTLY A REMODELING AND REBUILDING OF A PORTION THAT HAD BEEN DAMAGED BY FIRE (THE EAST END). SOLAR HEATING WAS APPLIED TO 2/3 OF THE STRUCTURE. THE LENGTH OF THE BUILDING IS 160'. THE WIDTHS OF THE THREE SEGMENTS ARE 36', 48', AND 56'. THE MAIN UPPER STORY CONTAINS A KITCHEN AND A DINING ROOM. THE LOWER STORY (BASEMENT) CONTAINS CLASSROOMS AND UTILITY ROOMS. THE TOTAL FLOOR AREA IS 14,000 SQ. FT.
TECHNIQUE	THE BUILDING FACES 27 DEGREES WEST OF SOUTH. SOME OF THE TRANSVERSE WALLS AND NORTH WALLS ARE 12" THICK AND MADE OF CONCRETE BLOCKS, HAVING ONLY 25% VOIDS. THE MASSIVE EXTERIOR WALLS ARE INSULATED ON THE EXTERIOR WITH 4" OF FIBERGLAS AND PROTECTED BY A LAYER OF STUCCO. THE MASONRY WALLS ARE REQUIRED BY THE FIRE CODE, BUT ALSO SERVE TO STORE THERMAL ENERGY. THE MAJOR PORTION OF THE VERTICAL SOUTH WALL CONSISTS OF WINDOWS FOR VIEW AND FOR THE COLLECTION OF SOLAR RADIATION. 40% OF THE WINDOW AREA IS DOUBLE GLAZED, 60% IS SINGLE GLAZED.
	THE SOLAR HEATING SYSTEM IS OF THE PASSIVE TYPE. THE COLLECTION IS VIA A 1,700 SQ. FT. SKYLIGHT SYSTEM ON THE 26 DEGREE-SLOPING SOUTH ROOF AREAS OF THE TWO WESTERN SEGMENTS OF THE BUILDING. THIS SYSTEM HAS A STAIRCASE-LIKE CROSS SECTION, WITH 18" WIDE "TREADS" AND 8-1/2" HIGH RISERS. THE RISERS ARE SINGLE GLAZED WITH DOUBLE STRENGTH GLASS. THE TREADS ARE OF SHINY ALUMINUM, AND, ESPECIALLY IN WINTER, GUIDE MUCH ADDITIONAL RADIATION TO THE TRANSPARENT RISERS, AND THEN TOWARD THE FLOORS, THE NORTH WALL, OR THE TRANSVERSE WALLS. THE SINGLE GLAZING, WHICH RESTS ON THE "STAIRCASE", IS OF CORRUGATED FILON. THE SKYLIGHT PRO-VIDES DAYLIGHT TO THE ROOMS, GREATLY REDUCING THE NEED FOR ELECTRIC LIGHTING.
	THE STORAGE IS PROVIDED BY THE BUILDING AS A WHOLE, AND ESPECI-ALLY BY THE MASSIVE TRANSVERSE WALLS AND THE MASSIVE LONGITUDI-NAL WALL STRUCK BY THE DIRECT RADIATION..
	THERE IS AN ELECTRICAL SENSING AND RECORDING SYSTEM, WHICH TAKES DATA ON TEMPERATURES AND IN SOLATION.
MATERIAL	12" THICK CONCRETE BLOCKS; CONCRETE; 4" OF FIBERGLAS; STUCCO; 40% OF THE WINDOW AREA IS DOUBLE GLAZED; 60% OF THE WINDOW AREA IS SINGLE GLAZED; SKYLIGHT; DOUBLE STRENGTH GLASS; SHINY ALUMINUM; CORRUGATED FILON (POLYESTER REINFORCED FIBERGLAS); ELECTRICAL SENSING AND RECORDING SYSTEM
AVAILABILITY	ADDITIONAL INFORMATION MAY BE OBTAINED FROM N.B. SAUNDERS, 15 ELLIS ROAD, WESTON, MASSACHUSETTS 02193. THE ERDA-WA-76-4947 REPORT MAY BE OBTAINED FROM ERDA.
REFERENCE	SHURCLIFF, W.A. SOLAR HEATED BUILDINGS, A BRIEF SURVEY CAMBRIDGE, MASSACHUSETTS, W.A. SHURCLIFF, MARCH 1976, 212 P

TITLE   CAPE COD SOLAR HOUSE

CODE

KEYWORDS   HOUSE WATER TYPE COLLECTOR STORAGE

AUTHOR   J. DEVRIES OF SOL-R-TECH

DATE & STATUS   THE HOUSE WAS COMPLETED IN LATE 1975.

LOCATION   QUECHEE, VERMONT

SCOPE   THE BUILDING IS A 2-STORY, 4-BEDROOM WOOD FRAME HOUSE WITH AN
                 ATTACHED 2-CAR GARAGE. THE LIVING ROOM IS TWO STORIES HIGH.

TECHNIQUE   THE COLLECTOR IS A 480 SQ. FT., WATER TYPE CLLECTOR, MOUNTED
                 ON THE ROOF SLOPING 45 DEGREES FROM THE HORIZONTAL. THERE
                 ARE 20 PANELS, EACH 8' X 3', MADE BY OLIN BRASS CO. AND
                 EMPLOYING A ROLL-BOND ALUMINUM SHEET WITH INTEGRAL PASSAGES
                 FOR A COOLANT. THE BLACK COATING IS NON-SELECTIVE. THE GLAZING IS
                 DOUBLE, WITH A FILON SHEET ON THE OUTSIDE AND TEDLAR FILM ON
                 THE INSIDE. THE COOLANT IS WATER THAT HAS BEEN SOFTENED. A
                 SODIUM CHROMATE INHIBITOR IS USED.

                 THE STORAGE SYSTEM CONTAINS 2,500 GALLONS OF WATER IN A 3,000-
                 GALLON RECTANGULAR, 10' X 10' X 4' HIGH, TANK OF STEEL-REINFORCED
                 CONCRETE. THE INNER SURFACE IS WATERPROOFED WITH THORO-SEAL;
                 THE OUTER SURFACE IS INSULATED WITH 2" OF STYROFOAM. TWO HEAT-
                 PUMPS ARE USED, YORK TRITON HEAT PUMPS DW-30H AND DW-40H.
                 THE ROOMS ARE HEATED BY A FAN-COIL-DUCT SYSTEM.

MATERIAL   ROLL-BOND ALUMINUM SHEET; NON-SELECTIVE BLACK COATING; DOUBLE
                 GLAZING; FILON SHEET; TEDLAR FILM; WATER; SODIUM CHROMATE
                 INHIBITOR; 2,500 GALLONS OF WATER; 3,000 GALLON RECTANGULAR TANK;
                 STEEL-REINFORCED CONCRETE; THORO-SEAL; 2" OF STYROFOAM; YORK
                 TRITON HEAT PUMPS DW-30H AND DW-40H

AVAILABILITY   ADDITIONAL INFORMATION MAY BE OBTAINED FROM SOL-R-TECH, MILL
                 ROAD, HARTFORD, VERMONT 05047.

REFERENCE   SHURCLIFF, W.A.
                 SOLAR HEATED BUILDINGS, A BRIEF SURVEY
                 CAMBRIDGE, MASSACHUSETTS, W.A. SHURCLIFF, MARCH 1976, 212 P

159a

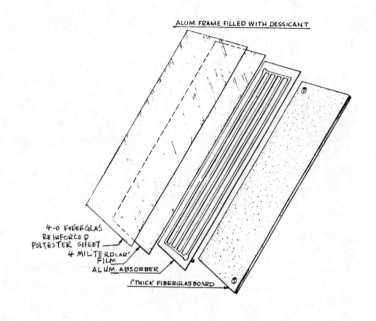

ALUM. FRAME FILLED WITH DESSICANT

4-0 FIBERGLAS REINFORCED POLYESTER SHEET

4 MIL TEDLAR FILM

ALUM. ABSORBER

1" THICK FIBERGLAS BOARD

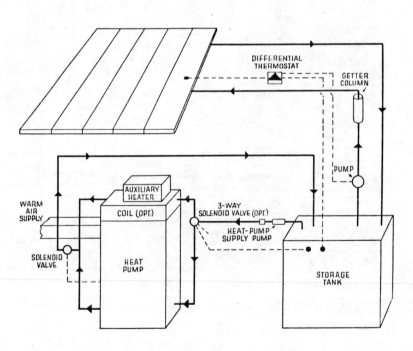

DIFFERENTIAL THERMOSTAT

GETTER COLUMN

PUMP

WARM AIR SUPPLY

AUXILIARY HEATER

COIL (OPT.)

3-WAY SOLENOID VALVE (OPT.)

HEAT-PUMP SUPPLY PUMP

SOLENOID VALVE

HEAT PUMP

STORAGE TANK

TITLE	CAPITAL ENERGY CONSERVING HOMES
CODE	
KEYWORDS	DOUBLE FRAMING INSULATION TRIPLE GLAZED WINDOWS DOORS FLOOR PLANS MINIMIZE EXTERIOR SURFACE
AUTHOR	KALEVI M. TURKIA, DIRECTOR OF RESEARCH AND DEVELOPMENT, UGI DEVELOPMENT CO.
DATE & STATUS	THE CAPITAL ENERGY CONSERVING HOMES ARE CURRENTLY AVAILABLE ON THE MARKET.
LOCATION	THE CAPITAL ENERGY CONSERVING HOMES ARE PREFABRICATED AND MAY BE CONSTRUCTED ON MANY DIFFERENT SITES.
SCOPE	CAPITAL HAS INTRODUCED THREE ENERGY-SAVING MODELS - A 1008 SQUARE FOOT HOME THAT SELLS FOR $28,200; A 1296 SQUARE FOOT UNIT PRICED AT $36,200; AND A 1600 SQUARE FOOT HOME PRICED AT $43,400. (ALL PRICES ARE EXCLUSIVE OF LAND.)
	IT IS ESTIMATED THAT THE ENERGY CONSERVING FEATURES WILL REDUCE ENERGY USE BY 33% TO 50%.
TECHNIQUE	SOME OF THE ENERGY-SAVING FEATURES ARE AN UNUSUAL "DOUBLE-FRAMING" ARRANGEMENT FOR SIDEWALLS THAT ACCOMMODATES TWO SEPARATE TRANSVERSE LAYERS OF BATT INSULATION. THE TECHNIQUE MINIMIZES HEAT CONDUCTION THROUGH 2X4 STUDS; SPECIALLY CONSTRUCTED AND SPECIALLY INSULATED CORNER JOISTS THAT CUT AIR INFILTRATION; TWO LAYERS OF CEILING INSULATION (EIGHT INCHES TOTAL), THE TOP TWO-INCH LAYER OF INSULATION COVERS ALL WOOD; SIX-INCH FLOOR INSULATION, PLUS AN INSULATION PACK BETWEEN HORIZONTAL MEMBERS AT THE PERIMETER OF THE SILL; TRIPLE-GLAZED WINDOWS; DOUBLE DOORS AT EACH ENTRY; SQUARED-OFF FLOOR PLANS THAT ALLOW FOR MAXIMUM LIVING AREA WITH THE LEAST POSSIBLE EXTERIOR WALL SURFACE.

THE "DOUBLE-FRAMING" TECHNIQUE TAKES TWO LAYERS OF INSULATION. THE FIRST LAYER - A STANDARD 3-1/2 INCH BATT - IS INSTALLED BETWEEN VERTICAL FRAMING MEMBERS, 2 X 4 STUDS PLACED 24" O.C. CREWS THEN BUILD WHAT IS ESSENTIALLY A SECOND WALL WITH 2 X 2S THAT ARE NAILED TO THE INTERIOR FACE OF THE 2 X 4S AND RUN HORIZONTALLY. A SECOND LAYER OF INSULATION - A TWO INCH BATT - IS THEN INSTALLED IN THE CAVITY BETWEEN THE 2 X 2S. THE TWO INCH BATT CRISSCROSSES THE 3-1/2 INCH BATT AND COVERS THE FACE OF THE 2 X 4S, THEREBY REDUCING HEAT LOSS THROUGH THE WOOD.

THE SPECIALLY CONSTRUCTED AND SPECIALLY INSULATED CORNER JOISTS STOP AIR INFILTRATION AT CORNER JOINTS AND STOP CONDUCTED HEAT LOSS THROUGH SIDEWALL STUDS. AT CORNER, A LAYER OF COMPRESSED FOAM INSULATION IS USED BETWEEN VERTICAL FRAMING MEMBERS (SEE DRAWING). THE STUDS ARE SECURED WITH INTERLOCKING SCREWS. INSULATION STUFFED BETWEEN THE CORNER AND EXTERIOR TRIM PROVIDES A SECOND SEAL AGAINST AIR INFILTRATION, AND MINIMIZES CONDUCTIVE LOSS.

THE TWO LAYERS OF CEILING INSULATION USE CRISSCROSSED LAYERS OF INSULATION. WORKMEN FIRST INSTALL A PRIMARY LAYER OF SIX INCH INSULATION BETWEEN THE 2 X 6 CEILING JOISTS (DRAWING). THEN A SECOND LAYER OF TWO INCH INSULATION IS LAID PERPENDICULAR TO AND ON TOP OF THE CEILING JOISTS, EFFECTIVELY COVERING THE WOOD SURFACES.

RECOGNIZING THAT AIR INFILTRATION AND HEAT LOSS AT FLOOR PERIMETERS CAUSE COLD SPOTS IN HOME INTERIORS, THE BUILDING SPECS CALL FOR INSTALLATION OF INSULATION BETWEEN HORIZONTAL MEMBERS AT THE SILL PERIMETER (DRAWING). IN ADDITION, SIX INCHES OF INSULATION IS USED BETWEEN FLOOR JOISTS.

AT ALL EXTERIOR ENTRIES, HOMES HAVE DOUBLE DOORS SEPARATED BY A TWO INCH INSULATING AIR SPACE. THE OUTER DOOR OF THIS DOUBLE-BLADED ASSEMBLY (DRAWING) IS AN INSULATED METAL DOOR. IT OPENS OUT. THE INNER DOOR IS A HOLLOW CORE, WEATHER-STRIPPED WOOD DOOR THAT CAN BE REMOVED FOR WARM WEATHER CONVENIENCE. THE INNER DOOR OPENS IN. DOOR KNOBS ON THE TWO DOORS ARE MOUNTED ONE ABOVE THE OTHER.

TRIPLE-GLAZED CASEMENT WINDOWS ARE USED. WINDOW EXPOSURE IS HELD
TO 10%, OR THE MINIMUM EXPOSURE PERMITTED BY FHA MINIMUM PROPERTY
STANDARDS. THE OUTER PANE OF THE WINDOWS CAN BE REMOVED.

MATERIAL   TWO SEPARATE TRANSVERSE LAYERS OF BATT INSULATION, 2 X 4 STUDS,
                 SPECIALLY CONSTRUCTED AND SPECIALLY INSULATED CORNER JOISTS,
                 TWO LAYERS OF CEILING INSULATION (EIGHT INCHES TOTAL), INTER-
                 LOCKING SCREWS, CEILING JOISTS, FLOOR JOISTS, INSULATED METAL
                 DOOR, HOLLOW CORE WEATHER-STRIPPED WOOD DOOR, TRIPLE-GLAZED
                 CASEMENT WINDOWS

AVAILABILITY   ADDITIONAL INFORMATION MAY BE OBTAINED FROM KALEVI M. TURKIA,
                 CAPITAL HOUSING INC., SOUTH AVIS ST., AVIS, PENNSYLVANIA.

REFERENCE   PROFESSIONAL BUILDER
                 DOUBLE FRAMING HELPS CUT ENERGY USE 33-50 PERCENT
                 PROFESSIONAL BUILDER, JULY 1976, PP 52, 56

TRIPLE PROTECTION AGAINST AIR INFLATION

One of the key energy-conservation features in the new Energy Saving House is triple-glazed casement windows. Each has an outer pane held in place by an easily loosened retainer to facilitate removal for cleaning. Window shown in open position is at the end of the living room module, with siding installed except for corner portion. Exposed portion of exterior on adjacent module (seen at right) reveals the additional horizontal layer (2") of insulation, further contributing to Capital's efficient use of energy.

CORNERING THE ENERGY SAVING MARKET

Two layers of fiberglass, with seams installed at right angles, as well as additional insulation and trim cover the innovative developed corner of the new Energy Saving House, introduced by UGI's Capital Housing Inc. This developed corner assures best possible structural integrity, provides high thermal resistance and minimizes unwanted air inside the house for maximum comfort and minimal heat loss.

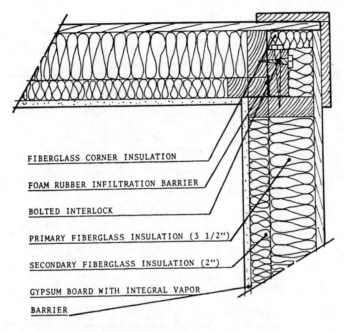

FIBERGLASS CORNER INSULATION

FOAM RUBBER INFILTRATION BARRIER

BOLTED INTERLOCK

PRIMARY FIBERGLASS INSULATION (3 1/2")

SECONDARY FIBERGLASS INSULATION (2")

GYPSUM BOARD WITH INTEGRAL VAPOR

BARRIER

UNIQUELY DEVELOPED CORNER OF CAPITOL'S

ENERGY SAVING HOUSE KEEPS OUT COLD

A uniquely developed corner joint distinguishes the Energy Saving House (ESH) introduced by UGI's Capital Housing Inc., of Avis, Pa. to prevent energy losses where traditionally built homes have no insulation. To overcome air infiltration through cracks between studs, Capital places a special layer of compressed foam insulation between vertical frame members, which are secured with bolts or lag screws to create a superior seal. In addition, the ESH uses two layers of 3½" and 2" high-resistance fiberglass insulation which are crisscrossed so that their seams are perpendicular. Finally, outside corner insulation is added before the corner trim is installed to overcome another source of heat loss and create a double layer of protection.

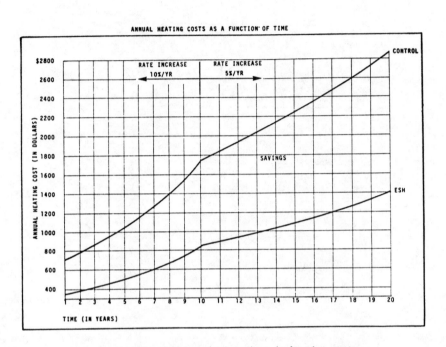

ANNUAL HEATING COSTS AS A FUNCTION OF TIME

RATE INCREASE 10%/YR   RATE INCREASE 5%/YR

CONTROL

SAVINGS

ESH

ANNUAL HEATING COST (IN DOLLARS)

TIME (IN YEARS)

Projection of theoretical savings in heating costs that can be achieved over a 20-year period by a typical Energy Saving House in comparison with a conventional home of similar design and size. The new concept was recently introduced by Capital Housing, Inc., Avis, Pa. a subsidiary of UGI Corp., the Valley Forge, Pa. utility. Calculations are based on assumption that electric rates will increase 10% in first ten years and 5% in next ten years.

TITLE   CARPENTER HOUSE

CODE

KEYWORDS   HOUSE COLLECTOR WATER TYPE INSULATION

AUTHOR   SOLAR HOMES, INC.

DATE & STATUS THE CARPENTER HOUSE WAS COMPLETED IN 1975.

LOCATION   SOUTH KINGSTOWN, RHODE ISLAND

SCOPE   THE BUILDING IS A ONE-STORY, 2-BEDROOM HOUSE. ITS OVERALL DIMENSIONS, INCLUDING A ONE-CAR GARAGE, ARE 24' X 64'. THE HEATED FLOOR AREA IS 1,200 SQ. FT. THERE IS A FULL BASEMENT AND AN ATTIC. THE BASEMENT IS HEATED IN FULL, AND CONTAINS A LABORATORY, WORKSHOP, OFFICE, UTILITY ROOM (HEAT PUMP AMD CONTROLS), AND A HEAT STORAGE BIN.

TECHNIQUE   THE WINDOWS ARE DOUBLE GLAZED. THE THICKNESS OF THE FIBERGLAS INSULATION IS 6" IN THE WALLS, AND 12" IN THE CEILING. THE BASEMENT-FLOOR CONCRETE SLAB RESTS ON A 2" LAYER OF STYRO-SM BOARD.

THE COLLECTOR IS A 360 SQ. FT., WATER TYPE COLLECTOR, MOUNTED ON THE ROOF, AND SLOPING 57 DEGREES FROM THE HORIZONTAL. THE COLLECTOR DIMENSIONS ARE 6' X 60'. THE COLLECTOR CONSISTS OF 15 PANELS, EACH 6' X 4', AND SPECIALLY MADE BY SOLAR HOMES, INC. THE HEART OF THE PANEL IS A 16-OZ. SHEET OF COPPER TO WHICH 1/2" DIAMETER COPPER TUBES, 6" APART ON CENTERS, HAVE BEEN SOLDERED. THE BLACK COATING IS OF NON-SELECTIVE, HEAT-RESISTANT RUSTOLEUM PAINT. THE GLAZING IS DOUBLE, OF 0.040" KALWALL SUN-LITE. THE BACKING CONSISTS OF 6" OF FIBERGLAS. NO ANTIFREEZE IS USED; THE WATER IS DRAINED FROM THE COLLECTOR BEFORE FREEZING CAN OCCUR. THE WATER IS CIRCULATED BY A 1/6 HP PUMP. NO HEAT EXCHANGER IS USED.

THE STORAGE SYSTEM CONSISTS OF 5,000 GALLONS OF WATER IN A POURED CONCRETE TANK WITH OUTSIDE DIMENSIONS OF 20' X 8' X 6' HIGH. THE TANK IS WATERPROOFED WITH THREE COATS OF PPG EPOXY. THE TANK WALLS ARE INSULATED ON THE EXTERIOR WITH 4" OF FIBERGLAS. THE TANK RESTS ON A 4" SLAB OF LOAD-BEARING STYRO-SM BOARD. THE TANK IS IN THE NORTHEAST CORNER OF THE BASEMENT. A VAUGHN EC-44, 44,000 BTU/H WATER-TO-AIR HEAT PUMP HEATS THE ROOMS WITH FORCED HOT AIR. IT EXTRACTS HEAT FROM THE TOP OF THE WATER TANK, WHICH IS NORMALLY IN THE RANGE OF 70 TO 100 DEGREES F AND NEVER BELOW 40 DEGREES F. THE OUTPUT TEMPERATURE OF THE HEAT PUMP IS 180 DEGREES F.

FOR COOLING IN SUMMER, THE HEAT PUMP IS OPERATED IN THE REVERSE MANNER TO EXTRACT HEAT FROM THE ROOM AIR AND DELIVER THIS HEAT TO THE DOMESTIC HOT WATER OR TO THE BIG STORAGE TANK.

MATERIAL   6" AND 12" FIBERGLAS INSULATION; CONCRETE SLAB; 2" LAYER OF STYRO-SM BOARD; 15 PANELS BY SOLAR HOMES, INC.; 16-OZ. SHEET OF COPPER; 1/2" DIAMETER COPPER TUBES, 6" APART ON CENTERS; NON-SELECTIVE, HEAT-RESISTANT RUSTOLEUM PAINT; 0.040" KALWALL SUN-LITE; 1/6 HP PUMP; 5,000 GALLONS OF WATER; POURED CONCRETE TANK; 4" OF FIBERGLAS; PPG EPOXY; 4" SLAB OF LOAD-BEARING STYRO-SM BOARD; VAUGHN EC-44, 44,000 BTU/H WATER-TO-AIR HEAT PUMP

AVAILABILITY ADDITIONAL INFORMATION MAY BE OBTAINED FROM SOLAR HOMES, INC., 2 NARRAGANSETT AVE., JAMESTOWN, RHODE ISLAND 02835.

REFERENCE   SHURCLIFF, W.A.
SOLAR HEATED BUILDINGS, A BRIEF SURVEY
CAMBRIDGE, MASSACHUSETTS, W.A. SHURCLIFF, MARCH 1976, 212 P

TITLE	CARR HOUSE
CODE	
KEYWORDS	HOUSE WATER TYPE COLLECTOR STORAGE
AUTHOR	L.H. CARR
DATE & STATUS	THE CARR HOUSE WAS RETROFITTED FOR SOLAR HEATING IN 1973.
LOCATION	SAN LUIS OBISPO, CALIFORNIA
SCOPE	THE BUILDING IS A ONE-STORY HOUSE, 20 FT. X 80 FT., WITH THE LONG AXIS EAST-WEST. THERE IS NO ATTIC OR BASEMENT. ABOUT 85% OF THE TOTAL COLLECTOR AREA IS DEVOTED TO HEATING THE BUILDING; THE REST IS FOR HEATING DOMESTIC HOT WATER. THERE IS NO COOLING IN SUMMER.
TECHNIQUE	THE FLOOR SLAB IS OF CONCRETE. IT IS 6" THICK AND RESTS ON SAND. THE WALLS HAVE LITTLE INSULATION. THE CEILING HAS 1" OF FIBER-BOARD UNDER THE COMPOSITION SHINGLES. THERE IS 300 SQ. FT. OF SINGLE GLAZED WINDOW AREA, AS WELL AS 3 DOUBLE-GLAZED, 2 FT. X 4 FT. SKYLIGHTS. THE PORCH ROOF EXTENDS 8 FT. SOUTH AND BLOCKS MUCH WINTER SUNLIGHT AT MIDDAY.

THE COLLECTOR AREA IS 800 SQ. FT. IT IS A WATER-TYPE COLLECTOR, MOUNTED 27 DEGREES FROM THE HORIZONTAL. IT IS INSTALLED 25 FT. TO THE NORTH ON A HIGH, SLOPING, BANK OF EARTH. THE COLLECTOR CONSISTS OF TWO LARGE PANELS, EACH 20 FT. X 20 FT. EACH EMPLOYS A LAYER OF NON-SELECTIVE BLACK PAINT ON A 3/4" THICK SHEET OF FIBERBOARD INSULATION BACKED BY WOODEN BOARDS. IMMEDIATELY IN FRONT OF THE BLACK FIBERBOARD, MORE OR LESS TOUCHING IT, IS AN ARRAY OF 1/2" DIAMETER BLACK POLYETHYLENE TUBING. THE ARRAY CONSISTS OF THREE SEPARATE LOOPS CONNECTED IN PARALLEL TO A 1-1/4" DIAMETER POLYETHYLENE-TUBING MANIFOLD. THE AGGREGATE LENGTH OF TUBING IS 1,800 FT. PER PANEL; I.E. 3,600 FT. IN ALL. THE TUBING IS HELD IN PLACE BY GUIDE POSTS WHICH ARE 3" LONG GALVANIZED IRON NAILS DRIVEN INTO THE BLACK FIBERBOARD. THE PANEL HAS A 3" HIGH EDGE OF WOOD. IN WINTER, A SINGLE GLAZING LAYER OF 0.006" FILM OF CLEAR POLYETHYLENE IS HELD 6" FROM THE BLACK FIBERBOARD BY A 5' X 8' MESH LATTICE OF SLENDER (1" X 2") WOODEN STRIPS AND SECURED TO THE PANEL EDGES BY STAPLES AND WOODEN BATTENS. ABOVE, THERE ARE CONFINING STRANDS OF CORD. A 1/2 HP PUMP CIRCULATES PLAIN WATER THROUGH THE TUBING AND TO THE STORAGE SYSTEM WHEN THE TEMPERATURE OF THE COLLECTOR EXCEEDS 110 DEGREES F. PRIOR TO MID-NOVEMBER AND AFTER MID-MARCH, THERE IS NO GLAZING. THIS ELIMINATES THE DANGER OF OVERHEATING THE POLYETHYLENE TUBING. COLLECTION PROCEEDS WITH MUCH LOWER EFFICIENCY AT THESE TIMES OF YEAR. EACH YEAR, IN MID-NOVEMBER, A FRESH POLYETHYLENE FILM IS INSTALLED IN 2 HOURS.

THE 6" THICK, 20' X 80' CONCRETE FLOOR SLAB, WHICH WEIGHS ABOUT 50 TONS, HAS EMBEDDED IN IT A SERPENTINE ARRAY OF 1/2" DIAMETER COPPER TUBING. THE TUBE SEGMENTS ARE 1' APART ON CENTERS. THE TOTAL LENGTH OF THE TUBING IS ABOUT 1,400 FT. IT IS DIVIDED INTO FOUR LOOPS. HOT WATER FROM THE COLLECTOR IS CIRCULATED DIRECTLY THROUGH THIS TUBING, AND THE ROOMS ARE HEATED BY RADIATION AND CONDUCTION FROM THE FLOOR. IF, IN SUNNY MILD WEATHER, THE ROOMS THREATEN TO BECOME TOO HOT, THE WATER CIRCU-LATION PUMP IS TURNED OFF MANUALLY.

MATERIAL	CONCRETE FLOOR SLAB; SAND; 1" FIBERBOARD; SINGLE GLAZED WINDOWS; DOUBLE GLAZED, 2' X 4' SKYLIGHTS; NON-SELECTIVE BLACK PAINT; FIBERBOARD INSULATION; WOODEN BOARDS; 1/2" DIAMETER BLACK POLY-ETHYLENE TUBING; 1-1/4" DIAMETER POLYETHYLENE TUBING; 3" LONG GALVANIZED IRON NAILS; 0.006" FILM OF CLEAR POLYETHYLENE; 5' X 8' MESH LATTICE; 1" X 2" WOODEN STRIPS; STAPLES; WOODEN BATTENS; 1/2 HP PUMP; PLAIN WATER
AVAILABILITY	ADDITIONAL INFORMATION MAY BE OBTAINED FROM LAURENCE H. CARR, SEE CANYON ROAD, BOX 170, RT. 1, SAN LUIS OBISPO, CALIFORNIA 93401.
REFERENCE	SHURCLIFF, W.A. SOLAR HEATED BUILDINGS, A BRIEF SURVEY CAMBRIDGE, MASSACHUSETTS, W.A. SHURCLIFF, MARCH 1976, 212 P

TITLE	CARROLL HOUSE
CODE	
KEYWORDS	HOUSE WINDOW AREA INSULATION WATER TYPE COLLECTOR STORAGE
AUTHOR	SUNSET ENGINEERING
DATE & STATUS	THE CARROLL HOUSE WAS COMPLETED IN 1975.
LOCATION	BANKS, OREGON
SCOPE	THE BUILDING IS A ONE-STORY, 3-BEDROOM, 2,100 SQ. FT., WOOD-FRAME HOUSE WITH A CRAWL SPACE, BUT NO ATTIC.
TECHNIQUE	THERE IS A SMALL AREA OF WINDOWS ON THE EAST, NORTH AND WEST. THEY ARE ALL DOUBLE GLAZED. THE TOTAL WINDOW AREA IS 370 SQ. FT. THE WALL AND CEILING INSULATION IS R-11 AND R-19. THERE IS MUCH USE OF FIBERGLAS BATTS.

THE COLLECTOR IS A 390 SQ. FT., WATER-TYPE COLLECTOR, MOUNTED ON AN ATTACHED STRUCTURE JUST TO THE SOUTH OF THE HOUSE. THE SLOPE OF THE COLLECTOR IS 60 DEGREES FROM THE HORIZONTAL. ITS DIMENSIONS ARE 33' X 13'. THE COLLECTOR INCLUDES 20 REVERE COPPER & BRASS, INC., 6-1/2' X 3' PANELS ARRANGED IN TWO ROWS OF 10 EACH. THE HEART OF THE PANEL IS A SHEET OF TYPE 110 COPPER (LAMINATED TO PLYWOOD) WITH A NON-SELECTIVE BLACK COATING. RECTANGULAR TUBES OF TYPE 122 COPPER ARE ATTACHED TO THE SHEET AT 6" INTERVALS BY MEANS OF CLIPS AND ADHESIVE. THE GLAZING CONSISTS OF A SINGLE LAYER OF PLATE GLASS. THE BACKING INCLUDES FOAM INSULATION. THE COOLANT IS WATER WITH NO ANTIFREEZE, AS THE SYSTEM IS DRAINED BEFORE FREEZE-UP CAN OCCUR. A RHO-SIGMA DIFFERENTIAL THERMOSTAT IS USED. THE WATER IS CIRCULATED DIRECTLY TO THE STORAGE SYSTEM AT 20 GPM.

THE STORAGE SYSTEM HAS 6,200 GALLONS OF WATER. THE WATER IS IN FIVE 1,250 GALLON VERTICAL CYLINDRICAL FIBERGLAS TANKS MADE BY THE HOFFMAN FIBERGLAS CO. THE TANK HEIGHT AND DIAMETER ARE 6-1/2' AND 6', RESPECTIVELY. EACH TANK IS INSULATED ON THE EXTERIOR WITH 2" FOAM. THE TANKS ARE IN AN INSULATED HOUSING BETWEEN THE COLLECTOR AND THE HOUSE. WHEN THE RESIDENCE NEEDS HEAT, A HEAT PUMP COMES INTO PLAY TO EXTRACT HEAT FROM THESE FIVE TANKS, OR FROM THE AUXILIARY-SYSTEM TANK, AND DELIVERS IT TO A FORCED-AIR DUCT SYSTEM. THE HEAT PUMP IS A SINGER, WATER-TO-AIR, TYPE CC520 PUMP WITH A NOMINAL CAPACITY OF 4 TONS COOLING.

THERE IS ALSO ONE ADDITIONAL 1,250 GALLON TANK THAT IS HEATED DURING "OFF-PEAK" HOURS BY AN 18 KW SIDEARM WATER HEATER. THIS PROVIDES THE AUXILIARY ENERGY USED TO SUPPLEMENT THE SOLAR CONTRIBUTION AS NEEDED. A 20 KW DUCT HEATER IS INSTALLED FOR EMERGENCY HEAT.

MATERIAL	DOUBLE GLAZED WINDOWS; FIBERGLAS BATTS; TYPE 110 COPPER (LAMINATED TO PLYWOOD); NON-SELECTIVE BLACK COATING; RECTANGULAR TUBES OF TYPE 122 COPPER; SINGLE LAYER OF PLATE GLASS; FOAM INSULATION; WATER; RHO-SIGMA DIFFERENTIAL THERMOSTAT; SIX 1,250 GALLON VERTICAL CYLINDRICAL FIBERGLAS TANKS; SINGER, WATER-TO-AIR, TYPE CC520 HEAT PUMP; 18 KW SIDEARM WATER HEATER; 20 KW DUCT HEATER
AVAILABILITY	ADDITIONAL INFORMATION MAY BE OBTAINED FROM JACK CARROLL, SUNSET ENGINEERING, 2944 SE POWELL BLVD., PORTLAND, OREGON 97202.
REFERENCE	SHURCLIFF, W.A. SOLAR HEATED BUILDINGS, A BRIEF SURVEY CAMBRIDGE, MASSACHUSETTS, W.A. SHURCLIFF, MARCH 1976, 212 P

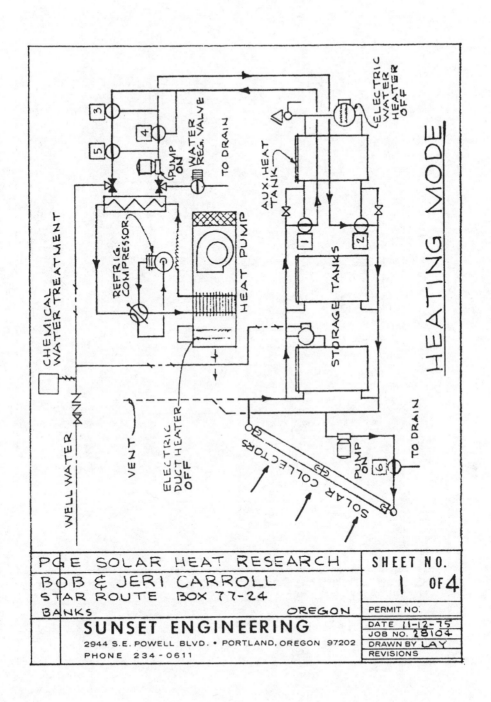

HEATING MODE

CHEMICAL WATER TREATMENT

WELL WATER

VENT

ELECTRIC DUCT HEATER OFF

REFRIG. COMPRESSOR

HEAT PUMP

PUMP ON

WATER REG. VALVE

TO DRAIN

AUX. HEAT TANK

ELECTRIC WATER HEATER OFF

STORAGE TANKS

SOLAR COLLECTORS

PUMP ON

TO DRAIN

3

5

4

1

2

6

PGE SOLAR HEAT RESEARCH	SHEET NO.
BOB & JERI CARROLL STAR ROUTE BOX 77-24 BANKS OREGON	1 OF 4

PERMIT NO.

SUNSET ENGINEERING
2944 S.E. POWELL BLVD. • PORTLAND, OREGON 97202
PHONE 234-0611

DATE 11-12-75
JOB NO. 25104
DRAWN BY LAY
REVISIONS

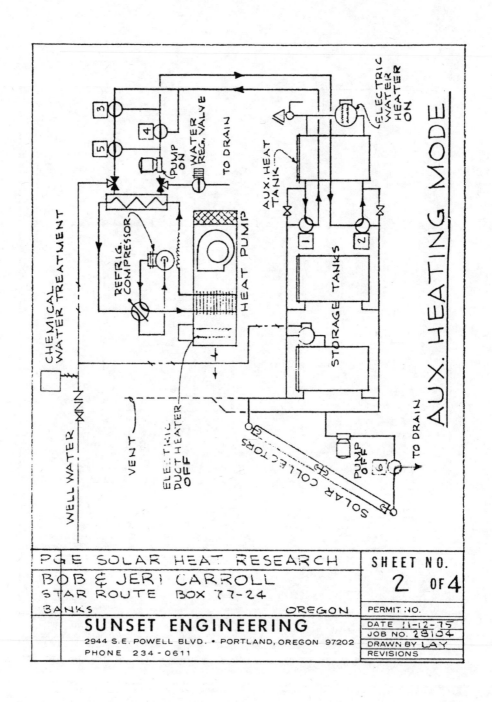

AUX. HEATING MODE

ELECTRIC WATER HEATER ON

AUX. HEAT TANK

STORAGE TANKS

ELECTRIC WATER HEATER ON

3

4

5

PUMP ON

WATER REG. VALVE

TO DRAIN

REFRIG. COMPRESSOR

HEAT PUMP

CHEMICAL WATER TREATMENT

WELL WATER

VENT

ELECTRIC DUCT HEATER OFF

SOLAR COLLECTORS

PUMP OFF

6

TO DRAIN

1

2

PGE SOLAR HEAT RESEARCH

BOB & JERI CARROLL
STAR ROUTE BOX 77-24

BANKS   OREGON

SUNSET ENGINEERING

2944 S.E. POWELL BLVD. • PORTLAND, OREGON 97202

PHONE 234-0611

SHEET NO.

2 OF 4

PERMIT NO.

DATE 11-12-75
JOB NO. 29104
DRAWN BY LAY
REVISIONS

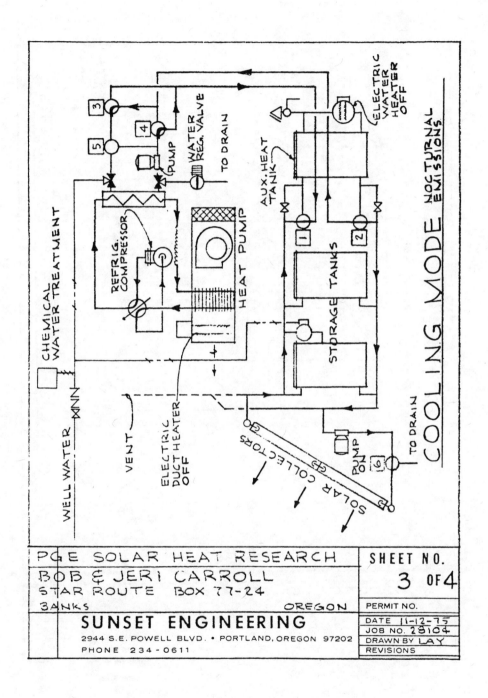

CHEMICAL WATER TREATMENT

WELL WATER

VENT

ELECTRIC DUCT HEATER OFF

REFRIG. COMPRESSOR

HEAT PUMP

3

4

5

PUMP

WATER REG. VALVE

TO DRAIN

AUX. HEAT TANK

ELECTRIC WATER HEATER OFF

1

2

STORAGE TANKS

SOLAR COLLECTORS

PUMP ON

6

TO DRAIN

COOLING MODE NOCTURNAL EMISSIONS

PGE SOLAR HEAT RESEARCH	SHEET NO.
BOB & JERI CARROLL	3 OF 4
STAR ROUTE BOX 77-24	
BANKS OREGON	PERMIT NO.

SUNSET ENGINEERING

2944 S.E. POWELL BLVD. • PORTLAND, OREGON 97202

PHONE 234 - 0611

DATE 11-12-75	
JOB NO. 28104	
DRAWN BY LAY	
REVISIONS	

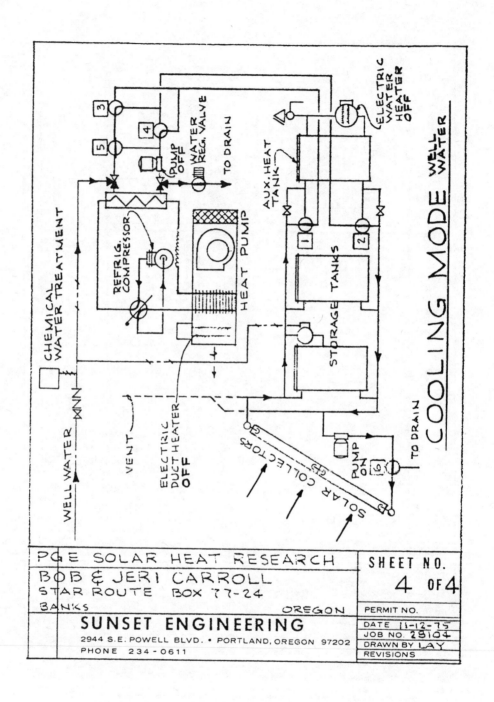

COOLING MODE

WELL WATER

ELECTRIC WATER HEATER OFF

AUX. HEAT TANK

STORAGE TANKS

ELECTRIC WATER HEATER OFF

WATER REG. VALVE

TO DRAIN

PUMP OFF

REFRIG. COMPRESSOR

HEAT PUMP

CHEMICAL WATER TREATMENT

WELL WATER

VENT

ELECTRIC DUCT HEATER OFF

SOLAR COLLECTORS

PUMP ON

TO DRAIN

PGE SOLAR HEAT RESEARCH

BOB & JERI CARROLL
STAR ROUTE BOX 77-24
BANKS   OREGON

SUNSET ENGINEERING

2944 S.E. POWELL BLVD. • PORTLAND, OREGON 97202
PHONE 234-0611

SHEET NO.
4 OF 4

PERMIT NO.

DATE 11-12-75
JOB NO. 29104
DRAWN BY LAY
REVISIONS

TITLE   CARY ARBORETUM ADMINISTRATION BUILDING

CODE

KEYWORDS   ADMINISTRATION BUILDING ARBORETUM SOLAR PANEL COLLECTOR POOL
                 RAINWATER HEAT GREENHOUSE SOD INSULATION

AUTHOR   MALCOLM WELLS AND FRED DUBIN

DATE & STATUS   THIS BUILDING WAS DESIGNED IN EARLY 1974 AND COMPLETED DURING
                 1976.

LOCATION   MILLBROOK, NEW YORK

SCOPE   THIS BUILDING IS A LARGE ADMINISTRATIVE CENTER TO BE LOCATED
                 ON THE GROUNDS OF THE CARY ARBORETUM. THERE WILL BE APPROXI-
                 MATELY 6,000 SQ. FT. OF SOLAR PANELS LOCATED ON THE ROOF OF
                 THIS 2-STORY, 35,000 SQ. FT. STRUCTURE. PEAK LOSS AND HEAT
                 GAIN FOR THE BUILDING WILL BE APPROXIMATELY 50% LESS THAN
                 CORRESPONDING STANDARD PRACTICE BUILDINGS, AND WATER CONSUMPTION
                 WILL BE REDUCED BY THE SAME AMOUNT. THE TOTAL AMOUNT OF ENERGY
                 USED BY THE FACILITY SHOULD BE UNDER 50,000 BTUS PER SQ. FT. PER
                 YEAR FOR ALL SERVICES, INCLUDING THE ELECTRICAL REQUIREMENTS.

TECHNIQUE   THIS BUILDING WILL USE EXTENSIVE ARRAYS OF SOLAR PANELS,
                 ARRANGED IN A SAWTOOTH PATTERN, FOR MOST OF ITS HEATING NEEDS.
                 A POOL IS INCLUDED ON THE SITE TO COLLECT RAINWATER FOR RECIRCU-
                 LATION. INSULATED TANKS ARE LOCATED UNDERGROUND FOR STORAGE OF
                 HEATED WATER. SOD IS BANKED AGAINST THE COLDEST SIDES OF THE
                 BUILDING FOR INSULATION. CARBON DIOXIDE FROM OFFICES WILL BE
                 RECIRCULATED THROUGH AN EXPERIMENTAL GREENHOUSE ATTACHED TO
                 THE BUILDING. DEEP WELLS ON THE PROPERTY WILL PROVIDE COOLING
                 DIRECTLY AND A SOURCE OF CONDENSOR WATER COOLING FOR THE SMALL
                 AMOUNT OF REFRIGERATION NEEDED FOR AIR CONDITIONING. THE
                 BUILDING WILL BE EQUIPPED WITH DOUBLE BUNDLE CONDENSOR HEAT
                 PUMPS WITH THE SOLAR COLLECTORS PROVIDING THE MAJORITY OF HEAT
                 REQUIRED.

                 THE WELL WATER WILL BE USED AS A SUPPLEMENTARY HEAT SOURCE FOR
                 THE HEAT PUMPS UNDER THE MOST SEVERE WINTER WEATHER CONDITIONS
                 WHEN THE SOLAR COLLECTORS CANNOT PROVIDE A HEAT SOURCE. THERE
                 WILL BE THREE STORAGE TANKS BELOW GRADE CONSTRUCTED OF CONCRETE
                 AND INSULATED ON THE EXTERIOR WITH URETHANE. THESE TANKS WILL
                 STORE HOT WATER FROM THE SOLAR COLLECTORS AND PROVIDE A HEAT
                 SOURCE FOR THE HEAT PUMPS IN THE WINTER TIME.

                 THE ENTIRE SYSTEM WILL BE MONITORED AND TESTED BY COMPUTER.

MATERIAL   SOLAR PANELS, STORAGE TANKS, INSULATED POOL, SOD, GREENHOUSE

AVAILABILITY   ADDITIONAL INFORMATION MAY BE OBTAINED FROM DUBIN/BLOOME ASSO-
                 CIATES, CONSULTING ENGINEERS, 42 W. 39TH STREET, NEW YORK,
                 NEW YORK 10018.

REFERENCE   BOOKMAN, GEORGE B.
                 BUILDING AN ARBORETUM: THE CARY CAMPUS AFTER THREE YEARS
                 GARDEN JOURNAL, AUGUST 1974, VOL. 24, NO. 4, 5 P

                 DUBIN/BLOOME ASSOCIATES
                 SCOPE OF CURRENT ENERGY CONSERVATION AND SOLAR ENERGY PROJECTS
                 NEW YORK, DUBIN/BLOOME ASSOCIATES PAMPHLET, MAY 1976, 12 P

TITLE	CASA DEL SOL OF THE FUTURE
CODE	
KEYWORDS	SOLAR DEMONSTRATION HOUSE ORIENTATION WATER TYPE COLLECTOR
AUTHOR	DEAN & HUNT ASSOCIATES
DATE & STATUS	THE CASA DEL SOL OF THE FUTURE WAS SCHEDULED FOR COMPLETION IN LATE 1976.
LOCATION	LAS CRUCES, NEW MEXICO
SCOPE	THE BUILDING IS A ONE-STORY, 2,000 SQ. FT., 3-BEDROOM HOUSE. THE COST IS SLIGHTLY GREATER THAN $100,000.
TECHNIQUE	THE WALLS INCLUDE 8" OF MASONRY WITH 2" OF POLYURETHANE FOAM SPRAYED ON. THE EXTERIOR FINISH IS STUCCO. THE ROOF INCLUDES 3" OF POLYURETHANE FOAM. THE WINDOWS ARE DOUBLE GLAZED. THE HOUSE FACES EXACTLY SOUTH.
	THE COLLECTOR IS AN 800 SQ. FT., WATER TYPE COLLECTOR, MOUNTED ON THE LARGE SOUTH FACE WHICH SLOPES 30 DEGREES FROM THE HORIZONTAL. THE COLLECTOR INCLUDES 46 PANELS. THESE WERE MADE BY REVERE COPPER & BRASS CO. THE PANELS HAVE A SELECTIVE COATING BY ENTHONE.
	THE STORAGE SYSTEM CONTAINS 2,000 GALLONS OF WATER IN A STEEL TANK COATED ON THE INTERIOR WITH EPOXY RESIN. THE HEAT IS DELIVERED TO THE TANK VIA A HEAT EXCHANGER. THE ROOMS ARE HEATED BY A FAN-COIL SYSTEM. A LIBR ABSORPTION SYSTEM (ARKLA, 3 TON COOLER) IS USED FOR SUMMER COOLING.
MATERIAL	8" OF MASONRY; 2" OF POLYURETHANE FOAM; DOUBLE GLAZED WINDOWS; 46 REVERE COPPER & BRASS CO. PANELS; SELECTIVE COATING BY ENTHONE; 2,000 GALLONS OF WATER; STEEL TANK; EPOXY RESIN; HEAT EXCHANGER; FAN-COIL; LIBR ABSORPTION SYSTEM (ARKLA, 3 TON COOLER)
AVAILABILITY	ADDITIONAL INFORMATION MAY BE OBTAINED FROM DEAN & HUNT ASSOCIA- TES, 210 LA VETA N.E., ALBUQUERQUE, NEW MEXICO 87108.
REFERENCE	SHURCLIFF, W.A. SOLAR HEATED BUILDINGS, A BRIEF SURVEY CAMBRIDGE, MASSACHUSETTS, W.A. SHURCLIFF, MARCH 1976, 212 P

179

180

TITLE	CENTER FOR APPLIED COMPUTER RESEARCH AND PROGRAMMING
CODE	
KEYWORDS	LAGOON WATER BASIN BARGE WALL SOLAR COLLECTOR ELECTRIC HELIO-VOLTAIC BATTERY COMPUTER CENTER FLOAT REST CLOUD EVAPORATION NOZZLE WINDMILL CONICAL TANK IRRIGATE PLANTING
AUTHOR	EMILIO AMBASZ
DATE & STATUS	THE BUILDING WAS CONCEIVED IN 1975, BUT HAS NOT YET BEEN BUILT.
LOCATION	MEXICO CITY, MEXICO
SCOPE	THE CLIENT NEEDS IN THIS PROJECT STIPULATED THE NEED FOR A FLEXIBLE PLAN, WHICH COULD BE CHANGED ACCORDING TO INTRINSIC REQUIREMENTS OF PROJECTS AT HAND.
TECHNIQUE	THE BUILDING IS SITUATED ON A LANDFILL LAGOON. INCORPORATED INTO THE PLAN IS AN EXTENSIVE WATER BASIN, WHICH IS CONSTANTLY DRAINING THE LAGOON TO SOLVE FOUNDATION PROBLEMS. THE OFFICE SPACES ARE DESIGNED AS BARGES THAT FLOAT UNTIL POSITIONED WHERE NEEDED. ONCE PLACED, A WATERTIGHT COMPARTMENT IS FILLED, AND THE BARGE COMES TO REST ON THE BASIN FLOOR. TWO LARGE WALLS SURROUND THE SITE, ONE OF WHICH IS INCLINED AND ACTS AS A SOLAR ENERGY GATHERING SURFACE. ELECTRIC ENERGY IS OBTAINED THROUGH A HELIOVOLTAIC PROCESS, AND STORED IN BATTERIES AT THE POWER PLANT TO SUPPLY POWER TO THE COMPUTER CENTER.
	PART OF THE CENTER IS HOUSED UNDER A HUGE PLATFORM, OVER WHICH HOVERS A CLOUD, CONCEIVED AS AN ARCHITECTURAL ELEMENT, WHICH SERVES TO REPLENISH THE BASIN'S EVAPORATED WATER AND TO COOL THE BUILDING'S EXTERIOR. THIS EVER CHANGING MASS OF COLD WATER MIST IS GENERATED BY HIGHLY PRESSURIZED WATER AS IT IS EXPELLED THROUGH ESPECIALLY DESIGNED NOZZLES. THE CLOUD IS FED BY WATER, PUMPED BY WINDMILLS AND STORED IN CONICAL TANKS THAT ARE TRADITIONALLY USED IN THE REGION AS SILOS. THIS WATER IS ALSO USED TO SATISFY ALL OF THE DEVELOPMENT'S WATER NEEDS, AND TO IRRIGATE ITS EXTENSIVE PLANTINGS.
MATERIAL	LAGOON, SOLAR ENERGY GATHERING SURFACE, WINDMILLS, WATER, CLOUD PRODUCING NOZZLE SYSTEM, EXTENSIVE PLANTINGS
AVAILABILITY	FOR FURTHER INFORMATION SEE REFERENCE BELOW.
REFERENCE	AMBASZ, EMILIO ULTIMATELY, A FLOWER BARGE PROGRESSIVE ARCHITECTURE, MARCH 1975, PP 76-79

Solar energy-gathering wall supplies electric power to center and some to community; windmills supply water to basin and to "cooling" cloud.

Offices barges can be rearranged or taken to dry dock (red square, facing page) as needed; they are connected by pedestrian-mechanical tubes.

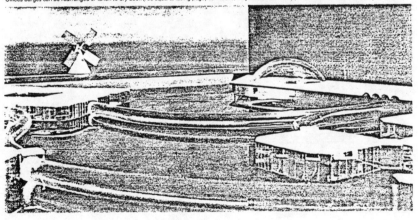

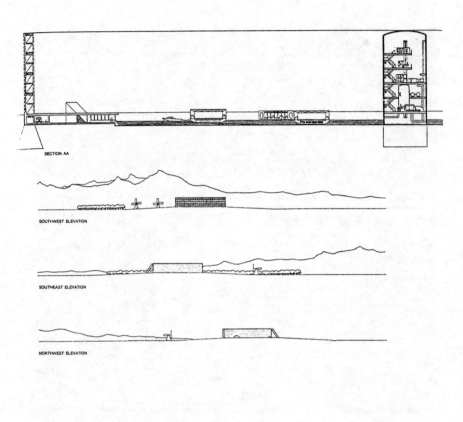

SECTION AA

SOUTHWEST ELEVATION

SOUTHEAST ELEVATION

NORTHWEST ELEVATION

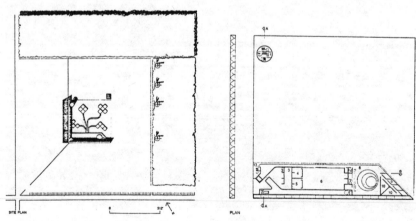

SITE PLAN

0   512'

PLAN

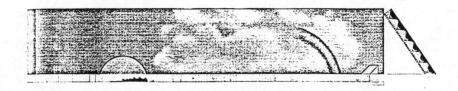

SECTION BB   Section (above) and drawings (below) show entrance where two walls meet.

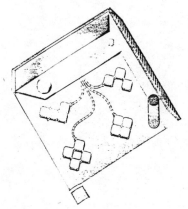
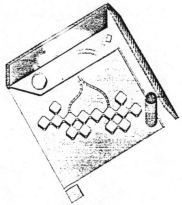

When offices are no longer needed, facility will become community park.

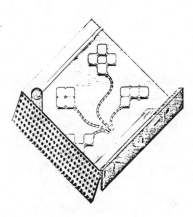
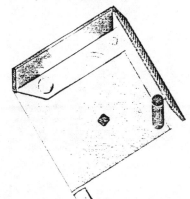

TITLE   CHARLESTON SOLAR HOME

CODE

KEYWORDS   L SHAPE AIR MOVEMENT TEMPERATURE HUMIDITY OVERHANG NOTCHED
               CORNER DEFLECT NEGATIVE PRESSURE ORIENTATION SUN EQUINOX
               COLLECTOR TILT STORAGE SPACE COOL THERMOSIPHON HOME

AUTHOR   DANIEL SCULLY

DATE & STATUS THE CHARLESTON SOLAR HOME WAS DESIGNED AS THE PROTOTYPE FOR
               A HOME WHICH CAN BE BUILT IN THE CHARLESTON, SOUTH CAROLINA
               AREA. IT WAS DESIGNED IN 1975, BUT IT HAS NOT YET BEEN BUILT.

LOCATION   CHARLESTON, SOUTH CAROLINA

SCOPE   THE TOTAL HEAT LOSS FOR THE CHARLESTON SOLAR HOUSE IS 16,893 BTU/
               DD OR 11.7 BTU/SQUARE FEET/DEGREE DAY. THE TOTAL NUMBER OF BTU'S
               LOST IN A HEATING SEASON IS 32,434,600. THE AMOUNT OF SOLAR HEAT
               COLLECTED BY THE SOLAR COLLECTION SYSTEM IS 13,250,000. THE
               AMOUNT OF HEAT GAINED THROUGH THE WINDOWS IS 15,000,000; OF
               THIS ONLY 13,500,000 IS USABLE. ADDITIONAL HEAT PROVIDED BY
               THE AUXILIARY OFF-PEAK HEATING SYSTEM IS 5,700,000 BTU/SEASON.
               ENERGY CONSUMPTION FOR HEATING THE HOUSE IS 1,700 KWH/SEASON.

TECHNIQUE   IN CHARLESTON, SOUTH CAROLINA, THE RELATIVE HUMIDITY DURING THE
               SUMMER MONTHS AVERAGES 80%. A 500 FT/MINUTE (6 MPH) AIR
               MOVEMENT IS NECESSARY TO MAKE THIS COMBINATION OF HIGH TEMPE-
               RATURE AND HUMIDITY COMFORTABLE. THE MONTHLY WIND SPEEDS
               AVERAGE ABOUT 10 MPH FROM THE SOUTHWEST. THE HOUSE IS DESIGNED
               TO TAKE MAXIMUM ADVANTAGE OF THESE WINDS. THE HOUSE IS INTENDED
               TO BE TRANSPARENT TO BREEZES FROM THE SOUTHWEST. IN ADDITION,
               BOTH THE "L" SHAPE OF THE PLAN AND THE DESIGN OF THE BEDROOM
               WING WALL SECTION ARE INTENDED TO COLLECT AND FOCUS WIND
               THROUGH THE HOUSE. THE ROOF OVERHANG PREVENTS THE WIND FROM
               BEING DEFLECTED UP OVER THE HOUSE. ALSO, THE EXTENDED WALLS
               TO THE NORTH AND SOUTH DEFLECT WINDS FROM THOSE DIRECTIONS
               INTO THE HOUSE. IN SEPTEMBER AND OCTOBER THE RELATIVE HUMIDITY
               STILL AVERAGES 80%, BUT THE WIND REVERSES ITSELF AND COMES FROM
               THE NORTHEAST. THE NOTCHED NORTHEAST CORNER OF THE HOUSE IS
               DESIGNED TO CAPTURE THESE BREEZES AND DEFLECT THEM THROUGHOUT
               THE HOUSE. ALSO THE NEGATIVE AIR PRESSURE CREATED ON THE
               NORTH SIDE OF THE HOUSE BY THE SUMMER BREEZES BLOWING UP
               AND OVER THE SOLAR COLLECTORS IS USED TO DRAW AIR UP THREE
               DUCTS AND OUT OF THE BATHROOM AND BEDROOMS. IF THE HOUSE WERE
               TO BE DESIGNED FOR ECONOMICAL AIR CONDITIONING, IT WOULD
               HAVE TO BE TIGHTENED UP, HAVE A MINIMUM SURFACE AREA, WHICH
               WOULD MAKE MORE DIFFICULT THE USE OF NATURAL VENTILATION WITH
               ITS BENEFITS.

               THE "L" SHAPE OF THE HOUSE, AND ITS ORIENTATION, ALLOWS MOST
               OF THE WINTER SUN AND MUCH LESS OF THE SUMMER SUN TO PENETRATE
               THE HOUSE. IN THE SUMMER, THE SUN SHINES ON THE SOUTH SIDE OF
               THE BUILDING FROM ABOUT 10 A.M. TO 1 P.M., WHILE IN THE WINTER,
               IT SHINES ON THIS SURFACE ALL DAY. AT THE FALL AND SPRING
               EQUINOXES, THE SOUTH WALLS ARE SHADED FROM ALL LATE AFTERNOON
               SUN. THE INTENTION IS TO MAXIMIZE HEAT GAIN IN THE WINTER AND
               MINIMIZE IT IN THE SUMMER.

               SINCE THE COLLECTOR IS USED FOR BOTH WINTER HEATING AND
               SUMMER COOLING, A RELATIVELY LOW COLLECTOR TILT ANGLE OF
               40 DEGREES IS USED. THE SYSTEM OPERATES AS FOLLOWS: AIR IS
               DRAWN INTO THE COLLECTOR FEED MANIFOLD AT THE BOTTOM OF THE
               COLLECTOR FROM THE BOTTOM OF BOTH THE STORAGE SPACES. COOL AIR
               FROM THE BOTTOM OF STORAGE TERMOSIPHONS UP THROUGH THE COLLEC-
               TOR. THIS AIR MOVEMENT IS AIDED BY A FAN BELOW THE RETURN
               MANIFOLD IN THE SINGLE DOWN FEED DUCT. ONLY ONE FAN IS USED
               FOR THE ENTIRE CLOSED HEAT COLLECTION SYSTEM. FROM THE FAN,
               'HOT' AIR IS DUCTED BACK INTO THE TOPS OF THE TWO STORAGE
               CHAMBERS. THE ENTIRE SYSTEM IS A CLOSED LOOP SYSTEM AND THE AIR
               WILL CONTINUE DOWN THROUGH THE STORAGE AND BACK UP THROUGH
               THE COLLECTOR AGAIN (PICKING UP MORE HEAT) BY THE COMBINED
               ACTION OF VERTICAL THERMOSIPHONING AND THE FAN. WATER USED
               AS THERMAL STORAGE IS HELD IN STACKED 20 GALLON MOLDED PLASTIC

CONTAINERS. THESE ARE DESIGNED TO INTERLOCK TOGETHER FORMING
ZIG-ZAG DIAGONAL AIR DUCTS THAT RUN FROM THE TOP TO THE BOTTOM
OF THE CHAMBER. EACH 20 GALLON CONTAINER CAN STORE 6,400 BTU
OVER THE 40 F TEMPERATURE GRADIENT EXPECTED IN THE THERMAL
STORAGE (75-115 DEGREES).

HOT AIR IS DISTRIBUTED TO THE HOUSE FROM THE TOP OF EACH
STORAGE ENCLOSURE. ONE DUCT FROM EACH ENCLOSURE REQUIRES ONE
DISTRIBUTION FAN PER STORAGE AREA. ALL HEATED AREAS ARE SUPPLIED
FROM THESE DISTRIBUTION DUCTS. THE BEDROOMS AND BATHROOM ARE
HEATED BY ONE STORAGE WALL, AND THE LIVING AND KITCHEN AREAS
BY THE OTHER. HEAT ENTERS AT THE CEILING IN EACH ROOM, AND AS
IT COOLS, FALLS DOWN AND INTO THE RETURNS WHICH ARE MOUNTED
IN THE FLOOR.

POSITIVE SUMMER VENTILATION IS PROVIDED BY SHORT CIRCUITING
THE HEATING SYSTEM AND USING IT PARTIALLY IN REVERSE. FOR
SUMMER COOLING, VENTS ARE OPENED TO THE OUTSIDE AT THE TOP OF
THE COLLECTOR FEED MANIFOLD. AS THE COLLECTOR OVERHEATS IN
SUMMER, HOT AIR RISES UP AND OUT THE OPEN VENT. THIS CHIMNEY
EFFECT IN TURN DRAWS HOT AIR FROM THE CEILING OF ALL THE ROOMS
AND THROUGH THE HEATING DUCTWORK, EXHAUSTING IT OUT THE TOP OF
THE COLLECTOR, THROUGH THE VENTS.

MATERIAL   RETURN MANIFOLD, FAN, STORAGE, COLLECTOR FEED, COLLECTOR,
                20 GALLON MOLDED PLASTIC CONTAINERS, BACK-UP ELECTRIC HEATER
                WITH FAN

AVAILABILITY   ADDITIONAL INFORMATION ON THE CHARLESTON SOLAR HOME MAY BE
                OBTAINED BY WRITING TO TOTAL ENVIRONMENTAL ACTION, HARRISVILLE,
                NEW HAMPSHIRE 03450.

REFERENCE   TOTAL ENVIRONMENTAL ACTION
                CHARLESTON SOLAR HOME
                HARRISVILLE, NEW HAMPSHIRE, IN SOLAR ENERGY HOME DESIGN IN
                FOUR CLIMATES, PUBLISHED BY TOTAL ENVIRONMENTAL ACTION, APRIL
                1975, PP 117-143

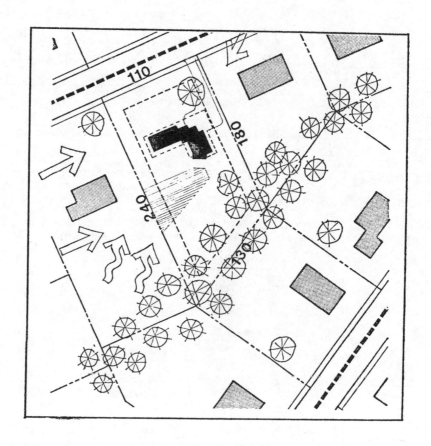

SITE PLAN OF CHARLESTON SOLAR HOME

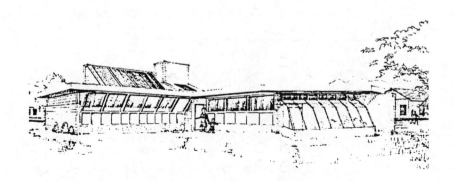

DESIGN INFORMATION - 1

Region: III   CHARLESTON, S.C. AREA   Climate: HOT-HUMID   Latitude: 32°-5' N. LAT.
Regions with similar climate characteristics: GULF COAST AREA - New Orleans, LA. - 30°-0' N. LAT.

CLIMATIC DATA

MONTH OF YEAR	HEATING DEGREE DAYS	COOLING DEGREE DAYS	% POSS. SUNSHINE	SUNSHINE HOURS	AVERAGE TEMPERATURE			WIND		RELATIVE HUMIDITY
					HIGH °F	DAY °F	NIGHT °F	SPEED mph	FROM dir.	%
	DD/MO	DD/MO	Decimal	HRS/MO	°F	°F	°F	mph	dir.	%
AUG	0		.66	281	80	84.5	75.5	8	SW	80
SEP	0		.67	244	76	81.5	71	8	NNE	80
OCT	59		.68	239	66	71.5	60.5	8	NNE	80
NOV	282		.68	210	56	62	50	8	NNE	75
DEC	471		.57	187	50	55.5	44.5	9	SW	79
JAN	487		.58	188	50	55.5	44	10	WSW	75
FEB	389		.60	189	52	57.5	46	11	NNE	71
MAR	291		.65	243	57	62.5	51	11	SSW	70
APR	54		.72	284	65	71	59	11	SSW	70
MAY	0		.73	323	73	78.5	67.5	9	S	71
JUN	0		.70	308	79	84	74	9	SW	72
JUL	0		.66	297	81	85	66.5	8	SW	80
TOTALS	2033		.66	2993	66	69.1	59.2	9	SW	75

(side note in Cooling Degree Days column: NO COOLING DEGREE DAY DATA IS AVAILABLE FROM N.A.H.B. INSULATION MANUAL GIVES 1250 HRS. ABOVE 80° F AT AVERAGE ANNUALLY.)

(left margin: HEATING SEASON spanning NOV through APR)

DD indicates Degree Days

Climatic Data for the Charleston Solar Home Design.

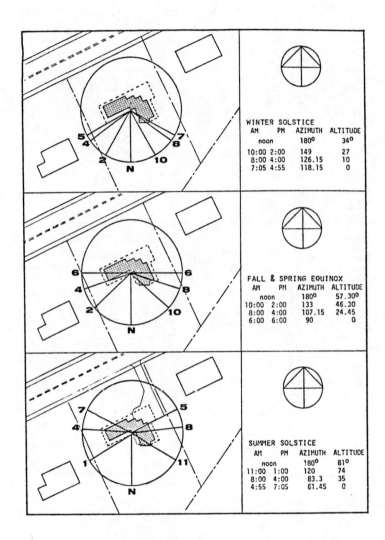

WINTER SOLSTICE

AM	PM	AZIMUTH	ALTITUDE
noon		180°	34°
10:00	2:00	149	27
8:00	4:00	126.15	10
7:05	4:55	118.15	0

FALL & SPRING EQUINOX

AM	PM	AZIMUTH	ALTITUDE
noon		180°	57.30°
10:00	2:00	133	46.30
8:00	4:00	107.15	24.45
6:00	6:00	90	0

SUMMER SOLSTICE

AM	PM	AZIMUTH	ALTITUDE
noon		180°	81°
11:00	1:00	120	74
8:00	4:00	83.3	35
4:55	7:05	61.45	0

SOLAR ANGLES SUPERIMPOSED ON PLAN OF CHARLESTON SOLAR HOME

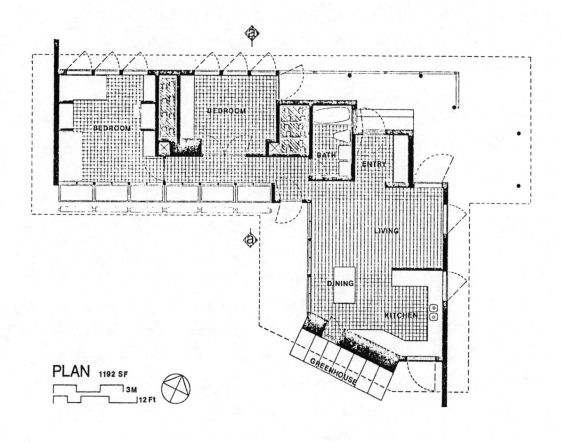

PLAN 1192 SF

3M

12 Ft

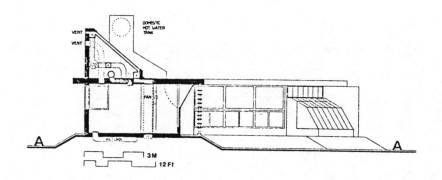

DOMESTIC
HOT WATER
TANK

VENT

VENT

FAN

A — A

3M

12 Ft

190

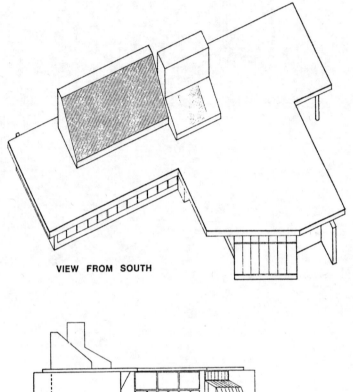

VIEW FROM SOUTH

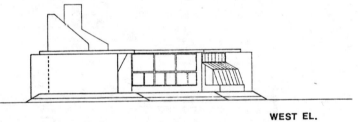

WEST EL.

WEST ELEVATION OF CHARLESTON SOLAR HOME AND VIEW FROM SOUTH OF ROOF

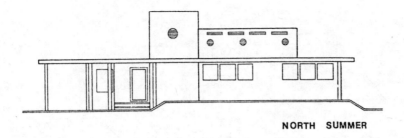

NORTH SUMMER

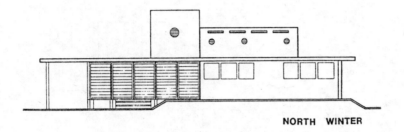

NORTH WINTER

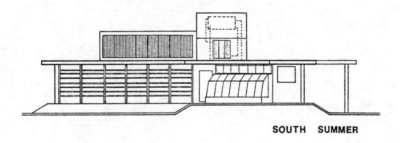

SOUTH SUMMER

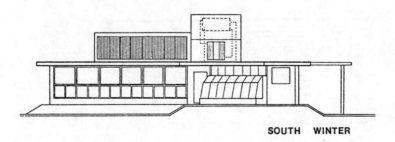

SOUTH WINTER

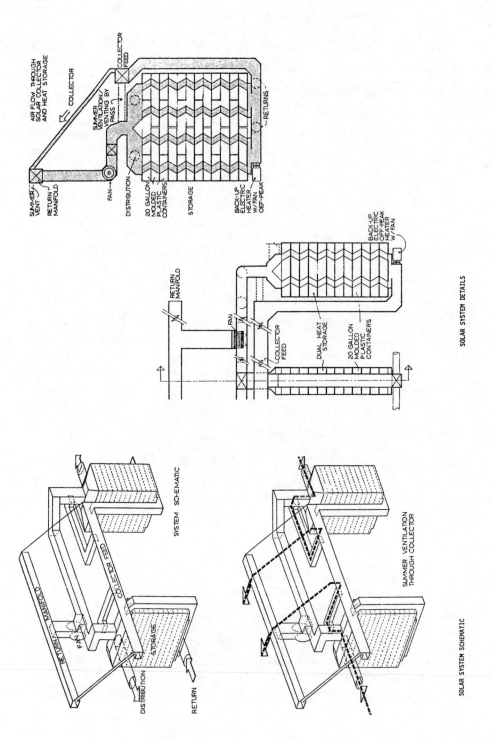

AIR FLOW THROUGH SOLAR COLLECTOR AND HEAT STORAGE

COLLECTOR

COLLECTOR FEED

SUMMER VENTILATION / VENTING BY PASS

RETURNS

SUMMER VENT

RETURN MANIFOLD

FAN

DISTRIBUTION

20 GALLON MOLDED PLASTIC CONTAINERS

STORAGE

BACK-UP ELECTRIC HEATER W/ FAN OFF-PEAK

RETURN MANIFOLD

FAN

COLLECTOR FEED

DUAL HEAT STORAGE

20 GALLON MOLDED PLASTIC CONTAINERS

BACK-UP ELECTRIC OFF-PEAK HEATER W/ FAN

SOLAR SYSTEM DETAILS

RETURN MANIFOLD

COLLECTOR FEED

FAN

STORAGE

DISTRIBUTION

RETURN

SYSTEM SCHEMATIC

SUMMER VENTILATION THROUGH COLLECTOR

SOLAR SYSTEM SCHEMATIC

193

TITLE	CHERRY CREEK OFFICE BUILDINGS
CODE	
KEYWORDS	OFFICE SOLAR COLLECTOR INSULATION HEATING COOLING BERM WIND TURBINE HEAT BOX PLENUM VENTILATION NEGATIVE ION WATER FOUNTAIN
AUTHOR	RICHARD L. CROWTHER, AIA
DATE & STATUS	THESE BUILDINGS WERE BUILT IN 1976.
LOCATION	THE NORTH BUILDING IS LOCATED AT 310 STEELE STREET, DENVER, COLORADO 80206. THE SOUTH BUILDING IS LOCATED AT 3201 EAST THIRD AVENUE, DENVER, COLORADO 80206.
SCOPE	EACH OF THESE BUILDINGS HAS 4,500' OF FLOOR SPACE ON TWO LEVELS. SOLAR ENERGY COMBINED WITH ENERGY OPTIMIZED ARCHITECTURAL FEATURES WILL PROVIDE APPROXIMATELY 75% OF THE HEATING AND 60% OF THE COOLING NEEDS.

SCOPE

THE BUILDING IS ALL ELECTRIC. FROM AN ANALYSIS OF PREVAILING UTILITY RATE STRUCTURES, IT BECAME APPARENT THAT ENERGY CONSERVATION COUPLED WITH TOTAL ENERGY MANAGEMENT WAS NECESSARY TO MINIMIZE UTILITY COSTS. INTERLOCKING SWITCHES WERE INSTALLED TO PREVENT THE SIMULTANEOUS OPERATION OF SELECTED ELECTRICAL DEVICES. THIS ARRANGEMENT LIMITS ELECTRICAL DEMAND AND MINIMIZES UTILITY COSTS UNDER A TWO PART DEMAND/ENERGY ELECTRICAL RATE STRUCTURE.

TECHNIQUE

THE PRIMARY BUILDING ENTRY FACES SOUTH FOR WINTER SUN EXPOSURE AND AS A PROTECTION FROM COLD WINTER WINDS. THE TOTAL GLASS AREA IS LIMITED TO LESS THAN 10% OF THE TOTAL FLOOR AREA. ALL WINDOWS ARE DOUBLE GLAZED WITH INSULATING GLASS TO MINIMIZE THERMAL TRANSFER. WALL CONSTRUCTION TYPE IS CONTINUOUS FROM THE WOODEN FOUNDATION PLATE TO THE TOP OF THE ROOF PARAPET. THIS SYSTEM PERMITS UNINTERRUPTED INSULATION TO BE PLACED IN THE WALLS FROM BELOW FLOOR SLABS TO ABOVE THE ROOF LINE. HEAT LOSS IS REDUCED BY HAVING THE LOWER LEVEL OF THE BUILDING SET INTO GRADE. EARTH BERMS ARE USED TO DIRECT WIND FLOW AWAY FROM BUILDING SURFACES. THE 175 SQ. FT. FLAT PLATE AIR TYPE SOLAR COLLECTOR IS SOUTH FACING AND IS TILTED AT A 45 DEGREE ANGLE TO THE HORIZON. AN UPPER MIRRORED OVERHANG REFLECTS WINTER SUN DIRECTLY INTO THIS SKYLIGHT, FOR INTENSIFIED INTERIOR ILLUMINATION AND INTERIOR HEAT GAINS.

NATURAL COOLING IS PROVIDED BY INDUCTIVE VENTILATION. ONCE THE WARM AIR REACHES THE HIGHEST POINT IN THE BUILDING, IT ENTERS DUCTS AND IS EITHER VENTED BY A WIND TURBINE OR IS INJECTED INTO THE BOTTOM OF THE SOLAR COLLECTOR AND VENTILATES THE COLLECTOR BEFORE BEING EXHAUSTED BY A FAN. COOL NOCTURNAL OR LATE AFTERNOON AIR IS DRAWN INTO THE BUILDING THROUGH GROUND LEVEL VENTS. THESE VENTS HAVE WEATHER HOODS, PREVENT DIRECT WIND FLOW, AND HAVE INSULATED PANELS WITH GASKETS TO BLOCK INFILTRATION WHEN CLOSED. HEAT PUMPS ARE PROVIDED TO REMOVE HEAT FROM THE INTERIOR WHEN THE OUTSIDE TEMPERATURES ARE TOO HIGH TO BE USED FOR NATURAL COOLING.

IN ADDITION TO THE ABOVE FEATURES, THE NORTH BUILDING HAS A GREENHOUSE ON THE LOWER LEVEL AND A NEGATIVE ION WATER FOUNTAIN. THE PLANTS IN THE GREENHOUSE WILL ABSORB CARBON DIOXIDE AND PRODUCE HUMIDITY. FRESHENING OF THE AIR WILL BE BY MEANS OF NEGATIVE IONIZATION. ADDITIONAL HUMIDITY IS PROVIDED BY THE FOUNTAIN. NEGATIVE IONS ARE ONE OF THE COMPONENTS OF FRESH AIR WHICH REVITALIZE AND INVIGORATE THE BODY. A NEGATIVE ION GENERATOR IS LOCATED IN THE POOL AT THE BASE OF THE FOUNTAIN. WATER FROM THE POOL IS PUMPED THROUGH A NOZZLE TO CREATE THE FOUNTAIN. SOME OF THE IONIZED WATER VAPOR IS CIRCULATED THROUGHOUT THE BUILDING BY AIR CURRENTS.

THE UNIQUE FEATURE OF THE SOUTH BUILDING IS THAT IT HAS A WEST FACING HEAT BOX PLENUM. THE PLENUM IS WARMED BY THE SUN IN THE AFTERNOON DURING SPRING, SUMMER AND FALL, AND COMBINED WITH THE BUILDING'S INTERNAL HEAT LOADS, INCREASES INDUCTIVE VENTILATION THROUGH A ROTARY WIND TURBINE. STACK ACTION IS PRODUCED WHICH IS USED TO VENTILATE THE BUILDING.

MATERIAL   DOUBLE GLAZING, UNINTERRUPTED INSULATION, EARTH BERMS, AIR
                  TYPE SOLAR COLLECTORS, MIRRORS, WIND TURBINES, HEAT PUMPS,
                  GREENHOUSE, HEAT BOX PLENUM, NEGATIVE ION WATER FOUNTAIN

AVAILABILITY   ADDITIONAL INFORMATON MAY BE OBTAINED FROM RICHARD L. CROWTHER,
                  AIA, 310 STEELE STREET, DENVER, COLORADO 80206.

REFERENCE   CROWTHER, RICHARD L.
                  CHERRY CREEK SOLAR OFFICE BUILDINGS
                  DENVER, COLORADO, CROWTHER/SOLAR GROUP PAMPHLET, 1976, 3 P

                  CROWTHER, RICHARD L.
                  CHERRY CREEK SOLAR OFFICE BUILDINGS: DENVER, COLORADO
                  DENVER, COLORADO, CROWTHER/SOLAR GROUP PAMPHLET, 1976, 15 P

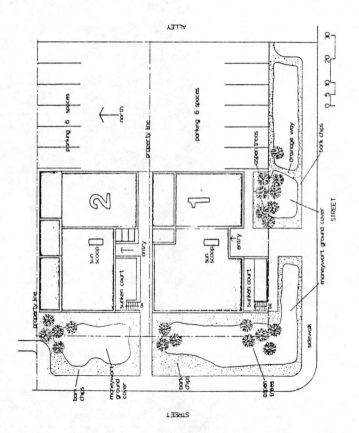

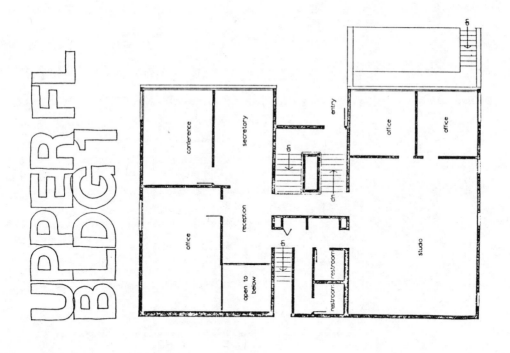

UPPER FL BLDG1

conference

secretary

entry

dn

dn

office

office

office

reception

dn

studio

open to below

dn

restroom restroom

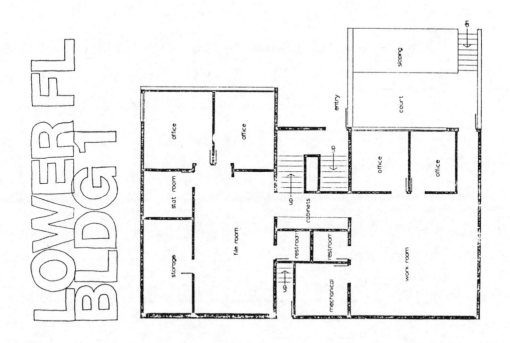

LOWER FL BLDG1

office

office

entry

sliding

court

stat room

up

office

office

file room

cabinets

up

storage

restroom

restroom

up

work room

mechanical

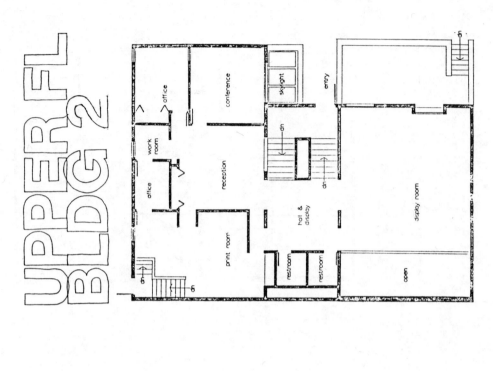

UPPER FL BLDG 2

office

work room

office

conference

reception

print room

skylight

entry

dn

dn

hall & display

restroom

restroom

display room

open

dn

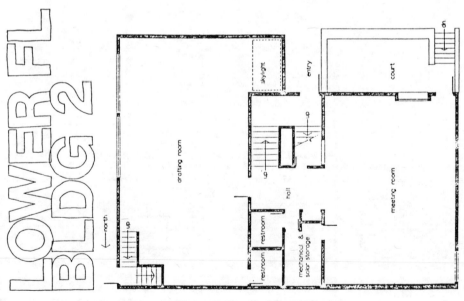

LOWER FL BLDG 2

← north

drafting room

skylight

entry

up

up

hall

restroom

restroom

restroom

mechanical & solar storage

meeting room

court

up

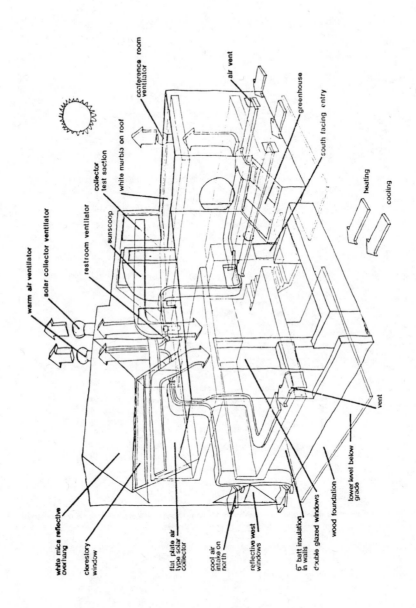

warm air ventilator

white mica reflective over hung

clerestory window

flat plate air type solar collector

cool air intake on north

reflective west windows

6" batt insulation in walls

double glazed windows

wood foundation

lower level below grade

vent

solar collector ventilator

restroom ventilator

sunscoop

collector test section

white marble on roof

conference room ventilator

air vent

greenhouse

south facing entry

heating

cooling

SECTION

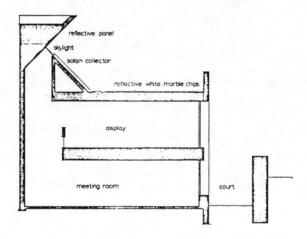

reflective panel

skylight

solar collector

reflective white marble chips

display

meeting room

court

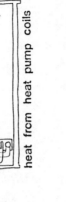

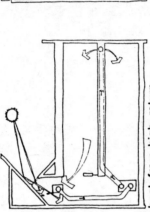

heat from heat pump coils

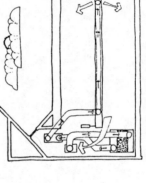

heat from resistance coils

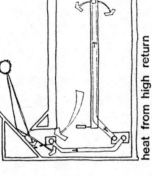

heat from high return

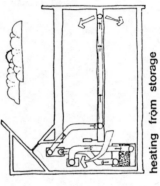

heating from storage

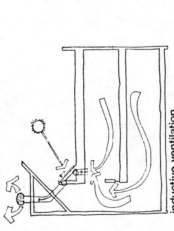

inductive ventilation

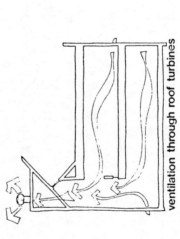

ventilation through roof turbines

TITLE	CHERRY CREEK SOLAR RESIDENCE - 419A ST. PAUL STREET
CODE	
KEYWORDS	SOLAR PROTOTYPE TOWNHOUSE CLUSTER HOUSING GREENHOUSE VAULTED CEILING MONITORING INSTRUMENTATION EVAPORATIVE WATER COOLING WIND TURBINE DOUBLE GLAZED INSULATED TANK COLLECTOR GAS FIRED FURNACE
AUTHOR	RICHARD L. CROWTHER
LOCATION	419A ST. PAUL STREET, DENVER, COLORADO
DATE & STATUS	419A ST. PAUL STREET WAS COMPLETED IN 1976.
SCOPE	419A ST. PAUL STREET IS A 2,000 SQUARE FOOT SOLAR RESIDENCE ON A MINIMAL SIZE LOT. IT IS A PROTOTYPE FOR EITHER ON-SITE OR MODULAR CONSTRUCTION FOR TOWNHOUSES, CLUSTER HOUSING, OR SINGLE FAMILY USE FOR INTENSIVE ENERGY CONSERVATION AND SOLAR COLLEC-TION.

THE HOUSE IS DESIGNED ON A 28 X 28 FOOT FOUNDATION. THE PLAN IS A SQUARE STACKING PLAN WITH BEDROOMS AND GREENHOUSE ON LOWER LEVEL, LIVING AND KITCHEN ON MAIN LEVEL, AND STUDIO AT THIRD LEVEL. THE ROOF CONFIGURATION WITH A SPACE BETWEEN THE INNER VAULTED CEILING AND THE OUTER ROOF LINE HAS EXTERNAL ACCESS FOR THE INSTALLATION, INSTRUMENTATION, AND MONITORING OF SOLAR COLLECTION WITHOUT DISTURBING THE OCCUPANTS.

TECHNIQUE THE LIVING ROOM SPACE HAS A LARGE OPEN VAULTED CEILING THAT CREATES A LARGE AIR VOLUME TO REDUCE THE DEMAND FOR EXTERNAL WINTER VENTILATION. IT ACTS AS A SPACE FOR UPWARD CONVECTION TO RETURN WARM AIR TO A DOWN-FLOW FURNACE LOCATED ON THE UPPER LEVEL. THE VAULTED CEILING AIDS THE PROCESS OF NOCTURNAL COOLING. IT ACTS AS AN EXHAUST SYSTEM, OPERATING ON A WIND TURBINE ABOVE THE ROOF, ASSISTED BY THE HOT AIR UPWARD FLOW THROUGH DUCTS FROM THE SOLAR RADIATION OF THE GREENHOUSE FROM THE LOWER LEVEL. A NORTH SIDE SCREENED INSULATED VENT WITH AN EXTERIOR HOOD ALLOWS FOR THE USE OF CONTROLLED NOCTURNAL AND TEMPERATURE DAYTIME OUTDOOR AIR.

A DROP CEILING OVER THE EAST PORTION OF THE LIVING ROOM PRO-VIDES FOR A WORK PLATFORM ABOVE THE CEILING FOR SOLAR ENERGY EQUIPMENT, AND SOLAR AND ENERGY CONSERVATION INSTRUMENTATION. IT SERVES AS A PLACE FOR DISTRIBUTION DUCTS FROM AN EXTERNAL 5,400 CFM EVAPORATIVE WATER COOLING UNIT THAT HAS AN IMPROVED EFFICIENCY FOR COOLING WHEN USED WITH THE NATURAL WIND TURBINE EXHAUST SYSTEM. A CONTROLLED DOUBLE GLAZED VIEW WINDOW TO THE EAST ADDS EARLY MORNING DAYLIGHT AND A PLEASANT VIEW OF TALL TREES, WHILE LIMITING ENERGY LOSSES AND GAINS.

AT THE SOUTHEAST CORNER OF THE LIVING ROOM IS A CONCEALED 1,000 GALLON INSULATED FIBERGLAS WATER TANK FOR SOLAR HEAT STORAGE. THE SOUTH WINDOW IS REGRESSED TO REDUCE AIR ENERGY LOSSES FROM THE INTERIOR, AND TO REDUCE THE EXTERNAL WIND EFFECT ON THE GLASS SURFACE. IT IS CALCULATED TO ALLOW DIRECT SOLAR RAYS TO PENETRATE IN AND WARM THE INTERIOR DURING LATE FALL AND THROUGH THE WINTER MONTHS. DURING LATE SPRING, SUMMER, AND EARLY FALL, THE HIGHER ANGLE OF THE SUN PREVENTS THE SUN'S RAYS FROM DIRECTLY ENTERING, THEREBY AVOIDING DIRECT INTERNAL SOLAR GAINS DURING THIS WARMER PERIOD. THE NORTHSIDE HORIZONTAL STRIP WINDOW IS REGRESSED THE FULL TEN INCHES OF THE WALL THICKNESS, AND EXCLUDES VIRTUALLY ALL DIRECT LATE NORTHWEST RAYS AND OVER-HEATING BY THE SETTING SUMMER SUN.

THE SOLAR COLLECTOR IS A FLAT PLATE WATER TYPE SOUTH FACING ROOF COLLECTOR SET AT A 53 DEGREE ANGLE. EACH DOUBLE GLASS COVER PLATE IS 1/8" THICK. THE INTERIOR COATING IS FLAT BLACK. A REINFORCED GUTTER AND TOP MEMBER IS PROVIDED TO SUPPORT A MOVEABLE LADDER FOR COLLECTOR ACCESS AND TO ACT AS A SNOW TRAP.

THERE IS A GAS FIRED FURNACE FOR AUXILIARY HEAT DURING SUSTAINED COLD AND CLOUDY WEATHER.

MATERIAL   WIND TURBINE, DUCTS, GREENHOUSE, EVAPORATIVE WATER COOLING UNIT,
                  DOUBLE GLAZED WINDOW, TALL TREES, 1,000 GALLON INSULATED FIBER-
                  GLAS WATER TANK, DOWNFLOW FURNACE, FLAT PLATE WATER TYPE
                  COLLECTOR, ELECTROSTATIC AND CHARCOAL AIR FILTERS

AVAILABILITY   ADDITIONAL INFORMATION MAY BE OBTAINED FROM RICHARD L. CROWTHER,
                  AIA, ARCHITECTS GROUP, 2830 EAST THIRD AVE., DENVER,
                  COLORADO 80206.

REFERENCE   CROWTHER/SOLAR GROUP
                  CHERRY CREEK SOLAR RESIDENCES
                  DENVER, COLORADO, CROWTHER/SOLAR GROUP, 1976, 4 P

BEFORE

AFTER

Solar Collectors
by Solaron Corp.

TITLE	CHERRY CREEK SOLAR RESIDENCE - 419B ST. PAUL STREET
CODE	
KEYWORDS	SOLAR COLLECTION THERMAL RESERVOIR HEATING COOLING DIRECT TRANSFER GRILLED FLOOR OPENING STORAGE COAL BIN INSULATED DOUBLE GLAZED WINDOW FLAT PLATE GRAVITY CONVECTION
AUTHOR	RICHARD L. CROWTHER
LOCATION	419B ST. PAUL STREET, DENVER, COLORADO
DATE & STATUS	419B ST. PAUL STREET WAS REMODELED IN 1976.
SCOPE	419 ST. PAUL STREET IS THE EXTENSIVE REMODELING OF AN OLDER 1,100 SQUARE FOOT RESIDENCE. THE RENOVATION INCLUDES EXTENSIVE USE OF INSULATION, AND THE APPLICATION OF A SOLAR COLLECTION SYSTEM.
TECHNIQUE	THE EXISTING FULL BASEMENT IS UTILIZED AS A THERMAL RESERVOIR TO MODERATE THE HEATING AND COOLING NEEDS OF THE HOUSE BY MEANS OF DIRECT TRANSFER GRILLED FLOOR OPENINGS. THE OLD COAL BIN IS RE-UTILIZED FOR SOLAR THERMAL STORAGE BY MEANS OF AN OPEN CONCRETE BLOCK FLOOR WITH AIR SPACE AND FILLED WITH 20 CUBIC YARDS OF LARGE GRAVEL, 3/4" TO 1-1/2" IN DIAMETER. A NEW INSULATED NORTH WALL WAS CONSTRUCTED THE LENGTH OF THE HOUSE THAT SHIELDS THE NORTH SIDE FROM WINTER STORMS AND WINDS, AS WELL AS CONVERTING THE OLD UNINSULATED 8" BRICK WALL TO AN INTERIOR THERMAL INERTIA WALL THAT CAN RETAIN HEAT OR COOLING. THIS NORTH WALL, EXTENDING OUT APPROXIMATELY 3', ALSO ALLOWS FOR RECESSED NORTHSIDE INSULATED AIR VENTS TO USE MODERATE DAY AND COOL NOCTURNAL TEMPERATURES, AND HAS DOUBLE GLAZED WINDOW OPENINGS. THE SOUTH ROOF ANGLE WAS ALTERED TO 53 DEGREES FOR EFFECTIVE USE OF SOLAR FLAT PLATE AIR TYPE COLLECTORS. 6" OF INSULATION WAS ADDED TO THE ROOF IN ADDITION TO FOUR INCHES OF EXISTING LOOSE FILL INSULATION.

THE COLLECTOR IS A SOLARON CORP. FLAT PLATE AIR TYPE SOUTH FACING SOLAR ROOF COLLECTOR SET AT A 53 DEGREE ANGLE. EACH DOUBLE GLASS COVER PLATE IS 1/8" THICK. THE INTERIOR COATING IS FLAT BLACK. A REINFORCED GUTTER AND TOP MEMBER IS PROVIDED TO SUPPORT A MOVEABLE LADDER FOR COLLECTOR ACCESS, AND TO ACT AS A SNOW TRAP.

THE SOLAR SYSTEM CONSISTS OF THE ROOFTOP COLLECTOR AND A DUCT SYSTEM FORCING SOLAR HEATED AIR TO A THERMAL GRAVEL STORAGE BIN (OLD COAL BIN) IN THE BASEMENT. DURING DAYTIME PERIODS OF FULL OR DIFFUSE SUNLIGHT, THE EXISTING BASEMENT FURNACE RECEIVES SOLAR HEATED AIR FROM THE COLLECTOR THROUGH THE GRAVEL STORAGE BIN. DURING NIGHT AND OVERCAST PERIODS, THE EXISTING BASEMENT FORCED-AIR SYSTEM WILL DRAW SOLAR HEATED AIR FROM THE GRAVEL STORAGE BIN. DURING SEASONS WHEN THE HOUSE NEEDS COOLING, COOL DAYTIME AND NOCTURNAL AIR WILL COOL THE GRAVEL BIN, AND COOL AIR WILL BE DELIVERED AS NEEDED THROUGH THE HOUSE BY MEANS OF THE EXISTING FORCED-AIR SYSTEM. |
MATERIAL	GRILLED FLOOR OPENINGS, COAL BIN, OPEN CONCRETE BLOCK FLOOR, AIR SPACE, GRAVEL, INSULATED WALL, 8" BRICK WALL, DOUBLE GLAZED WINDOWS, SOLAR FLAT PLATE AIR TYPE COLLECTOR, INSULATION, DUCT SYSTEM
AVAILABILITY	ADDITIONAL INFORMATION MAY BE OBTAINED FROM RICHARD L. CROWTHER, AIA, ARCHITECTS GROUP, 2830 EAST THIRD AVE., DENVER, COLORADO 80206.
REFERENCE	CROWTHER/SOLAR GROUP CHERRY CREEK SOLAR RESIDENCES DENVER, COLORADO, CROWTHER/SOLAR GROUP, 1976, 4 P

TITLE	CHESTER, CONNECTICUT HOME
CODE	
KEYWORDS	CLERESTORY SINGLE GLAZED WINDOWS FLOOR VENTILATION SUNLIGHT
AUTHOR	DONALD WATSON
DATE & STATUS	THE CHESTER, CONNECTICUT HOME WAS REMODELED IN 1975.
LOCATION	CHESTER, CONNECTICUT
SCOPE	THE GOAL OF THIS REMODELING PROJECT WAS TO MAKE A YEAR-ROUND VACATION HOUSE MORE LIVABLE WHILE REDUCING THE HEATING BILL. THE TOP LEVEL OF THE HOUSE HAD A CLERESTORY CEILING AND LARGE SINGLE-GLAZED WINDOWS, WHICH CAUSED HEATING DIFFICULTIES. ON THE TOP LEVEL WERE A KITCHEN, DINING AREA, BEDROOM, AND LIVING ROOM. THE LOWER LEVEL WAS A LARGE UNFINISHED ROOM WITH A CONCRETE FLOOR. THE GROUND LEVEL RECEIVED NO VENTILATION AND LITTLE SUNLIGHT; HIGH HUMIDITY WAS A PROBLEM IN THE SUMMER.
	THE GROUND LEVEL WAS REMODELED TO CONTAIN A DINING AREA AND A KITCHEN. A WELL OPENS INTO BOTH TOP AND GROUND LEVEL; OVER THIS WELL IS A 20' X 4' COMBINATION SKYLIGHT AND VENTILATING CLERESTORY.
	THE REMODELING REDUCED THE HEATING BILL BY 25% COMPARED TO PREVIOUS YEAR, WHILE INCREASING THE FLOOR AREA BY 50%.
TECHNIQUE	TO DEAL WITH THE DOUBLE-HEIGHT CEILING AT ONE END OF THE TOP LEVEL (THE RESULT OF THE SLOPING ROOF), THE CEILING WAS LOWERED, REDUCING THE HEATED AIR VOLUME AND CUTTING HEAT LOSS THROUGH THE ROOF. THE LOWERED CEILING PERMITTED AN INCREASE IN INSULATION WHICH WAS PLACED WITHIN THE NEW CEILING CONSTRUCTION.
	AN INSULATING PANEL, SLIDING ON HORIZONTAL TRACKS, CAN BE CLOSED DURING WINTER SUNLESS HOURS, OR DURING SUMMER DAYS WHEN TREE SHADE IS NOT SUFFICIENT. WHEN THE PANEL IS OPEN, SOLAR HEAT IS ADMITTED TO THE INTERIOR AND DISTRIBUTED TO THE HEATING SYSTEM THROUGH AN AIR RETURN PLACED HIGH IN THE CLERESTORY. A VENT FAN AUGMENTS NATURAL COOLING DURING THE SUMMER (THE HOUSE IS NOT AIR CONDITIONED).
	INSULATING GLASS WAS INSTALLED AT ALL WINDOWS, AND FOLDING INSULATING SHUTTERS (U VALUE OF .1) WERE INSTALLED TO SLIDE OVER THE WINDOW WALLS. THE WHOLE HOUSE WAS ALSO RE-INSULATED TO REDUCE HEAT FLOW. THE FLOORS BENEATH THE SKYLIGHT ARE QUARRY TILE, WHICH HELPS TO RETAIN THE SOLAR HEAT AFTER THE SUN HAS GONE DOWN.
MATERIAL	LARGE SINGLE-GLAZED WINDOWS, CONCRETE FLOOR, INSULATION, A 20' X 4' COMBINATION SKYLIGHT, VENTILATING CLERESTORY, INSULATING PANEL, AIR RETURN, VENT FAN, INSULATING GLASS, FOLDING INSULATING SHUTTERS, QUARRY TILE
AVAILABILITY	ADDITIONAL INFORMATION MAY BE OBTAINED FROM DONALD WATSON, AIA, BOX 401, GUILFORD, CONN. 06437.
REFERENCE	AMERICAN INSTITUTE OF ARCHITECTS CASE STUDY 3: PRIVATE RESIDENCE, CHESTER, CONNECTICUT WASHINGTON, D.C., IN AIA ENERGY NOTEBOOK, PUBLISHED BY AMERICAN INSTITUTE OF ARCHITECTS, 1975, PP CS-11 TO CS-12

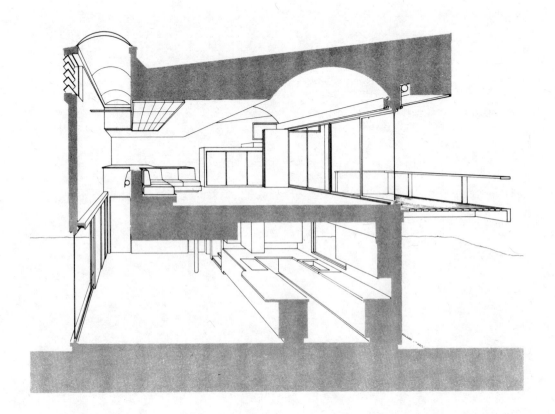

TITLE	CHESTERFIELD-MARLBORO TECHNICAL COLLEGE SOLAR HEATED LABORATORY
CODE	
KEYWORDS	LABORATORY
AUTHOR	R. THOMAS AND D. SMITH
DATE & STATUS	THE CHESTERFIELD-MARLBORO TECHNICAL COLLEGE SOLAR HEATED LABORATORY WAS COMPLETED IN THE SPRING OF 1975.
LOCATION	CHERAW, SOUTH CAROLINA
SCOPE	THE PARTIALLY SOLAR HEATED BUILDING IS A ONE-STORY, 40' X 40' LABORATORY BUILDING, CALLED THE AIR CONDITIONING, REFRIGERATION, AND HEATING LAB. THE COLLECTOR IS ON A NEARBY ONE-STORY SHED, 14' X 12' IN PLAN, WITH A RIDGE ROOF. THERE IS NO ATTIC OR BASEMENT.
TECHNIQUE	THE COLLECTOR AREA IS 125 SQ. FT. THE COLLECTOR IS OF THE AIR TYPE, MOUNTED ON THE ROOF, AND SLOPING 45 DEGREES FROM THE HORIZONTAL. IT CONSISTS OF FIVE SIDE-BY-SIDE PANELS, EACH CONSISTING OF A 3' X 8' BLACK-PAINTED METAL SHEET, OR SHALLOW TROUGH, WHICH IS SINGLE GLAZED WITH 1/8" GLASS. A 1/3 HP BLOWER CIRCULATES THE AIR TO THE STORAGE SYSTEM.
	THE STORAGE SYSTEM CONSISTS OF 20 TONS OF 3" DIAMETER STONES IN AN UNINSULATED RECTANGULAR BIN, 4' X 10' IN PLAN, AND 2' TO 5' HIGH. THE BIN OCCUPIES MOST OF THE SPACE WITHIN THE SHED. HOT AIR IS SENT, WITH THE AID OF THE BLOWER AND VIA AN INSULATED, BELOW-GROUND DUCT WITH A 16" X 16" CROSS-SECTION, TO THE LAB. THE DUCT IS MADE OF FIBERGLAS AND PLASTIC.
MATERIAL	3' X 8' BLACK-PAINTED METAL SHEET; 1/8" GLASS; 1/3 HP BLOWER; 20 TONS OF 3" DIAMETER STONES; UNINSULATED RECTANGULAR BIN; FIBERGLAS; PLASTIC
AVAILABILITY	ADDITIONAL INFORMATION MAY BE OBTAINED FROM RAYFORD THOMAS, TEC SCHOOL, CHERAW, SOUTH CAROLINA 29520.
REFERENCE	SHURCLIFF, W.A. SOLAR HEATED BUILDINGS, A BRIEF SURVEY CAMBRIDGE, MASSACHUSETTS, W.A. SHURCLIFF, MARCH 1976, 212 P

TITLE	CHILDREN'S HOSPITAL / NATIONAL MEDICAL CENTER
CODE	
KEYWORDS	GLASS HEAT LIGHT REFLECTIVE BUFFER ZONE INTERSTITIAL AIR SOLAR HOSPITAL CIRCULATION CHILDREN
AUTHOR	LEE WINDHEIM, AIA, EXECUTIVE ARCHITECT, OF LEO A. DALY CO.
DATE & STATUS	THIS HOSPITAL IS TO BE OPENED IN THE SUMMER OF 1977.
LOCATION	WASHINGTON, D.C.
SCOPE	THIS HOSPITAL HAS 4 STORIES OF INTERSTITIAL SERVED USE SPACE, ALLOWING FOR 250 BEDS, A COMPLETE CLINIC, LABORATORY AND SURGICAL FACILITIES, AND AN ADDITIONAL TRI-LEVEL UNDERGROUND PARKING GARAGE. THERE IS NO FORMAL GRADE LEVEL ENTRANCE. PATIENTS AND STAFF DRIVE INTO THE PARKING GARAGE AND USE ESCALATORS TO REACH THE HOSPITAL'S 90' HIGH CENTRAL, SKY-LIGHTED, ENTRY COURT. CONSTRUCTION COSTS ARE EXPECTED TO BE ABOUT $70.5 MILLION, WITH EXPANSION TO 500 BEDS SLATED FOR THE FUTURE.
TECHNIQUE	HERE A DOUBLE EXTERIOR WALL SYSTEM, WITH A BUFFER AIR ZONE EN-CLOSED BETWEEN A SINGLE PANE HEAT AND LIGHT REFLECTIVE GLASS OUTER WALL AND A COMBINATION SINGLE PANE GLASS AND DRY WALL INNER WALL, PROVIDES LOW COST INSULATION AND SUBSTANTIAL OUTWARD VISION. THE BUFFER ZONE, WHICH VARIES FROM 3' TO 5' IN DEPTH, IS SERVICED BY AIR HANDLING UNITS SEPARATE FROM THOSE THAT SERVICE THE REMAINDER OF THE BUILDING. SUMMER SUN STRIKES PRIMARILY THE OUTWARD SLOPING PANELS, WHICH ARE LOCATED ON INTERSTITIAL SERVICE FLOORS. THUS, MOST SOLAR LOADING OCCURS ON NON-PATIENT FLOORS. THE OUTWARD SLOPING PANELS ALSO SHADE THE PATIENT FLOORS, THUS THE SOLAR LOAD ON THE PATIENT FLOORS IS SUBSTANTIALLY REDUCED. DURING WINTER, SOLAR HEATED AIR FROM THE INTERSTITIAL LEVELS CAN BE CIRCULATED THROUGHOUT THE BUFFER ZONE TO HELP MAINTAIN A RELATIVELY WARM PERIPHERAL TEMPERATURE. THE SLOPE OF THE EXTERIOR WALLS IS SUCH THAT IT MAXIMIZES SOLAR HEAT GAIN DURING THE WINTER.
MATERIAL	HEAT AND LIGHT REFLECTIVE GLASS, SLOPING WALL CONSTRUCTION
AVAILABILITY	ADDITIONAL INFORMATION MAY BE OBTAINED FROM LEE WINDHEIM, LEO A. DALY CO., 45 MAIDEN LANE, SAN FRANCISCO, CALIFORNIA 94108.
REFERENCE	BUILDING DESIGN AND CONSTRUCTION CHILDREN'S HOSPITAL: UNIQUE LOOKING AND FUNCTIONAL BUILDING DESIGN AND CONSTRUCTION, JANUARY 1976, P 48 PROGRESSIVE ARCHITECTURE NEW RULES FOR THE GAME PROGRESSIVE ARCHITECTURE, JULY 1972, PP 84-92

Detail of hospital exterior shows inward-and-outward sloping glass.

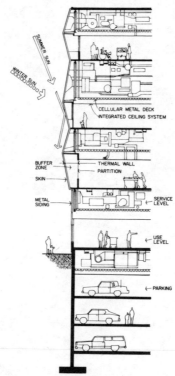

Wall section of hospital.

212

TITLE	CHILDREN'S HOSPITAL OF PHILADELPHIA
CODE	CHOP
KEYWORDS	HOSPITAL TINTED GLASS ATRIUM CONSTANT VOLUME TERMINAL REHEAT SYSTEM CUBE PERIMETER HEAT GAIN LOSS LIGHTING REUSE INDIVIDUAL TEMPERATURE CONTROL COURTYARD PLENUM OUTDOOR PLAYDECK VENETIAN BLIND CALIBRATED SUN ANGLE
AUTHOR	HARBESON HOUGH LIVINGSTON & LARSON, AND WILLIAM A. AMENTA, LEONARD WEGER ASSOC., CONSULTING ENGINEERS
DATE & STATUS	THE CHILDREN'S HOSPITAL OF PHILADELPHIA HAS BEEN BUILT AND IS FUNCTIONING (1975).
LOCATION	PHILADELPHIA, PENNSYLVANIA
SCOPE	A HUGE, ONE MILLION CUBIC FOOT GLASS-TOPPED ATRIUM RISES THROUGH 8 ABOVE-GRADE LEVELS OF THE 11 STORY STRUCTURE. ON GROUND LEVEL, THE ATRIUM ACTS AS THE MAIN "LOBBY" OF THE BUILDING. HERE, PLAY, WAITING, AND DINING AREAS ARE PROVIDED AMIDST A VERITABLE FOREST OF FIG TREES. A BANK, SNACK BAR, AUDITORIUM, AND GIFT SHOP - ALL AVAILABLE TO THE PUBLIC - AND A GYMNASIUM, DINING ROOMS, ADMINISTRATIVE OFFICES, EXAM AND TREATMENT ROOMS, AND PATIENT LOBBIES ENCIRCLE THE SPACE. AS THE ATRIUM RISES THROUGH THE BUILDING, IT IS SURROUNDED BY OPEN, BALCONY-TYPE CORRIDORS ON EACH LEVEL THAT ARE THE MAIN CIRCULATION PATHS TO THE PATIENT ROOMS. THEY ARE ALSO THE MAJOR ORIENTING DEVICE OF THE BUILDING. THERE ARE GLASS-ENCLOSED ELEVATORS RISING THROUGH LOWER LEVELS OF THE SPACE, SERVING BELOW GRADE PARKING AND THREE FLOOR LEVELS OF OUT-PATIENT DEPARTMENTS.

IT IS ESTIMATED THAT THE SAVINGS IN OPERATING COSTS THROUGH THIS ALL-AIR CONSTANT VOLUME TERMINAL REHEAT SYSTEM (IT CUTS FUEL BILLS $55,000 ANNUALLY) WILL PAY FOR THE INITIAL COST OF THE EQUIPMENT IN THE FIRST FIVE TO SIX YEARS.

TECHNIQUE	UNLESS A SPHERICAL FORM CAN BE JUSTIFIED, THE CUBE PRESENTS THE LEAST EXTERIOR SURFACE IN RELATION TO THE VOLUME OF THE INTERIOR. AND, THE LARGER THE CUBE, THE MORE EFFICIENT THE RATIO BECOMES. CONSEQUENTLY, THE HOSPITAL HAS A SHAPE THAT WAS CLOSEST, PRACTICALLY, TO THAT OF A CUBE. THIS MEANT THAT PERIMETER HEAT GAINS AND LOSSES COULD BE HELD TO A MINIMUM, BUT IT ALSO MEANT THAT SPACES DEEP WITHIN THE BUILDING, WHERE THERE IS A HEAT BUILD-UP FROM EQUIPMENT, LIGHTING, AND PEOPLE, WOULD REQUIRE AIR-CONDITIONING EVEN DURING THE WINTER. THE INTERNALLY GENERATED HEAT, HOWEVER, AS WELL AS HEAT FROM THE SUN, COULD BE REUSED, RATHER THAN REJECTED.

IN WINTER, WHEN THE REFRIGERATION MACHINES OPERATE AT A 1,200 TON CAPACITY, THE HEAT THEY WOULD OTHERWISE REJECT IS REDIRECTED TO THE PERIMETER OF THE BUILDING TO OFFSET THE HEAT LOSS THAT OCCURS THERE. NO ADDITIONAL PURCHASED STEAM OR ENERGY IS REQUIRED FOR ENVIRONMENTAL HEATING AS LONG AS THE OUTSIDE TEMPERATURE DOES NOT FALL BELOW 29 F. THIS PAYS OFF HANDSOMELY IN PHILADELPHIA, WHERE THE WINTER TEMPERATURE AVERAGES A MODERATE 40 F. THROUGHOUT THE ENTIRE YEAR, THIS NORMALLY REJECTED HEAT IS USED TO HEAT DOMESTIC HOT WATER AND TO TEMPER AIR. IN THE SUMMER, IF THE OUTSIDE TEMPERATURE DROPS, SOME AREAS OF THE BUILDING MAY NOT REQUIRE AIR AS COLD AS THAT EMANATING FROM THE DUCTS. IN ADDITION, DIFFERENT AREAS OF THE HOSPITAL, SUCH AS OPERATING ROOMS, RESEARCH DEPARTMENTS, PATIENT ROOMS, PRIVATE OFFICES, ANIMAL RESEARCH AREAS, ETC., REQUIRE DIFFERENT TEMPERATURES BECAUSE OF THEIR DIFFERENT FUNCTIONS. CONSEQUENTLY, THE BUILDING IS DESIGNED SO THAT ALL AREAS ARE PERMITTED INDIVIDUAL TEMPERATURE CONTROL, AND THIS CONTROL IS ACHIEVED THROUGH REHEATING WITH THE NORMALLY REJECTED HEAT, THE AIR AT ANY TERMINAL OUTLET. THE IDEA OF REHEATING PREVIOUSLY REFRIGERATED AIR MAY SOUND PECULIAR, BUT THE SYSTEM HAS CERTAIN DISTINCT ADVANTAGES. FIRST OF ALL, IT IS A REHEAT SYSTEM WITH A FREE HEAT SOURCE. SECOND, THE SYSTEM ALLOWS AIR VOLUME TO REMAIN CONSTANT, WHILE PERMITTING ANY VARIATION IN TEMPERATURE IN ANY AREA WITHIN THE BUILDING.

THE COURTYARD GREATLY ENHANCES THE SYSTEM'S EFFICIENCY. THE
GROUND FLOOR SPACE, WHICH IS ACTUALLY THE THIRD LEVEL OF THE
BUILDING, IS AIR-CONDITIONED WITH TEMPERATURE AND HUMIDITY
APPROPRIATE FOR THE MANY TREES AND PLANTS THAT FILL THE AREA.
FROM THE SECOND FLOOR THROUGH THE EIGHTH, THE COURTYARD IS USED
AS A RETURN AIR PLENUM. RETURN AIR IS DISCHARGED THROUGH THE
SLOTS IN THE FLOORS OF THE CORRIDOR-BALCONIES THAT SURROUND THE
INTERIOR OF THE COURT ON SIX OF THESE LEVELS, THUS ELIMINATING
THE NEED FOR LARGE VERTICAL DUCT RUNS TO THE AIR HANDLING UNITS
ON TOP OF THE BUILDING. THIS RESULTED IN A SAVING OF OVER
$100,000 IN DUCT CONSTRUCTION.

WHILE THE COURTYEARD IS SET INTO THE BUILDING, IT IS EXPOSED
TO THE EXTERIOR ON ITS SOUTH SIDE AND, OF COURSE, THROUGH ITS
TRUSS-SUPPORTED GLAZED ROOF. THE CONCRETE OUTDOOR PLAYDECKS ON
THE SOUTH SIDE OF THE COURTYARD ACT AS A LARGE VENETIAN BLIND
IN A FIXED POSITION. THEY ARE CALIBRATED TO SUN ANGLES TO PRO-
VIDE MAXIMUM INTERVENTION IN SUMMER TO REDUCE HEAT AND GLARE
FROM DIRECT RAYS, AND TO PRESENT MINIMAL OBSTACLES TO WARMTH
AND ILLUMINATION FROM THE SUN IN THE WINTER. ALTHOUGH THE ROOF
OF THE COURTYARD IS OF CLEAR GLASS, IT IS DESIGNED AS A SERIES
OF PARALLEL, SAW-TOOTH FORMS ANGLED TO REFLECT THE SUMMER SUN.
(REFLECTIVE MIRROR GLASS WAS NOT USED, AS CALCULATIONS INDI-
CATED THAT IT WOULD NOT REDUCE HEAT GAIN ENOUGH TO OFFSET COST;
TINTED GLASS WOULD HAVE DEFEATED THE PURPOSE.) IN WINTER, THE
ROOF HAS LITTLE EFFECT AS A SOLAR REFLECTOR DUE TO THE LOW ANGLE
OF THE SUN. THE HIGHEST PERCENTAGE OF SOLAR HEAT GAIN THEN
COMES THROUGH THE SPECIALLY ANGLED SOUTH BALCONIES.

MATERIAL   ONE MILLION CUBIC FOOT GLASS-TOPPED ATRIUM, FIG TREES,
                ALL-AIR CONSTANT VOLUME TERMINAL REHEAT SYSTEM, REFRIGERATION
                MACHINES, COURTYARD, TRUSS-SUPPORTED GLAZED ROOF, OUTDOOR
                CONCRETE PLAYDECKS, CLEAR GLASS

AVAILABILITY   ADDITIONAL INFORMATION MAY BE OBTAINED FROM THE ARCHITECTS,
                HARBESON HOUGH LIVINGSTON & LARSON, OR WILLIAM A. AMENTA,
                1500 ARCHITECTS BLDG., PHILADELPHIA, PENNSYLVANIA 19103.

REFERENCE   PROGRESSIVE ARCHITECTURE
                CHOPPING ENERGY COSTS
                PROGRESSIVE ARCHITECTURE, MAY 1975, PP 46-53

Outdoor playdecks on south side (above) shield huge inner atrium

Energy concern and tight site made form cubical. Main entry is on east.

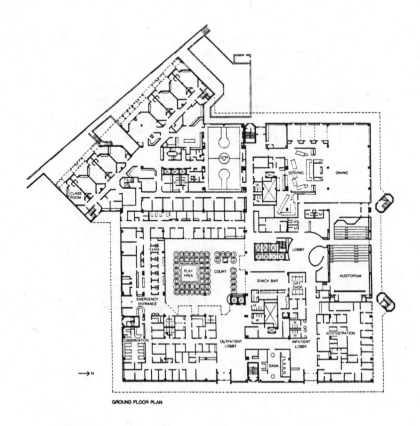

GROUND FLOOR PLAN

→ N

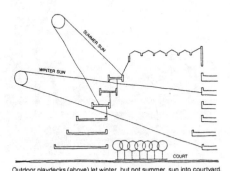

Outdoor playdecks (above) let winter, but not summer, sun into courtyard.

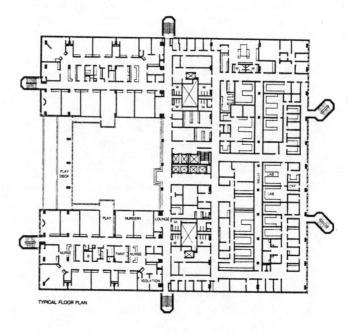

TYPICAL FLOOR PLAN

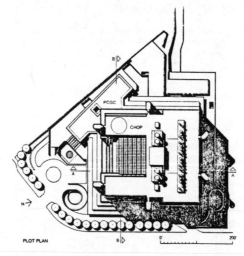

PLOT PLAN

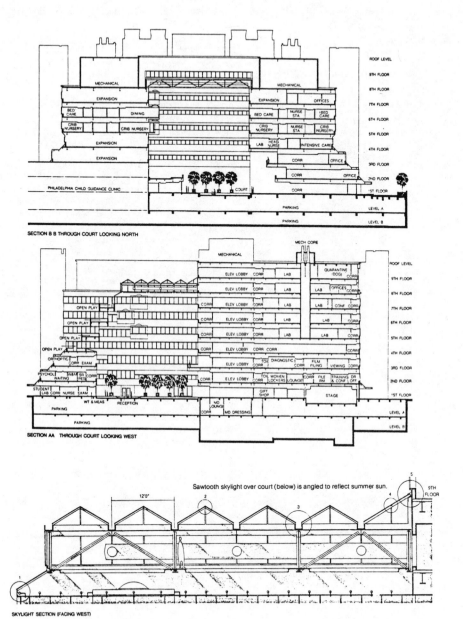

SECTION B B THROUGH COURT LOOKING NORTH

SECTION AA THROUGH COURT LOOKING WEST

Sawtooth skylight over court (below) is angled to reflect summer sun.

SKYLIGHT SECTION (FACING WEST)

219

TITLE	CITICORP CENTER
CODE	
KEYWORDS	SOLAR ENERGY DEHUMIDIFICATION SYSTEM FLAT PLATE COLLECTOR LIQUID HEAT RECLAMATION TRIETHYLENE GLYCOL OFFICE
AUTHOR	THE ARCHITECT IS HUGH STUBBINS & ASSOCIATES, INC.; THE OWNER IS FIRST NATIONAL CITY BANK.
DATE & STATUS	CITICORP CENTER IS SCHEDULED FOR COMPLETION IN THE FALL OF 1976. HOWEVER, THE SOLAR ENERGY DEHUMIDIFICATION SYSTEM WILL NOT BE USED WHEN THE PROJECT IS BUILT. IT IS INCLUDED HERE FOR ITS EDUCATIONAL VALUE.
LOCATION	LEXINGTON AVE. AND 53RD ST., NEW YORK, NEW YORK
SCOPE	THE CITICORP CENTER FEATURES A 46-STORY TOWER, RESTING ON A PLATFORM 112 FEET HIGH, WHICH WILL CONTAIN MORE THAN 1 MILLION SQUARE FEET OF OFFICE SPACE. THE CENTER WILL ALSO INCLUDE A NEW FREE-STANDING ST. PETER'S LUTHERAN CHURCH ON A CORNER ADJACENT TO THE TOWER, A LOWRISE BUILDING. PART OF THE COMPLEX WILL HAVE EIGHT STORIES OF TERRACED LEVELS. A SHOPPING AREA IN THE CENTER OF THE BLOCK WILL HAVE THREE LEVELS OF RESTAURANTS, STORES, AND OTHER RETAIL ESTABLISHMENTS THAT WILL FORM A U AROUND A 70 X 80 FOOT GALLERIA OR COURTYARD. A LARGE SUNKEN PLAZA WILL PROVIDE A PEDESTRIAN AREA WITH SCULPTURE, FOUNTAINS AND LANDSCAPING. THE COMPLEX HAS BEEN DESIGNED TO CONSERVE ENERGY. THE BUILDING'S EXTERIOR WALLS WILL BE INSULATED AND WINDOWS DOUBLE GLAZED WITH REFLECTIVE GLASS TO REDUCE DEMANDS ON AIR CONDITIONING AND HEATING. IT IS ESTIMATED THAT ONLY ABOUT 20 PERCENT OF SOLAR RADIATION WILL BE TRANSMITTED INTO THE BUILDING. NEWLY DEVELOPED LIGHTING FIXTURES ARE DESIGNED TO REDUCE ENERGY REQUIREMENTS. AIR FIBERS WILL REDUCE THE AMOUNT OF AIR TAKEN INTO THE BUILDING TO DIMINISH THE NEED TO COOL OR HEAT OUTSIDE AIR AND LET THE AIR INSIDE THE STRUCTURE BE CLEANER AND FREE OF POLLUTANTS.
TECHNIQUE	THE MAIN ENERGY CONSERVING TECHNIQUE USED IN THE CITICORP CENTER IS A SOLAR ENERGY DEHUMIDIFICATION SYSTEM. THE TWO COMPONENTS ARE A DEHUMIDIFICATION SYSTEM AND "STATE OF THE ART" FLAT PLATE COLLECTORS (APPROXIMATELY 20,000 FEET) MOUNTED ON THE SLOPING SOUTHERN FACE OF THE TOWER. ENERGY FROM THE COLLECTOR WILL OPERATE THE DEHUMIDIFICATION SYSTEM. MOISTURE FROM INCOMING AIR WILL BE REMOVED BY A LIQUID DESSICANT SPRAY. WATER ABSORBED BY THE DESSICANT WILL BE DISSIPATED BY HEATING THE DESSICENT SOLUTION TO HIGH TEMPERATURES. IN WINTER, THE SOLAR SYSTEM IS USED FOR DOMESTIC HOT WATER HEATING AND THE DEHUMIDIFICATION CYCLE IS REVERSED TO HUMIDIFY THE INCOMING FRESH AIR.
MATERIAL	LIQUID DESSICANT SPRAY, FLAT-PLATE COLLECTOR, TRIETMYLENE GLYCOL, REFLECTIVE GLASS
AVAILABILITY	THE READER IS ADVISED TO REFER TO REFERENCES BELOW.
REFERENCE	AIA JOURNAL MANHATTAN'S FIFTH TALLEST BUILDING IS DESIGNED FOR ENERGY CONSERVATION AIA JOURNAL, OCTOBER 1973, PP 11, 61 OWENS-CORNING FIBERGLAS CORPORATION OWENS-CORNING FIBERGLAS CORPORATION - ENERGY CONSERVATION AWARDS PROGRAM, CITICORP CENTER PROJECT TOLEDO, OHIO, OWENS-CORNING FIBERGLAS CORPORATION BROCHURE, AUGUST 30, 1974, 8 P ARCHITECTURAL RECORD MIT TEAM WILL LEAD MAJOR SOLAR STUDY ARCHITECTURAL RECORD, MARCH 1975, P 35

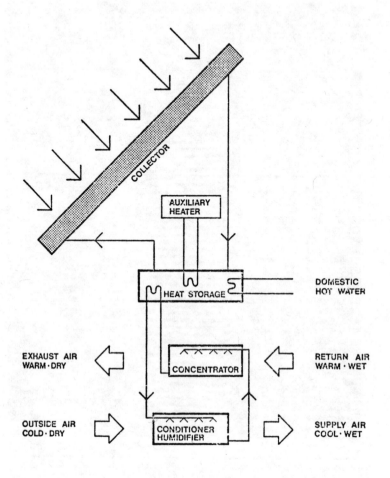

COLLECTOR

AUXILIARY HEATER

HEAT STORAGE

DOMESTIC HOT WATER

EXHAUST AIR
WARM · DRY

CONCENTRATOR

RETURN AIR
WARM · WET

OUTSIDE AIR
COLD · DRY

CONDITIONER HUMIDIFIER

SUPPLY AIR
COOL · WET

WINTER HUMIDIFICATION

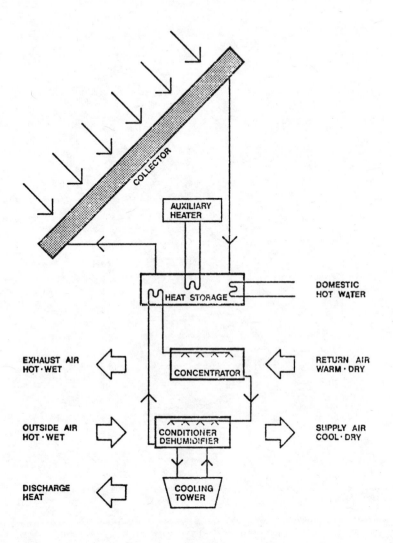

COLLECTOR

AUXILIARY
HEATER

HEAT STORAGE

DOMESTIC
HOT WATER

EXHAUST AIR
HOT · WET

CONCENTRATOR

RETURN AIR
WARM · DRY

OUTSIDE AIR
HOT · WET

CONDITIONER
DEHUMIDIFIER

SUPPLY AIR
COOL · DRY

DISCHARGE
HEAT

COOLING
TOWER

SUMMER DEHUMIDIFICATION

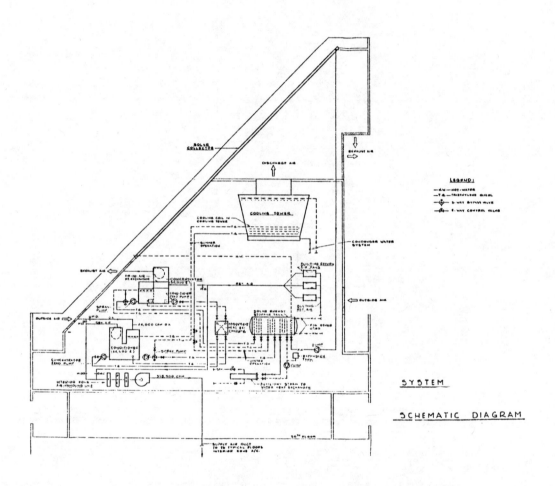

SYSTEM

SCHEMATIC DIAGRAM

224

TITLE	CITY SAVINGS BANK
CODE	
KEYWORDS	SOLAR COLLECTOR PANEL HEATING COOLING HOT WATER
AUTHOR	KENNETH F. PARRY ASSOCIATES; CA CROWLEY ENGINEERING CO.
DATE & STATUS	THE CITY SAVINGS BANK IS SCHEDULED FOR COMPLETION IN THE SUMMER OF 1977.
LOCATION	MIDDLETON, CONNECTICUT
SCOPE	
TECHNIQUE	ALL HOT WATER WILL BE HEATED BY SOLAR COLLECTORS IN THIS BUILDING, AND 48% OF THE HEATING AND 25% OF THE COOLING NEEDS WILL ALSO BE MET BY SOLAR POWER.
MATERIAL	SOLAR COLLECTOR
AVAILABILITY	ADDITIONAL INFORMATION MAY BE OBTAINED FROM DAYSTAR CORP., 90 CAMBRIDGE STREET, BURLINGTON, MASSACHUSETTS 01803.
REFERENCE	DAYSTAR CORPORATION ALL-COPPER SOLAR COLLECTORS FOR THREE NEW ENGLAND BANKS BURLINGTON, MASSACHUSETTS, DAYSTAR CORP. PAMPHLET, 1976, 1 P

TITLE	CLUSTER DESIGN FOR A COOL CLIMATE
CODE	
KEYWORDS	HOUSING WINTER WINDS ROOF MOUNTED FLAT PLATE SOLAR COLLECTOR SKYLIGHTED ENTRY COURT CLUSTER CONIFEROUS VEGETATION EARTH BERM BUFFER BAY WINDOW INSULATION LIGHT AIR CENTRAL FIREPLACE ORIENTATION NATURAL VENTILATION HEAT EXCHANGER
AUTHOR	GIFFELS ASSOCIATES, INC.
DATE & STATUS	THE LIVING UNITS AND CLUSTER DESIGN FOR A COOL CLIMATE HAVE BEEN DESIGNED, BUT NOT BUILT AS OF 1975.
LOCATION	DENVER, COLORADO
SCOPE	THE BASIC UNIT OF THE CLUSTER IS A TWO-BEDROOM, TWO STORY RESIDENCE, ADAPTABLE TO OCCUPANCY BY SHARING ADULTS OR SMALL FAMILIES. THE TWO STORY UNIT IS USED TO ALLOW A MAXIMUM OF PRIVACY WITHIN PARTS OF THE UNIT ITSELF, AS WELL AS FOR ENERGY CONSERVATION REASONS. THE UTILITY SPACE, SECOND BATHROOM (ON THE MAIN FLOOR), AND HALLWAY TO THE KITCHEN ARE DESIGNED TO PROVIDE A WORKING AREA FOR HOUSEKEEPING ACTIVITIES WITHIN THE UNIT ITSELF. THERE ARE ALSO ONE-BEDROOM UNITS AND AN EFFICIENCY UNIT. THERE ARE 31 UNITS IN 4 BUILDINGS. NINETEEN OF THESE UNITS ARE TWO-BEDROOM, ELEVEN ARE ONE-BEDROOM AND ONE IS AN EFFICIENCY UNIT. THERE IS A COMMUNITY BUILDING AND PARKING GARAGE SPACE FOR 50 CARS.
TECHNIQUE	GOOD INSULATION FROM CLIMATIC EXTREMES IS PROVIDED BY COVERING BOTH EAST AND WEST WALLS OF THE LIVING UNIT WITH BUFFER SPACES. THE MASSING OF THE BUILDINGS MAY THEN CONCENTRATE ON REDUCING WIND EXPOSURE OF EXTERIOR SPACES, INCLUDING ENTRY WAYS.

THE SOLAR ENERGY SYSTEM FOR THE COOL REGION HAS A PROVISION FOR SOLAR HEATING ONLY. A HEATING ONLY SYSTEM IS CHOSEN BECAUSE THE AMOUNT OF COOLING REQUIRED IN THIS REGION AND THE PERCENTAGE OF SUNSHINE AVAILABLE DURING THE COOLING SEASON ARE RELATIVELY SMALL WHEN COMPARED TO THE DEGREE OF SOPHISTICATION AND COSTS OF SOLAR COOLING SYSTEMS. WITH PROPER CONSIDERATIONS FOR ENERGY CONSERVATION AND NATURAL VENTILATION MEASURES, SPACE COOLING SHOULD NOT BE NECESSARY OVER MOST OF THE COOLING SEASON.

SOLAR ENERGY COLLECTION EMPLOYS CLOSED LOOP, ALUMINUM, FLAT PLATE, COLLECTOR UNITS. THE ENERGY TRANSPORT FLUID MEDIUM CONSISTS OF A WATER AND ETHYLENE GLYCOL (ANTI-FREEZE) SOLUTION. THIS SOLAR HEATED FLUID IS STORED IN AN INSULATED TANK FOR USE WHEN HEATING IS REQUIRED, AND DELIVERED TO INDIVIDUAL FAN-COIL UNITS LOCATED WITHIN EACH LIVING UNIT. DELIVERY OF SPACE HEATING IS AUGMENTED BY DUCTWORK INTEGRAL WITH THE FIREPLACE. A DOMESTIC HOT WATER HEAT EXCHANGER UNIT IS ALSO INTEGRATED WITH THE SOLAR ENERGY STORAGE SYSTEM TO PROVIDE DOMESTIC HOT WATER NEEDS. THE BACK-UP HEATING UNITS ARE GAS FIRED FLUID HEAT EXCHANGERS.

THE SOLAR ENERGY FLAT PLATE COLLECTORS ARE WELDED AND AIR EXPANDED 2-PLY ALUMINUM, WITH INTERNALIZED FLUID CHANNELS. WITH A SELECTIVE COATED SURFACE, BACK INSULATION AND TWO PANE LOW IRON CONTENT GLASS, THE ATTAINABLE FLUID TEMPERATURES GENERALLY RANGE BETWEEN 100 F TO 180 F. TYPICAL ASSEMBLED UNIT DIMENSIONS ARE APPROXIMATELY 36"X84"X4". THE SOLAR EFFICIENCY RANGES FROM 32% TO 48%, WITH A YEARLY AVERAGE EFFICIENCY OF AROUND 40%. THE ANGULAR TILT (ATTITUDE) OF THE PLATE COLLECTORS DEPENDS UPON THE LATITUDE OF THE BUILDING. THE OPTIMUM FOR WINTER HEATING IS THE LATITUDE PLUS 10 DEGREES TO 15 DEGREES, AND FOR SUMMER COOLING, THE LATITUDE MINUS 10 DEGREES TO 15 DEGREES. FOR COMBINED HEATING AND COOLING, THE LATITUDE IS USED. THE ORIENTATION RANGE IS FROM 170 DEGREES TO 200 DEGREES FOR GOOD SOLAR COLLECTOR PERFORMANCE. THIS INSTALLATION INCLUDES COLLECTORS LOCATED AT AN ATTITUDE OF 45 DEGREES (WITH A WINTER OPTIMUM ATTITUDE OF APPROXIMATELY 50 DEGREES TO 55 DEGREES).

THE HEAT ABSORBING FLUID MEDIUM IS A MIXTURE OF WATER (50%) AND ETHYLENE GLYCOL (50%) TO PREVENT FREEZING OF THE FLUID MEDIUM WITHIN THE COLLECTOR UNITS DURING THE WINTER NIGHTS.

USING AN ANTI-FREEZE IN WATER SOLUTION LOWERS THE HEAT CAPACITY OF WATER, REQUIRING A GREATER FLUID CIRCULATION RATE TO COLLECT AN EQUIVALENT AMOUNT OF AVAILABLE SOLAR ENERGY.

THE SOLAR ENERGY STORAGE SYSTEM CONSISTS OF TWO SEPARATE INSULATED STORAGE TANKS: THE DOMESTIC HOT WATER HEAT EXCHANGER/ STORAGE TANK AND THE PRIMARY SOLAR HEATED FLUID STORAGE TANK.

AS THE SOLAR HEATED FLUID FIRST ARRIVES AT THE STORAGE SYSTEM, IT IS USED TO HEAT THE DOMESTIC HOT WATER, HEAT EXCHANGER/ STORAGE TANK. THIS IMMERSED TANK SHOULD BE TIGHTLY SEALED TO PREVENT ANY POSSIBLE CROSS CONTAMINATION BETWEEN THESE TWO FLUIDS.

HAVING TRANSFERRED PART OF ITS HEAT CONTENT TO THE DOMESTIC HOT WATER SYSTEM, THE SOLAR HEATED FLUID THEN FLOWS INTO THE MAIN SOLAR HEATED STORAGE TANK. HERE, THE FLUID IS STORED UNTIL THERE IS A HEATING DEMAND IN ONE OF THE LIVING UNITS. UPON INITIATION OF THIS DEMAND, PART OF THE STORED SOLAR HEATED FLUID IS PUMPED UP TO THE UNIT'S INDIVIDUAL FAN COIL HEATING AND HUMIDIFYING EQUIPMENT WHERE THE HEAT IS TRANSFERRED, VIA THE FAN-COIL SYSTEM, TO THE HOT AIR DUCTWORK DISTRIBUTION SYSTEM WITHIN THE LIVING UNIT. THE MAIN DUCT LEADING FROM THE EQUIPMENT IS INTEGRATED WITH THE FIREPLACE FOR MORE EFFICIENT COLLECTION AND DISTRIBUTION OF THE HEAT GENERATED WHEN THE FIREPLACE IS IN USE.

THE BACK-UP DOMESTIC HOT WATER AND MAIN STORAGE TANK HEATING UNITS ARE GAS FIRED HOT FLUID SYSTEMS, WHERE THE AUXILIARY HEAT IS TRANSFERRED TO THE SOLAR STORAGE SYSTEMS, VIA THE USE OF FLUID-TO-FLUID HEAT EXCHANGERS.

MATERIAL   FLAT-PLATE SOLAR COLLECTOR, LOW THERMAL CONDUCTIVITY INSULATING MATERIALS, CENTRAL FIREPLACE WITH INTEGRAL DUCTWORK, LIGHT AND AIR IMPERVIOUS CURTAINS, SKYLIGHT, BAY WINDOW, ALUMINUM, WATER, ETHYLENE GLYCOL, FAN-COIL, DOMESTIC HOT WATER HEAT EXCHANGER, GAS FIRED FLUID HEAT EXCHANGER, SELECTIVE COATED SURFACE, TWO PANE, LOW IRON CONTENT GLASS

AVAILABILITY   ADDITIONAL INFORMATION ON THE CLUSTER DESIGN FOR A COOL CLIMATE MAY BE OBTAINED BY WRITING TO GIFFELS ASSOCIATES, INC., 1000 MARQUETTE BLDG., DETROIT, MICHIGAN 48226.

REFERENCE   GIFFELS ASSOCIATES, INC. DESIGN CONCEPT NO. 1: COOL REGION DETROIT, MICHIGAN, IN SOLAR ENERGY AND HOUSING DESIGN CONCEPTS, PUBLISHED BY GIFFELS ASSOCIATES, INC., JANUARY 1975, PP 17-35

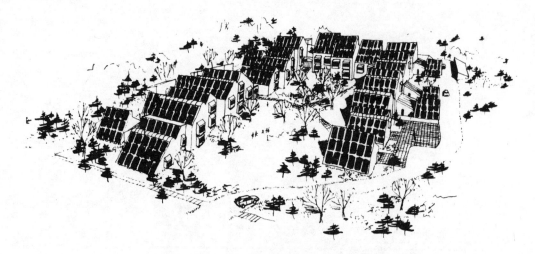

PRIMARY ORIENTATION: SOLAR
SECONDARY ORIENTATION: WIND, PRECIPITATION

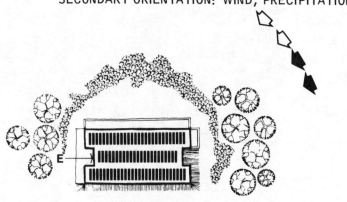

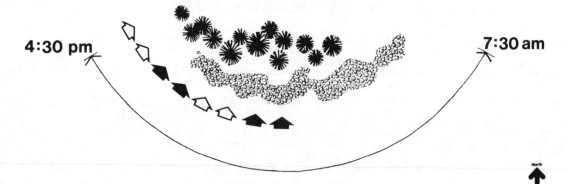

4:30 pm   7:30 am

North

EFFECTIVE SOLAR HEATING IN COOL MONTHS

SUN PROTECTION IN WARM MONTHS
WITHOUT SHADE ON SOLAR COLLECTORS

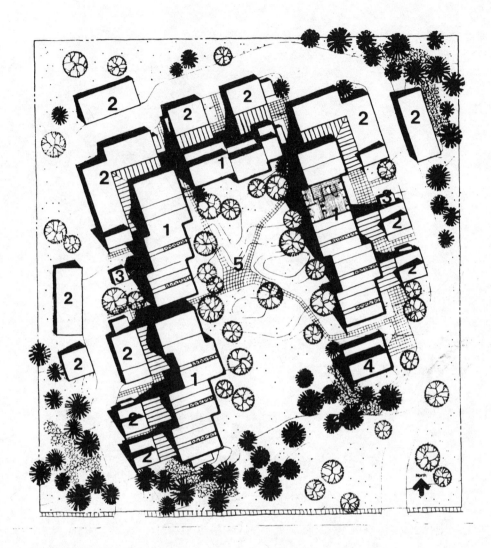

SITE PLAN

1. LIVING UNITS
2. GARAGE PARKING
3. BUILDING SERVICES (STORAGE, TRASH COLLECTION, RECREATION
   AND EQUIPMENT)
4. COMMUNITY BUILDING (OFFICE)
5. COMMUNITY SPACE

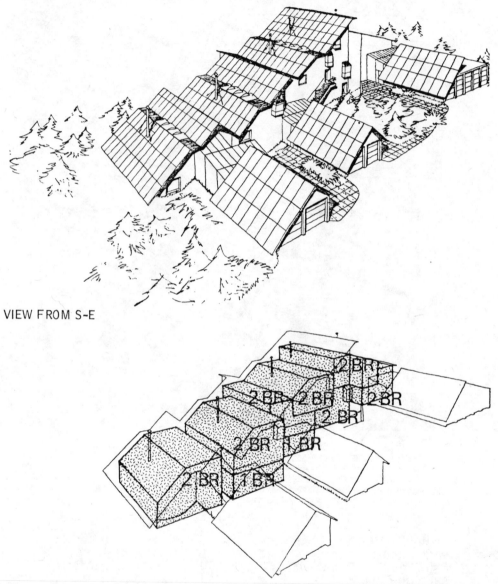

VIEW FROM S-E

SECTION VIEW FROM S-E

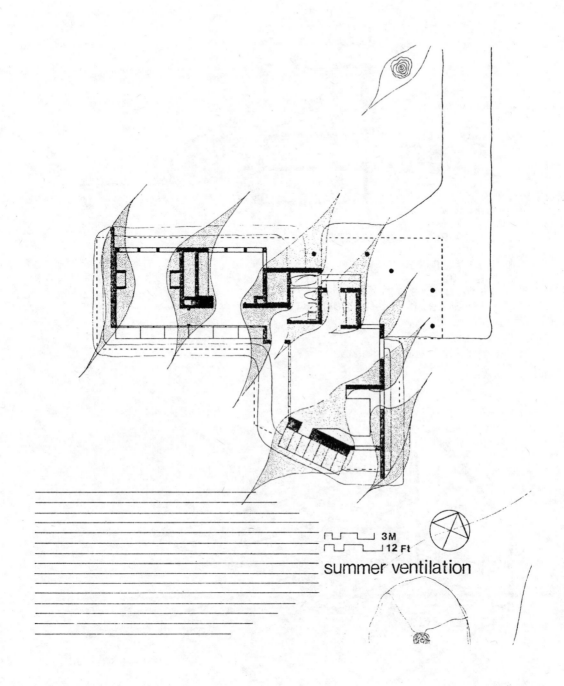

3M
12 Ft

summer ventilation

VENTILATING AIR FLOW THROUGH CHARLESTON SOLAR HOME

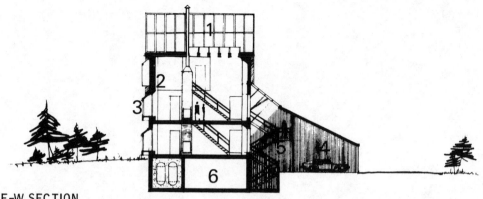

E–W SECTION

1. FLAT PLATE COLLECTORS
2. BUFFER SPACE
3. SOLAR BAY WINDOW
4. GARAGE
5. ENCLOSED ENTRY COURT
6. BASEMENT (STORAGE, LAUNDRY, ETC.)

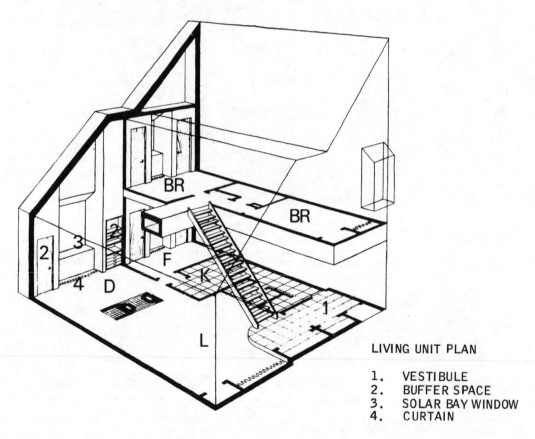

LIVING UNIT PLAN

1. VESTIBULE
2. BUFFER SPACE
3. SOLAR BAY WINDOW
4. CURTAIN

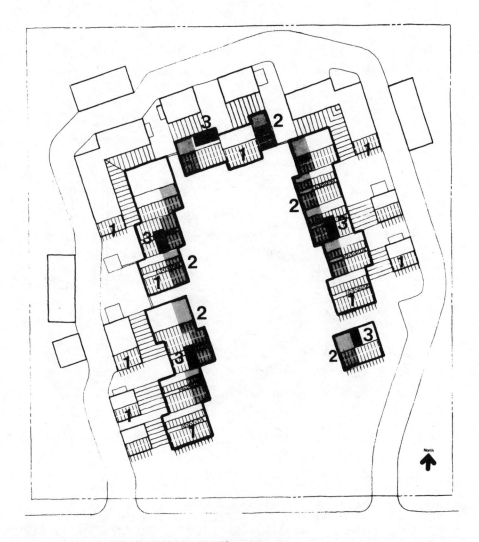

SOLAR ENERGY SYSTEM PLAN (BUILDING)

1. ROOF MOUNTED FLAT PLATE COLLECTORS
2. HEATED FLUID STORAGE TANKS (BASEMENT)
3. BUILDING SERVICES (PUMPS, BACK-UP ENERGY SOURCE, ETC.)

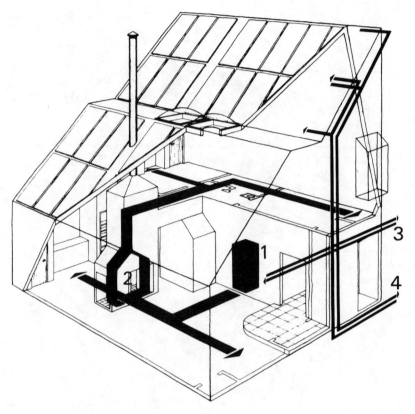

SOLAR ENERGY SYSTEM PLAN   (LIVING UNIT)

1.   HEATING AND HUMIDIFYING EQUIPMENT
2.   CENTRAL FIREPLACE WITH INTEGRAL DUCTWORK
3.   HOT FLUID DISTRIBUTION FROM CENTRAL ENERGY STORAGE TANKS
4.   HOT FLUID SUPPLY TO CENTRAL ENERGY STORAGE TANKS

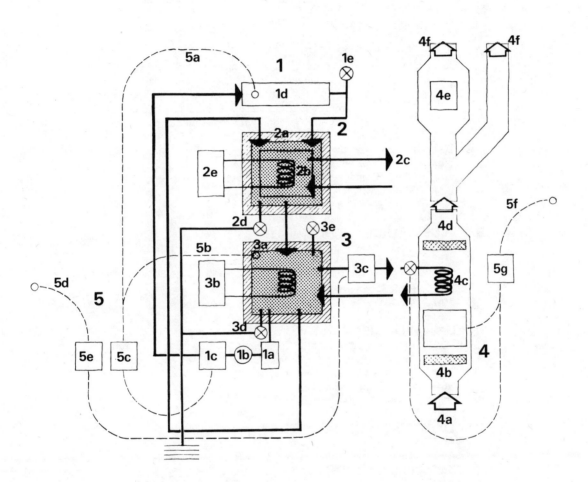

SOLAR SYSTEM SCHEMATIC
COOL AND TEMPERATE REGIONS

1) **SOLAR COLLECTION**
   Description: Expanded aluminum-internalized fluid channels,
                       with selective coating, typ. unit dimension 36"
                       x 84" x 4" aluminum frame with low iron con-
                       tent insulating glass. Fiberglass insulation @
                       4" - 6" integral with roof structured.

   1(a)   Fluid Filter
   1(b)   Fluid Flow Meter
   1(c)   Fluid Pump To Solar Collectors/Storage Tanks
   1(d)   Solar Collector
   1(e)   Pressure Release Valve

2) **DOMESTIC HOT WATER STORAGE**
   2(a)   Low Temperature Solar Heat Storage Tank (Insulated)
   2(b)   Domestic Hot Water Heat Exchanger/Storage Tank
   2(c)   Domestic Hot Water Supply To Living Units
   2(d)   Fluid Drain and Valve
   2(e)   Auxiliary Gas Heater (Immersion Coil-Closed Fluid Loop)

3) **MAIN SOLAR HEAT STORAGE**
   3(a)   Low Temperature Main Solar Heat Storage Tank (Insulated)
   3(b)   Auxiliary Gas Heater (Immersion Coil-Closed Fluid Loop)
   3(c)   Fluid Pump To Living Units
   3(d)   Fluid Drain and Valve
   3(e)   Pressure Release Valve

4) **LIVING UNIT FAN-COIL HEATER**
   4(a)   Air Supply
   4(b)   Air Filter
   4(c)   Unit Fan-Coil Heater
   4(d)   Humidifier
   4(e)   Fireplace With Integral Ductwork
   4(f)   Hot Air Distribution

5) **THERMISTOR SENSORS AND ELECTRONIC SWITCHING**
   5(a)   Solar Collector Monitoring Thermistor
   5(b)   Storage Tank Monitoring Thermistor
   5(c)   Fluid Pump Switching Monitor
   5(d)   Inside Temperature Monitoring Thermistor
   5(e)   Fan/Valve Switching Monitor
   5(f)   Fluid Pump Switching Monitor

TITLE	CLUSTER DESIGN FOR A HOT-ARID CLIMATE
CODE	
KEYWORDS	HOUSING SHADING DEVICES VEGETATION BREEZES ORIENTATION SUN SCREEN LOUVERED BALCONIES AWNING GREENHOUSE CROSS VENTILATION FLAT PLATE SOLAR COLLECTOR TRANSPORT FLUID HEATING COOLING INSULATED TANK HEAT EXCHANGER ANGULAR TILT
AUTHOR	GIFFELS ASSOCIATES, INC.
DATE & STATUS	THE LIVING UNITS AND THE CLUSTER PATTERN FOR A HOT, ARID CLIMATE HAVE BEEN DESIGNED, BUT NOT BUILT, AS OF 1975.
LOCATION	SOUTHWESTERN UNITED STATES
SCOPE	THE LIVING UNITS ARE ARRANGED IN TWO BUILDINGS ALONG A CENTRAL COMMUNITY SPACE, ORIENTED TO EXPOSE EACH LIVING UNIT TO PENETRATION BY COOLING EAST-WEST BREEZES. THE BUILDINGS ARE THREE STORIES HIGH, WITH PARTIAL BASEMENT AREAS BELOW AND DOUBLE RAINWATER COLLECTOR ROOF CONSTRUCTION ABOVE. THE PLAN INVOLVES SUN SCREENS, BALCONIES, EXTERIOR ENTRY STAIRS, EXTERIOR WALKWAYS, AND OTHER EXTERIOR LIVING SPACES (BOTH AT GRADE AND ABOVE PARKING AREAS). VEGETATION IS CLUSTERED ABOUT THE NORTH AND SOUTH SIDES TO CHANNEL COOLING BREEZES INTO THE LIVING UNITS. AN OFFICE/COMMUNITY BUILDING AND POOL ARE INCLUDED WITHIN THE CENTRAL COMMUNITY SPACE.

THE UNIT WHICH IS ILLUSTRATED, IS A TWO-BEDROOM, TWO STORY RESIDENCE, WHICH IS ADAPTABLE TO OCCUPANCY BY SHARING ADULTS OR SMALL FAMILIES. THE LIVING AND DINING SPACES, AND THE TWO SLEEPING SPACES ARE EACH ARRANGED ALONG THE PREDOMINANT WIND AXIS, WITH EXPANSIVE WINDOWS AT THE EXTERIOR. THIS PROVIDES A MAXIMUM CROSS VENTILATION. THE BALCONIES, PATIOS, AND EXTERIOR LIVING SPACES ARE ARRANGED ALONG SIDE LIVING, DINING, AND SLEEPING SPACES. THIS USE OF THE EXTERIOR SPACES IN LIEU OF THE INTERIOR SPACES LESSENS THE HEATING AND COOLING REQUIREMENTS.

TECHNIQUE	BOTH EXPANSIVE EAST AND WEST FACES OF THE BUILDINGS ARE FULLY PROTECTED FROM DIRECT SUNLIGHT BY EXTENSIVE EGG CRATE SUN SCREENS, VERTICAL LOUVERED SUN SCREENS, OPEN BALCONIES, COVERED EXTERIOR LIVING SPACES, AWNINGS, GREENHOUSES, ETC. THESE ELEMENTS COMBINE WITH THE SCREEN WALL AND CISTERNS AT THE SOUTHERN EDGES, AND WITH THE RAINWATER COLLECTION ROOF TO GIVE ALMOST COMPLETE ISOLATION OF THE BUILDING FROM DIRECT SUNLIGHT. EAST-WEST CROSS VENTILATION THROUGH LARGE WINDOW AND CLERESTORY OPENINGS IS ALSO MAXIMIZED. THE SOLAR ENERGY COLLECTORS, HEATED FLUID AND MECHANICAL EQUIPMENT ARE DETAILED TO BE SEPARATE FROM THE BUILDING TO MINIMIZE THE TRANSMISSION OF HEAT TO LIVING SPACES DURING THE PREDOMINANT COOLING SEASON.

THE SOLAR ENERGY SYSTEM FOR THE HOT-ARID REGION HAS PROVISIONS FOR BOTH SOLAR HEATING AND COOLING TO PROVIDE RELIEF FROM THE EXTENSIVE COOLING SEASON AND SOME MEANS OF HEATING DURING THE RELATIVELY SHORT HEATING SEASON.

SOLAR COOLING BECOMES ECONOMICAL IN THIS REGION DUE TO THE FAVORABLE AMOUNT OF CLEAR DAYS OVER THE COOLING SEASON. SOLAR ENERGY COLLECTION EMPLOYS CLOSED LOOP, ALUMINUM, FLAT PLATE COLLECTOR UNITS. THE ENERGY TRANSPORT FLUID MEDIUM CONSISTS OF A WATER AND ETHYLENE GLYCOL (ANTI-FREEZE) SOLUTION. THIS SOLAR HEATED FLUID IS STORED IN AN INSULATED TANK FOR USE WHEN HEATING OR COOLING IS REQUIRED, AND DELIVERED TO INDIVIDUAL FAN-COIL UNITS LOCATED WITHIN EACH LIVING UNIT. DELIVERY OF THE SPACE HEAT IS THROUGH DUCTWORK, WHICH IS INTEGRATED WITH THE FIREPLACE FOR INCREASED HEAT COLLECTION AND DISTRIBUTION WHEN THE FIREPLACE IS IN USE. A DOMESTIC HOT WATER HEAT EXCHANGER UNIT IS ALSO INTEGRATED WITH THE SOLAR ENERGY STORAGE SYSTEM, TO PROVIDE DOMESTIC HOT WATER NEEDS.

THE SOLAR ENERGY FLAT PLATE COLLECTORS ARE WELDED AND AIR EXPANDED 2 PLY ALUMINUM, WITH INTERNALIZED FLUID CHANNELS, A SELECTIVE COATED SURFACE, BACK INSULATION AND TWO PANE, LOW IRON CONTENT GLASS. THE ATTAINABLE FLUID TEMPERATURES GENERALLY RANGE BETWEEN 100 F TO 180 F. TYPICAL ASSEMBLED UNIT DIMENSIONS ARE APPROXIMATELY 36X84X4 INCHES.

THE ORIENTATION (AZIMUTH) RANGE IS FROM 170 TO 200 DEGREES
FOR GOOD SOLAR COLLECTOR PERFORMANCE. THE COLLECTOR'S ANGULAR
TILT IS ADJUSTABLE AND VARIES BETWEEN 10 TO 45 DEGREES (WITH
A SEASONAL OPTIMUM ATTITUDE RANGE OF APPROXIMATELY 15 TO 48
DEGREES).

THE HEAT ABSORBING FLUID MEDIUM IS A MIXTURE OF WATER (80%)
AND ETHYLENE GLYCOL (20%) TO PREVENT FREEZING OF THE FLUID
MEDIUM WITHIN THE COLLECTOR UNITS DURING WINTER NIGHTS.

THE SOLAR ENERGY STORAGE SYSTEM CONSISTS OF THREE SEPARATE
INSULATED STORAGE TANKS: THE DOMESTIC HOT WATER HEAT EXCHANGER/
STORAGE TANK, THE MAIN SOLAR HEATED FLUID STORAGE TANK AND THE
HIGH TEMPERATURE SOLAR HEATED FLUID STORAGE TANK (180 F TO
220 F) FOR THE ABSORPTION COOLING MACHINE.

AS THE SOLAR HEATED FLUID FIRST ARRIVES AT THE STORAGE SYSTEM,
IT IS USED TO HEAT THE DOMESTIC HOT WATER HEAT EXCHANGER/STORAGE
TANK. THIS IMMERSED TANK SHOULD BE TIGHTLY SEALED TO PREVENT ANY
POSSIBLE CROSS-CONTAMINATION BETWEEN THESE TWO FLUIDS.

HAVING TRANSFERRED PART OF ITS HEAT CONTENT TO THE DOMESTIC
HOT WATER SYSTEM, THE SOLAR HEATED FLUID THEN FLOWS INTO THE
MAIN SOLAR HEATED FLUID STORAGE TANK. DURING THE HEATING
SEASON, THE FLUID IS STORED HERE UNTIL THERE IS A HEATING
DEMAND IN ONE OF THE LIVING UNITS. UPON INITIATION OF THIS
DEMAND, PART OF THE STORED SOLAR HEATED FLUID IS PUMPED UP TO THE
UNIT'S INDIVIDUAL FAN-COIL AIR CONDITIONER WHERE THE HEAT IS
TRANSFERRED, VIA THE FAN-COIL SYSTEM, TO THE HOT AIR DUCTWORK
DISTRIBUTION SYSTEM WITHIN THE DWELLING UNIT. THE MAIN DUCT,
LEADING FROM THE AIR CONDITIONER, IS INTEGRATED WITH THE
DWELLING UNIT'S FIREPLACE FOR MORE EFFICIENT COLLECTION AND
DISTRIBUTION OF THE HEAT GENERATED WHEN THE FIREPLACE IS IN USE.

DURING THE COOLING SEASON, PART OF THE SOLAR HEATED FLUID IS
PUMPED TO A SEPARATE STORAGE TANK WHERE IT IS ELEVATED IN
TEMPERATURE BY A GAS FIRED BOOSTER HEATER TO PROVIDE HIGH GRADE
HEATED FLUID TO THE ABSORPTION COOLING MACHINE. THE BUILDING'S
CENTRAL ABSORPTION COOLING MACHINE GENERATES THE CHILLED WATER
FOR THE INDIVIDUAL-DWELLING UNIT'S FAN-COIL ABSORPTION AIR
CONDITIONERS. THE FLUID COOLED AIR IS DISTRIBUTED BY THE
LIVING UNITS' DUCTWORK.

THE DOMESTIC HOT WATER, MAIN SOLAR HEATED FLUID STORAGE TANK,
AND THE HIGH TEMPERATURE FLUID STORAGE TANK ARE SUPPLEMENTED
BY GAS FIRED BACK-UP HEATING UNITS, WHERE THE AUXILIARY HEAT
IS TRANSFERRED TO THE SOLAR STORAGE SYSTEMS, VIA THE USE OF
FLUID-TO-FLUID HEAT EXCHANGERS.

MATERIAL   EARTH BERMS, HORIZONTAL LOUVERED SUNSCREENS, ROOFTOP RAINWATER
                COLLECTORS, CLOSED LOOP ALUMINUM FLAT PLATE COLLECTORS, WATER,
                ETHYLENE GLYCOL, INSULATED TANKS, FAN-COIL UNITS, DUCTWORK
                INTEGRATED WITH FIREPLACE, DOMESTIC HOT WATER HEAT EXCHANGER,
                SELECTIVE COATED SURFACE, BACK INSULATION, TWO PANE, LOW IRON
                CONTENT GLASS, HEAT ABSORPTION COOLING MACHINE

AVAILABILITY   ADDITIONAL INFORMATION ON THE CLUSTER DESIGN FOR A HOT-ARID
                CLIMATE MAY BE OBTAINED BY WRITING TO GIFFELS ASSOCIATES, INC.,
                1000 MARQUETTE BLDG., DETROIT, MICHIGAN 48226.

REFERENCE   GIFFELS ASSOCIATES, INC.
                DESIGN CONCEPT NO. 4: HOT ARID REGION
                DETROIT, MICHIGAN, IN SOLAR ENERGY AND HOUSING DESIGN CONCEPTS,
                PUBLISHED BY GIFFELS ASSOCIATES, INC., JANUARY 1975, PP 75-91

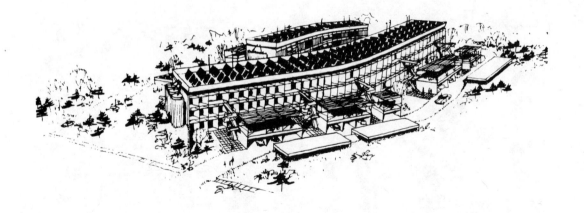

PRIMARY ORIENTATION: SOLAR, WIND

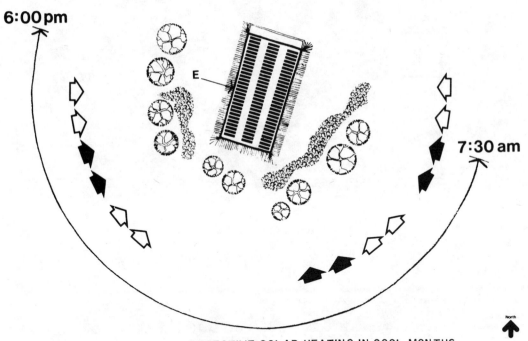

6:00 pm

E

7:30 am

North

EFFECTIVE SOLAR HEATING IN COOL MONTHS
AND SUMMER NIGHTS

SUN PROTECTION IN WARM MONTHS
WITHOUT SHADE ON SOLAR COLLECTORS
OR INTERFERENCE WITH SUMMER BREEZE

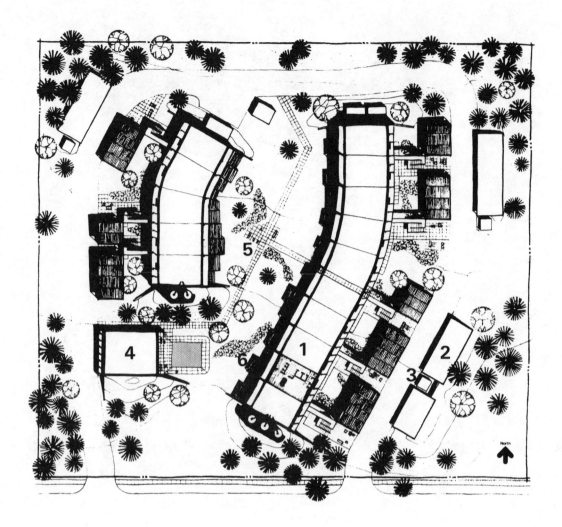

SITE PLAN

1.   LIVING UNITS
2.   COVERED PARKING
3.   BUILDING SERVICES (STORAGE, TRASH COLLECTION, RECREATION
        AND EQUIPMENT)
4.   COMMUNITY BUILDING (OFFICE AND POOL)
5.   COMMUNITY SPACE
6.   EXTERIOR LIVING SPACE (PATIO, BALCONIES ABOVE)
7.   EXTERIOR LIVING SPACE (OVER PARKING)

VIEW FROM S-E

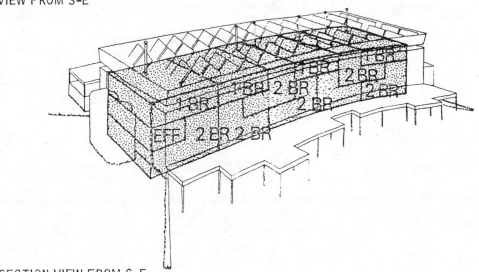

SECTION VIEW FROM S-E

COMPONENT	CONSTRUCTION
FOUNDATIONS & STRUCTURE	Concrete footings and foundation walls.
	Masonry bearing walls, separating individual units precast concrete floor deck and concrete rain water collection structure above roof.
EXTERIOR WALLS	Masonry.
ROOFING	Built up asphaultic roofing material.
EXTERIOR WINDOW OPENINGS	Metal or wood frame operating windows, with double pane insulating glass, tinted in color.
DOORS & FRAMES	VARIES ACCORDING TO USER PREFERENCE.
FLOORS	VARIES ACCORDING TO USER PREFERENCE.
CEILINGS	VARIES ACCORDING TO USER PREFERENCE.
INTERIOR PARTITIONS	VARIES ACCORDING TO USER PREFERENCE.
INTERIOR FINISHES	VARIES ACCORDING TO USER PREFERENCE.
	Note: Paint color and type specifically relates to the heating/ cooling requirements of the space and should be considered in detail.
MOISTURE PROTECTION	COMPLETE IMPERVIOUS VAPOR BARRIER.
INSULATION	Extensive application of very low thermal conductivity insulating materials.
CABINETWORK	VARIES ACCORDING TO USER PREFERENCE.
VERTICAL CIRCULATION	Wood or metal stairway open to the exterior.

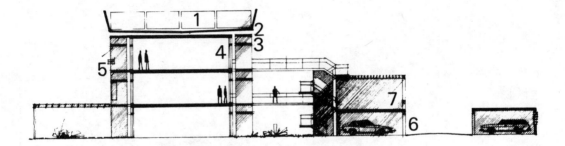

E-W SECTION

1. ADJUSTABLE FLAT PLATE COLLECTORS
2. DOUBLE ROOF AND RAINWATER COLLECTOR
3. SUN SCREEN
4. FULL HEIGHT WINDOW AND CLEARSTORY
5. BALCONY WITH LOUVERED HAND RAIL AND ADJUSTABLE SUN SCREEN
6. GARAGE
7. EXTERIOR LIVING AREA

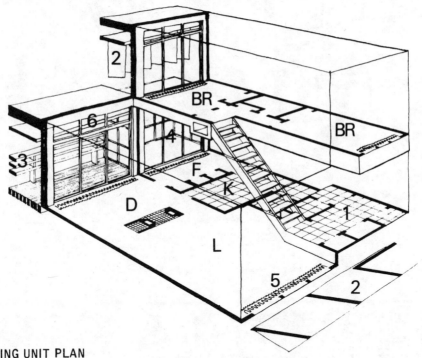

LIVING UNIT PLAN

1. VESTIBULE
2. FIXED EGG CRATE SUN SCREEN
3. BALCONY WITH LOUVERED HAND RAIL AND SUN SCREEN
4. GREENHOUSE
5. CURTAIN(S)
6. CLEARSTORY WINDOW OPENINGS

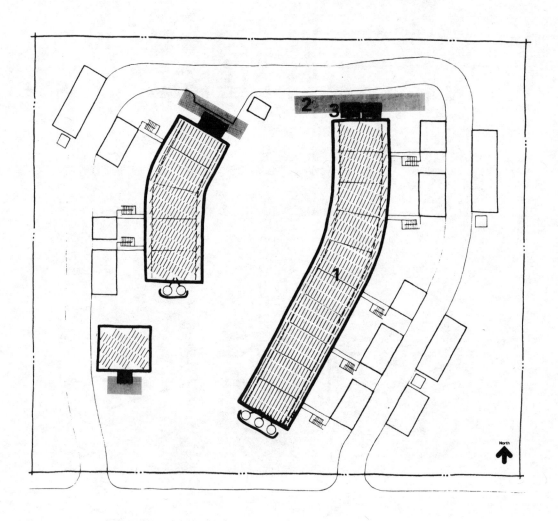

SOLAR ENERGY SYSTEM PLAN (BUILDING)

1. ADJUSTABLE FLAT PLATE COLLECTORS
2. HEATED FLUID STORAGE TANKS (BELOW GRADE)
3. BUILDING SERVICES (PUMPS, CHILLER, BACK-UP ENERGY SOURCE, ETC.)

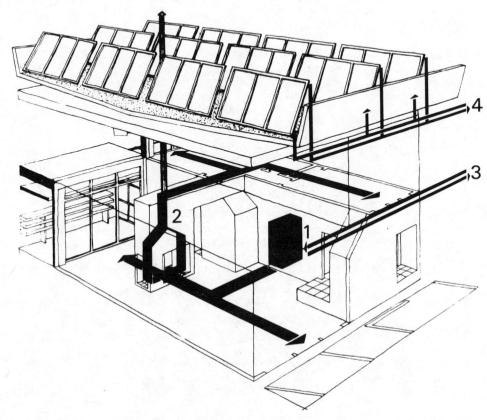

SOLAR ENERGY SYSTEM PLAN (LIVING UNIT)

1. HEATING/AIR CONDITIONING AND HUMIFYING EQUIPMENT
2. CENTRAL FIREPLACE WITH INTEGRAL DUCTWORK
3. HOT FLUID DISTRIBUTION FROM CENTRAL ENERGY STORAGE TANKS
4. HOT FLUID SUPPLY TO CENTRAL ENERGY STORAGE TANKS

TITLE	CLUSTER DESIGN FOR A HOT-HUMID CLIMATE
CODE	
KEYWORDS	HOUSING HORIZONTAL LOUVERED ECO ROOF CROSS VENTILATION LARGE WINDOW CLERESTORY FLAT PLATE SOLAR COLLECTOR STORAGE HOT HUMID HEATING COOLING INSULATED TANK HEAT EXCHANGER INTERNALIZED CHANNEL ABSORPTION AIR CONDITIONER
AUTHOR	GIFFELS ASSOCIATES, INC.
DATE & STATUS	THE LIVING UNITS AND THE CLUSTER DESIGN FOR A HOT-HUMID CLIMATE HAVE BEEN DESIGNED, BUT NOT BUILT AS OF 1975.
LOCATION	GULF COAST
SCOPE	THE LIVING UNITS ARE ARRANGED IN FOUR BUILDINGS ALONG A CENTRAL COMMUNITY SPACE, ORIENTED TO EXPOSE EACH LIVING UNIT TO PENETRATION BY COOLING NORTH-SOUTH BREEZES. THE BUILDINGS ARE THREE STORIES IN HEIGHT, RAISED ABOVE PARTIAL BASEMENT AREAS BELOW AND WITH DOUBLE ECO-ROOF CONSTRUCTION ABOVE. PARKING SHELTER ROOFS ARE USED TO HOUSE THE SOLAR ENERGY COLLECTORS, AND BERM CONSTRUCTION BEYOND THE PARKING COVERS THE HEATED FLUID STORAGE.

THE PLAN INVOLVES EXTENSIVE EXTERIOR SURFACE AREA, INCLUDING BALCONIES, EXTERIOR ENTRY STAIRS, EXTERIOR WALKWAYS, AND OTHER EXTERIOR LIVING SPACES. THESE FEATURES ARE ARRANGED TO PROVIDE SUNSCREEN PROTECTION.

VEGETATION IS CLUSTER ABOUT THE EAST AND WEST SIDES TO CHANNEL COOLING BREEZES INTO THE LIVING UNITS. BERM AND VEGETATION PROTECTION ARE PROVIDED ALONG THE EXPOSED EAST AND WEST HEDGES.

AN OFFICE/COMMUNITY BUILDING ARE INCLUDED WITHIN THE CENTRAL COMMUNITY SPACE.

TECHNIQUE	THE EXPANSIVE SOUTHERN FACES OF THE BUILDINGS ARE FULLY PROTECTED FROM DIRECT SUNLIGHT BY EXTENSIVE HORIZONTAL LOUVERED SUNSCREENS, OPEN BALCONIES, EXTERIOR LIVING SPACES, EXTERIOR ENTRYWAYS. THESE ELEMENTS COMBINE WITH THE ECO-ROOF TO GIVE ALMOST COMPLETE ISOLATION OF THE BUILDING FROM THE DIRECT SUNLIGHT. NORTH-SOUTH CROSS VENTILATION THROUGH LARGE WINDOW AND CLERESTORY OPENINGS IS ALSO MAXIMIZED.

THE SOLAR ENERGY COLLECTORS, HEATED FLUID STORAGE, AND MECHANICAL EQUIPMENT ARE LOCATED AWAY FROM THE BUILDING TO MINIMIZE THE TRANSMISSION OF HEAT TO LIVING SPACES DURING THE PREDOMINANT COOLING SEASON.

THE SOLAR ENERGY SYSTEM FOR THE HOT-HUMID REGION HAS PROVISIONS FOR BOTH SOLAR HEATING AND COOLING, TO PROVIDE BOTH RELIEF FROM THE EXTENSIVE COOLING SEASON AND SOME MEANS OF HEATING DURING THE RELATIVELY SHORT HEATING SEASON.

SOLAR ENERGY COLLECTION EMPLOYS CLOSED LOOP, ALUMINUM, FLAT PLATE COLLECTOR UNITS. THE ENERGY TRANSPORT FLUID MEDIUM CONSISTS OF A WATER AND ETHYLENE GLYCOL SOLUTION. THIS SOLAR HEATED FLUID IS STORED IN AN INSULATED TANK FOR USE WHEN HEATING OR COOLING IS REQUIRED AND DELIVERED TO INDIVIDUAL FAN COIL UNITS LOCATED WITHIN EACH LIVING UNIT. DELIVERY OF THE SPACE HEAT IS THROUGH DUCTWORK, WHICH IS INTEGRATED WITH THE FIREPLACE FOR INCREASED HEAT COLLECTION AND DISTRIBUTION WHEN THE FIREPLACE IS IN USE. A DOMESTIC HOT WATER HEAT EXCHANGER UNIT IS ALSO INTEGRATED WITH THE SOLAR ENERGY STORAGE SYSTEM TO PROVIDE FOR DOMESTIC HOT WATER NEEDS.

THE SOLAR ENERGY FLAT PLATE COLLECTORS ARE WELDED AND AIR EXPANDED 2-PLY ALUMINUM, WITH INTERNALIZED FLUID CHANNELS. WITH A SELECTIVE COATED SURFACE, BACK INSULATION AND TWO PANE, LOW IRON CONTENT GLASS, THE ATTAINABLE FLUID TEMPERATURE GENERALLY RANGES BETWEEN 100 F TO 180 F. TYPICAL ASSEMBLED UNIT DIMENSIONS ARE APPROXIMATELY 36"X84"X4". THE SOLAR CONVERSION EFFICIENCY RANGES FROM 32% TO 48% WITH A YEARLY EFFICIENCY OF AROUND 40%.

THE ORIENTATION (AZIMUTH) RANGE IS FROM 170 DEGREES TO
200 DEGREES FOR GOOD SOLAR COLLECTOR PERFORMANCE. THIS
INSTALLATION INCLUDES COLLECTORS ADJUSTABLE FROM AN ATTITUDE
OF 10 DEGREES TO 45 DEGREES (WITH A SEASONAL OPTIMUM ATTITUDE
RANGE OF APPROXIMATELY 12 F TO 45 F.

THE HEAT ABSORBING FLUID MEDIUM IS A MIXTURE OF WATER (80%) AND
ETHYLENE GLYCOL (20%) TO PREVENT FREEZING OF THE FLUID MEDIUM
WITHIN THE COLLECTOR UNITS DURING THE WINTER NIGHTS.

THE SOLAR ENERGY STORAGE SYSTEM CONSISTS OF THREE SEPARATE
INSULATED STORAGE TANKS: THE DOMESTIC HOT WATER HEAT EXCHANGER/
STORAGE TANK, THE MAIN SOLAR HEATED FLUID STORAGE TANK AND THE
HIGH TEMPERATURE SOLAR HEATED FLUID STORAGE TANK (180 F - 220 F)
FOR THE ABSORPTION COOLING MACHINE.

AS THE SOLAR HEATED FLUID FIRST ARRIVES AT THE STORAGE SYSTEM,
IT IS USED TO HEAT THE DOMESTIC HOT WATER HEAT EXCHANGER/STORAGE
TANK. THIS IMMERSED TANK SHOULD BE TIGHTLY SEALED TO PREVENT
ANY POSSIBLE CROSS-CONTAMINATION BETWEEN THESE TWO FLUIDS.

HAVING TRANSFORMED PART OF ITS HEAT CONTENT TO THE DOMESTIC HOT
WATER SYSTEM, THE SOLAR HEATED FLUID THEN FLOWS INTO THE MAIN
SOLAR HEATED FLUID STORAGE TANK. DURING THE HEATING SEASON, THE
FLUID IS STORED HERE UNTIL THERE IS A HEATING DEMAND IN ONE OF
THE LIVING UNITS. UPON INITIATION OF THIS DEMAND, PART OF THE
STORED SOLAR HEATED FLUID IS PUMPED UP TO THE UNIT'S INDIVIDUAL
FAN COIL AIR CONDITIONER WHERE THE HEAT IS TRANSFERRED, VIA
THE FAN-COIL SYSTEM, TO THE HOT AIR DUCTWORK DISTRIBUTION
SYSTEM WITHIN THE DWELLING UNIT'S FIREPLACE FOR MORE EFFICIENT
COLLECTION AND DISTRIBUTION OF THE HEAT GENERATED WHEN THE
FIREPLACE IS IN USE.

DURING THE COOLING SEASON, PART OF THE SOLAR HEATED FLUID IS
PUMPED TO A SEPARATE STORAGE TANK WHERE IT IS ELEVATED IN
TEMPERATURE BY A GAS FIRED BOOSTER HEATER TO PROVIDE HIGH GRADE
HEATED FLUID TO THE ABSORPTION COOLING MACHINE. THE BUILDING'S
CENTRAL ABSORPTION COOLING MACHINE GENERATES THE CHILLED WATER
FOR THE INDIVIDUAL DWELLING UNIT'S FAN-COIL ABSORPTION AIR
CONDITIONERS. THE FLUID COOLED AIR IS DISTRIBUTED BY THE
LIVING UNIT'S DUCTWORK.

THE DOMESTIC HOT WATER, MAIN SOLAR HEATED FLUID STORAGE TANK,
AND THE HIGH TEMPERATURE FLUID STORAGE TANK ARE SUPPLEMENTED BY
GAS BACK-UP HEATING UNITS, WHERE THE AUXILIARY HEAT IS TRANS-
FERRED TO THE SOLAR STORAGE SYSTEMS, VIA THE USE OF FLUID-TO-
FLUID HEAT EXCHANGERS.

MATERIAL   EARTH BERMS, HORIZONTAL LOUVERED SUNSCREENS, ECO-ROOF, LARGE
                WINDOW, CLERESTORY, CLOSED LOOP ALUMINUM FLAT PLATE COLLECTORS,
                WATER, ETHYLENE GLYCOL, INSULATED TANK, FAN-COIL UNITS,
                DUCTWORK INTEGRATED WITH THE FIREPLACE, DOMESTIC HOT WATER
                HEAT EXCHANGER, SELECTIVE COATED SURFACE, BACK INSULATION,
                TWO PANE LOW IRON CONTENT GLASS, HEAT ABSORPTION COOLING MACHINE

AVAILABILITY   ADDITIONAL INFORMATION ON THE CLUSTER DESIGN FOR A HOT-HUMID
                CLIMATE MAY BE OBTAINED BY WRITING TO GIFFELS ASSOCIATES, INC.,
                1000 MARQUETTE BLDG., DETROIT, MICHIGAN 48226.

REFERENCE   GIFFELS ASSOCIATES, INC.
                DESIGN CONCEPT NO. 3: HOT-HUMID REGION
                DETROIT, MICHIGAN, IN SOLAR ENERGY AND HOUSING DESIGN CONCEPTS,
                PUBLISHED BY GIFFELS ASSOCIATES, INC., JANUARY 1975, PP 55-73

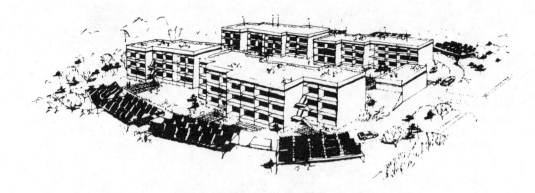

PRIMARY ORIENTATION: SOLAR, WIND
SECONDARY ORIENTATION: PRECIPITATION, WATER TABLE

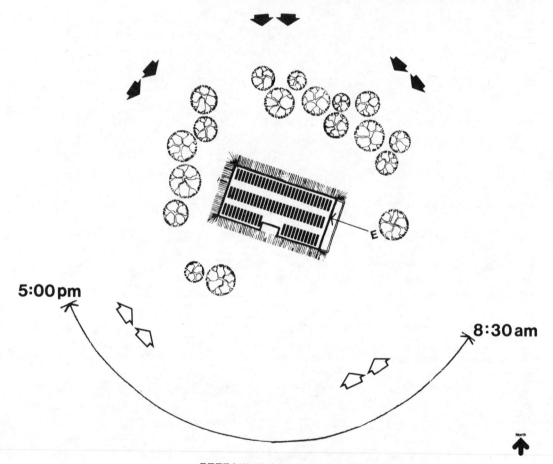

5:00 pm

8:30 am

E

North

EFFECTIVE SOLAR HEATING IN COOL MONTHS

SUN PROTECTION IN WARM MONTHS
WITHOUT SHADE ON SOLAR COLLECTORS
OR INTERFERENCE WITH SUMMER BREEZE

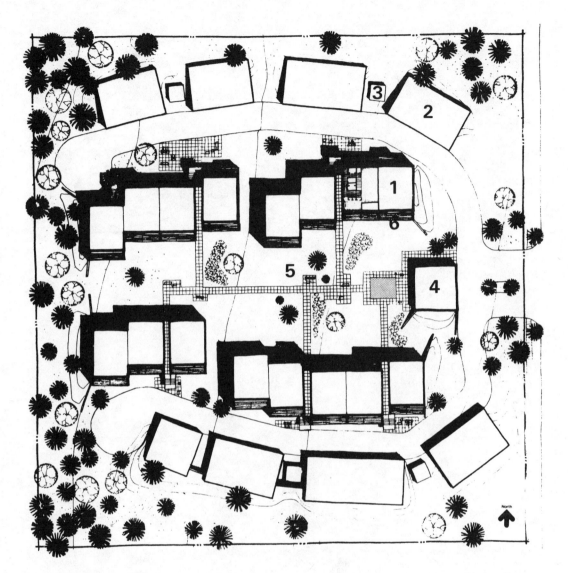

SITE PLAN

1. LIVING UNITS
2. COVERED PARKING
3. BUILDING SERVICES (STORAGE, TRASH COLLECTION, RECREATION
   AND EQUIPMENT)
4. COMMUNITY BUILDING (OFFICE AND POOL)
5. COMMUNITY SPACE
6. EXTERIOR LIVING SPACES (PATIO, BALCONIES ABOVE)

VIEW FROM S-W

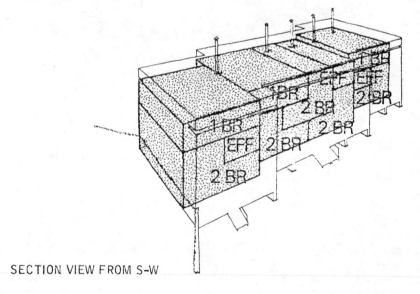

SECTION VIEW FROM S-W

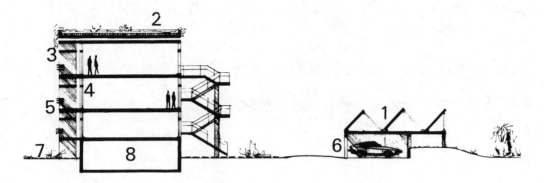

N-S SECTION

1.   ADJUSTABLE FLAT PLATE COLLECTORS
2.   DOUBLE ROOF AND ECO-ROOF
3.   SUN SCREEN
4.   FULL HEIGHT WINDOW AND CLEARSTORY
5.   BALCONY WITH LOUVERED HAND RAIL AND ADJUSTABLE SUN SCREEN
6.   COVERED PARKING
7.   EXTERIOR LIVING AREA
8.   BASEMENT (STORAGE, LAUNDRY, ETC.)

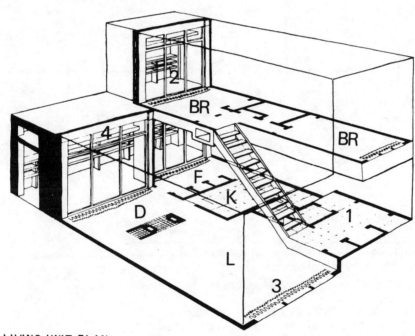

LIVING UNIT PLAN

1.   VESTIBULE
2.   BALCONY WITH LOUVERED HANDRAIL AND SUNSCREEN
3.   CURTAIN(S)
4.   CLEARSTORY WINDOW OPENINGS

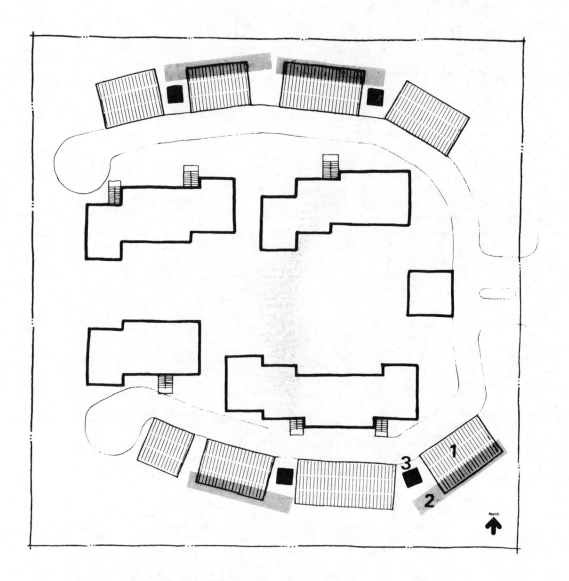

SOLAR ENERGY SYSTEM PLAN (BUILDING)

1.   ADJUSTABLE FLAT PLATE COLLECTORS
2.   HEATED FLUID STORAGE TANKS (BELOW GRADE)
3.   BUILDING SERVICES (PUMPS, CHILLER, BACK-UP ENERGY SOURCE, ETC.)

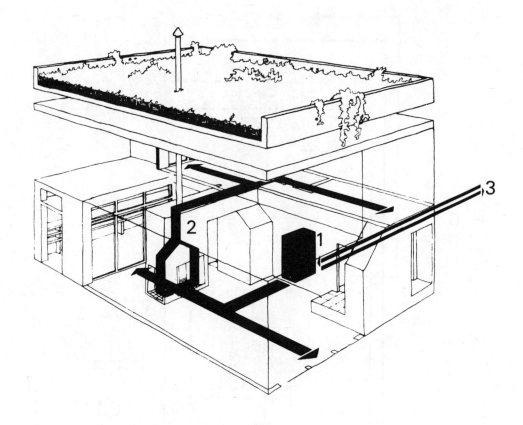

SOLAR ENERGY SYSTEM PLAN (LIVING UNIT)

1.   HEATING/AIR CONDITIONING EQUIPMENT
2.   CENTRAL FIREPLACE WITH INTEGRAL DUCTWORK
3.   HOT FLUID DISTRIBUTION FROM CENTRAL ENERGY STORAGE TANKS

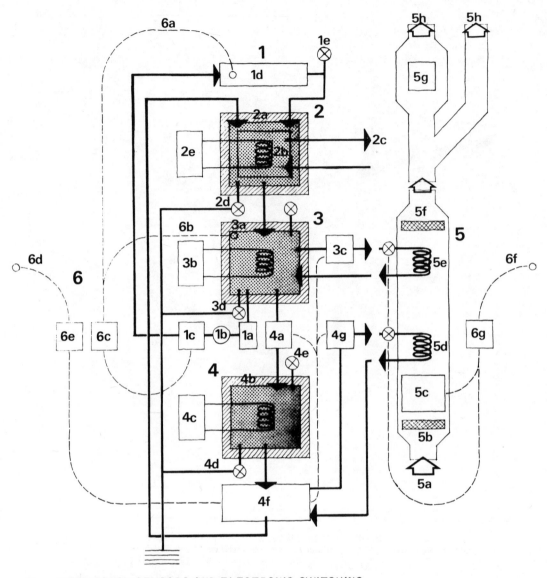

6) THERMISTOR SENSORS AND ELECTRONIC SWITCHING
6(a) Solar Collector Monitoring Thermistor
6(b) Storage Tank Monitoring Thermistor
6(c) Fluid Pump Switching Monitor
6(d) Outside Temperature Monitoring Thermistor
6(e) Air Conditioner and Fluid Pumps (3) Switching Monitor
6(f) Inside Temperature Monitoring Thermistor
6(g) Fan/Valves (2) Switching Monitor

TITLE	CLUSTER DESIGN FOR A TEMPERATE CLIMATE
CODE	
KEYWORDS	HOUSING CONIFEROUS FLAT PLATE SOLAR COLLECTOR TRANSPORT FLUID INSULATED TANK FAN COIL UNIT DUCTWORK FIREPLACE HEAT EXCHANGER DOMESTIC HOT TWO PLY WELDED INTERNALIZED FLUID CHANNEL SELECTIVE COATED SURFACE PANE LOW IRON GLASS CONVERSION EFFICIENCY STORAGE IMMERSED HUMIDIFYING AUXILIARY GAS FIRED
AUTHOR	GIFFELS ASSOCIATES, INC.
DATE & STATUS	THE LIVING UNITS AND CLUSTER DESIGN FOR A TEMPERATE CLIMATE HAVE BEEN DESIGNED, BUT NOT BUILT AS OF 1975.
LOCATION	BOSTON, MASSACHUSETTS
SCOPE	THE LIVING UNITS ARE CLUSTERED WITHIN SIX BUILDINGS, ORIENTED TO PROVIDE PROTECTION OF EXTERIOR SPACES FROM COLD EAST-WEST WINDS. THE BUILDINGS INCLUDE HEIGHTS FROM ONE TO THREE STORIES, WITH ROOFS, AND ROOF MOUNTED SOLAR ENERGY COLLECTORS ORIENTED TO THE SOUTH-WEST. TO PROVIDE THE LARGE SOLAR ENERGY COLLECTOR AREA REQUIRED, GARAGE ROOFS ARE ALSO USED TO HOUSE COLLECTORS. GARAGES ARE ATTACHED ON THE NORTH SIDES OF THE BUILDINGS AS BUFFER SPACES. FENCES, CONIFEROUS VEGETATION, AND HEDGES AUGMENT THIS PROTECTION AS WELL AS SHELTERING THE ENTRY POINTS.

THE BASIC UNIT IN THE CLUSTER IS A TWO-BEDROOM, TWO STORY RESIDENCE, WHICH IS ADAPTABLE TO OCCUPANCY BY SHARING ADULTS OR SMALL FAMILIES. THE TWO STORY UNIT IS USED TO ALLOW A MAXIMUM OF PRIVACY WITHIN ITS PARTS, AS WELL AS FOR ENERGY CONSERVATION REASONS. THE UTILITY SPACE, SECOND BATHROOM (ON THE MAIN FLOOR), AND HALLWAY TO THE KITCHEN ARE DESIGNED TO PROVIDE A WORKING AREA FOR HOUSEKEEPING ACTIVITIES WITHIN THE UNIT ITSELF. THE RAISED AND TILED ENTRY VESTIBULE IS DESIGNED TO PROVIDE A WEATHER LOCK FOR INTERIOR LIVING AREAS AND A MORE FORMAL ENTRY. THE TOTAL AREA OF THE LIVING UNIT IS 1,200 SQUARE FEET. |
| TECHNIQUE | THE SOLAR ENERGY SYSTEM FOR THE TEMPERATE REGION HAS A PROVISION FOR SOLAR HEATING ONLY. A HEATING ONLY SYSTEM IS CHOSEN BECAUSE THE AMOUNT OF COOLING REQUIRED IN THIS REGION AND THE PERCENTAGE OF SUNSHINE AVAILABLE DURING THE COOLING SEASON ARE RELATIVELY SMALL WHEN COMPARED TO THE DEGREE OF SOPHISTICATION AND COSTS OF THE SOLAR COOLING SYSTEMS.

SOLAR ENERGY COLLECTION EMPLOYS CLOSED LOOP, ALUMINUM, FLAT PLATE, COLLECTOR UNITS. THE ENERGY TRANSPORT FLUID MEDIUM CONSISTS OF A WATER AND ETHYLENE GLYCOL (ANTI-FREEZE) SOLUTION. THIS SOLAR HEATED FLUID IS STORED IN AN INSULATED TANK FOR USE WHEN HEATING IS REQUIRED, AND DELIVERED TO INDIVIDUAL FAN-COIL UNITS LOCATED WITHIN EACH LIVING UNIT. DELIVERY OF SPACE HEATING IS AUGMENTED BY DUCTWORK INTEGRAL WITH THE FIREPLACE. A DOMESTIC HOT WATER HEAT EXCHANGER UNIT IS ALSO INTEGRATED WITH THE SOLAR ENERGY STORAGE SYSTEM, TO PROVIDE THE DOMESTIC HOT WATER NEEDS.

THE SOLAR ENERGY FLAT PLATE COLLECTORS ARE WELDED AND AIR EXPANDED 2-PLY ALUMINUM, WITH INTERNALIZED FLUID CHANNELS. WITH A SELECTIVE COATED SURFACE, BACK INSULATION AND TWO PANE, LOW IRON CONTENT GLASS, THE ATTAINABLE FLUID TEMPERATURE GENERALLY RANGES BETWEEN 100 F TO 180 F. TYPICALLY ASSEMBLED UNIT DIMESIONS ARE APPROXIMATELY 36"X84"X4".

THE HEAT ABSORBING FLUID MEDIUM IS A MIXTURE OF WATER (50%) AND ETHYLENE GLYCOL (50%) TO PREVENT FREEZING OF THE FLUID MEDIUM WITHIN THE COLLECTOR UNITS DURING THE WINTER NIGHTS.

THE SOLAR STORAGE SYSTEM CONSISTS OF TWO SEPARATE, INSULATED STORAGE TANKS: THE DOMESTIC HOT WATER HEAT EXCHANGER/STORAGE TANK AND THE PRIMARY SOLAR HEATED FLUID STORAGE TANK. AS THE SOLAR HEATED FLUID FIRST ARRIVES AT THE STORAGE SYSTEM, IT IS USED TO HEAT THE DOMESTIC HOT WATER, HEAT EXCHANGER/ STORAGE TANK. THIS IMMERSED TANK SHOULD BE TIGHTLY SEALED |

TO PREVENT ANY CROSS CONTAMINATION BETWEEN THESE TWO FLUIDS. HAVING TRANSFERRED PART OF ITS HEAT CONTENT TO THE DOMESTIC HOT WATER SYSTEM, THE SOLAR HEATED FLUID THEN FLOWS INTO THE MAIN SOLAR HEATED STORAGE TANK. HERE, THE FLUID IS STORED UNTIL THERE IS A HEATING DEMAND IN ONE OF THE LIVING UNITS. UPON INITIATION OF THIS DEMAND, PART OF THE STORED SOLAR HEATED FLUID IS PUMPED UP TO THE UNIT'S INDIVIDUAL FAN-COIL HEATING AND HUMIDIFYING EQUIPMENT WHERE THE HEAT IS TRANSFERRED VIA THE FAN-COIL SYSTEM, THE HOT AIR DUCTWORK DISTRIBUTION SYSTEM WITHIN THE LIVING UNIT. THE MAIN DUCT LEADING FROM THE EQUIPMENT IS INTEGRATED WITH THE FIREPLACE FOR MORE EFFICIENT COLLECTION AND DISTRIBUTION OF THE HEAT GENERATED WHEN THE FIREPLACE IS IN USE.

THE BACK-UP DOMESTIC HOT WATER AND MAIN STORAGE TANK HEATING UNITS ARE GAS FIRED HOT FLUID SYSTEMS, WHERE THE AUXILIARY HEAT IS TRANSFERRED TO THE SOLAR STORAGE SYSTEMS VIA THE USE OF FLUID-TO-FLUID HEAT EXCHANGERS.

MATERIAL   CONIFEROUS VEGETATION, HEDGES, INSULATION, FIREPLACE, CLOSED LOOP, ALUMINUM, FLAT PLATE, COLLECTOR UNITS, WATER AND ETHYLENE GLYCOL SOLUTION, INSULATED TANK, DUCTWORK, DOMESTIC HOT WATER HEAT EXCHANGER, INDIVIDUAL FAN COIL UNITS, SELECTIVE COATED SURFACE, LOW IRON CONTENT GLASS

AVAILABILITY   ADDITIONAL INFORMATION ON THE CLUSTER DESIGN FOR A TEMPERATE CLIMATE MAY BE OBTAINED BY WRITING TO GIFFELS ASSOCIATES, INC., 1000 MARQUETTE BLDG., DETROIT, MICHIGAN 48226.

REFERENCE   GIFFELS ASSOCIATES, INC.
                   DESIGN CONCEPT NO. 2: TEMPERATE REGION
                   DETROIT, MICHIGAN, IN SOLAR ENERGY AND HOUSING DESIGN CONCEPTS,
                   PUBLISHED BY GIFFELS ASSOCIATES, INC., JANUARY 1975, PP 37-53

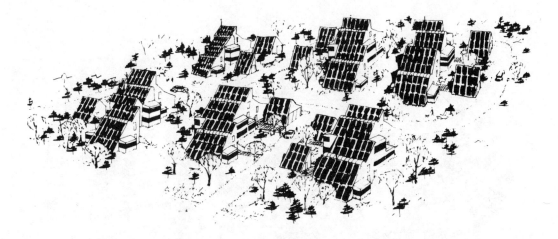

PRIMARY ORIENTATION: SOLAR, WIND
SECONDARY ORIENTATION: PRECIPITATION

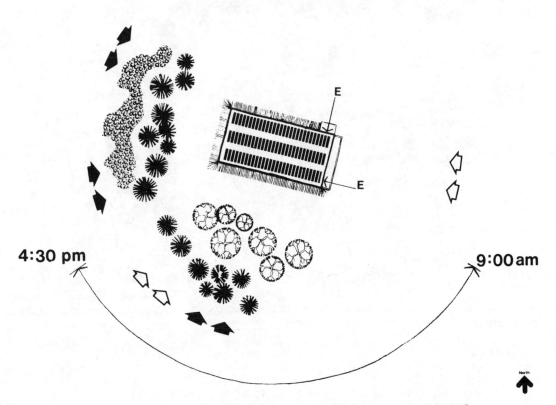

4:30 pm

9:00 am

North

EFFECTIVE SOLAR HEATING IN COOL MONTHS

SUN PROTECTION IN WARM MONTHS
WITHOUT SHADE ON SOLAR COLLECTORS

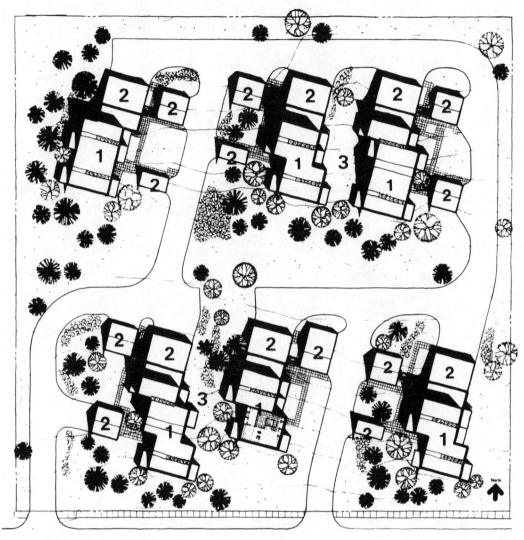

SITE PLAN

1.   LIVING UNITS
2.   GARAGE PARKING
3.   COMMUNITY SPACE
3.   BUILDING SERVICES (PUMPS, CHILLER, MAINTENANCE EQUIPMENT, ETC.)

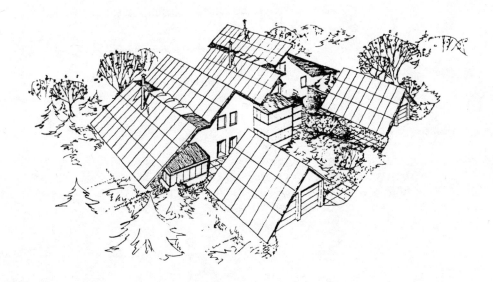

VIEW FROM S-E

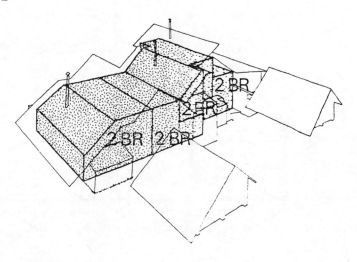

SECTION VIEW FROM S-E

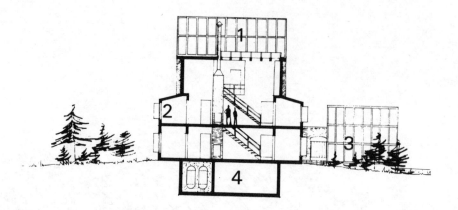

E-W SECTION

1. FLAT PLATE COLLECTORS
2. PORCH WITH ADJUSTABLE WINDOWS
3. GARAGE
4. BASEMENT (STORAGE, LAUNDRY, ETC.)

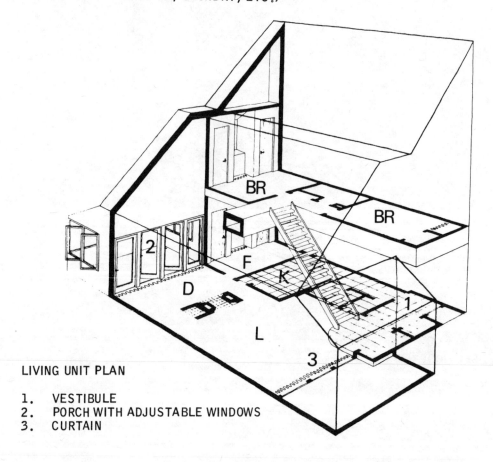

LIVING UNIT PLAN

1. VESTIBULE
2. PORCH WITH ADJUSTABLE WINDOWS
3. CURTAIN

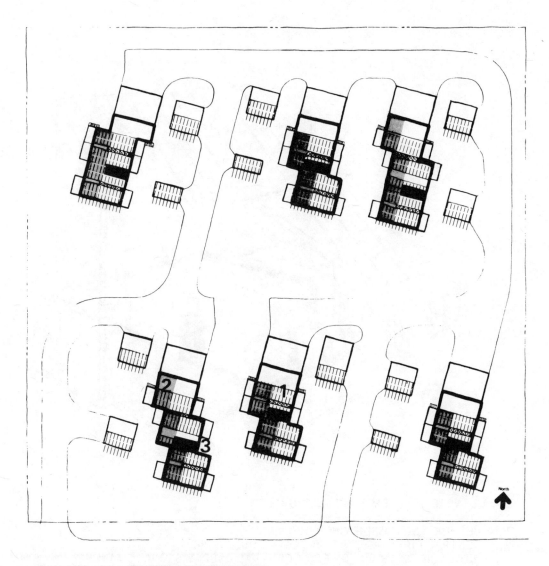

SOLAR ENERGY SYSTEM PLAN (BUILDING)

1.   ROOF MOUNTED FLAT PLATE COLLECTORS
2.   HEATED FLUID STORAGE TANKS (BASEMENT)
3.   BUILDING SERVICES (PUMPS, BACK-UP ENERGY SOURCE, ETC.)

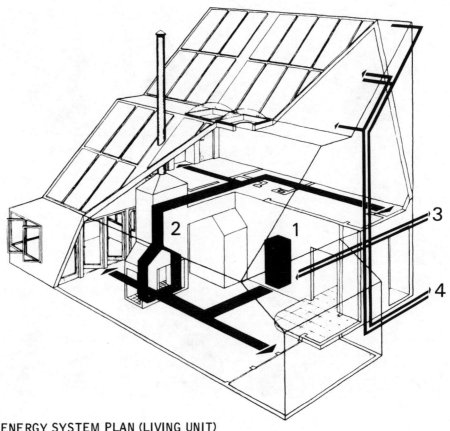

SOLAR ENERGY SYSTEM PLAN (LIVING UNIT)

1. HEATING AND HUMIDIFYING EQUIPMENT
2. CENTRAL FIREPLACE WITH INTEGRAL DUCTWORK
3. HOT FLUID DISTRIBUTION FROM CENTRAL ENERGY STORAGE TANKS
4. HOT FLUID SUPPLY TO CENTRAL ENERGY STORAGE TANKS

TITLE	COLORADO STATE UNIVERSITY HOUSE
CODE	
KEYWORDS	COLLECTOR COMPUTER PROGRAM HEAT EXCHANGER HOUSE LABORATORY SOLAR HEATING COOLING AUXILIARY FUEL THERMAL STORAGE INSULATION DOUBLE GLAZED AIR LOCKED WEATHER TIGHT PRESSURIZED TANK BOILER PUMP
AUTHOR	GEORGE LOF, DAN WARD
DATE & STATUS	THE COLORADO STATE UNIVERSITY HOUSE IS FUNCTIONING AS A SOLAR ENERGY LABORATORY IN 1975.
LOCATION	FORT COLLINS, COLORADO
SCOPE	THE COLORADO STATE UNIVERSITY HOUSE IS AN EXPERIMENTAL RESIDENTIAL SIZED LABORATORY, EQUIPPED WITH AN INTEGRATED SYSTEM UTILIZING SOLAR ENERGY TO DRIVE HEATING, COOLING AND HOT WATER SUBSYSTEMS, COMBINED WITH AUXILIARY FUEL AND THERMAL STORAGE. THE INTERIOR OF THE HOUSE CAN BE CONVERTED TO RESIDENTIAL USE, ALTHOUGH MOST OF THE ROOMS ARE CURRENTLY (WINTER 1974 - 1975) BEING USED AS OFFICES FOR THE RESEARCH STAFF WHICH IS MONITORING THE PERFORMANCE OF THE BUILDING. THE BUILDING HAS BEEN DESIGNED TO PERMIT SPEEDY REPLACEMENT OF THE VARIOUS SUBSYSTEMS SO THAT OTHER SOLAR COMPONENTS CAN BE INSTALLED AND EVALUATED. A COMPUTER PROGRAM DEVELOPED BY THE UNIVERSITY OF WISCONSIN SHOWED THAT, DURING AN AVERAGE YEAR, A DOUBLE-GLAZED COLLECTOR, 800 SQUARE FEET IN AREA, USING A NON-SELECTIVE SURFACE, SHOULD BE ABLE TO PROVIDE APPROXIMATELY 82% OF THE HEATING LOAD, AND 88% OF THE COOLING LOAD. THE ACTUAL AREA OF THE COLLECTORS AS INSTALLED IS 768 SQUARE FEET, SO IT IS EXPECTED THAT ABOUT 83% OF THE HEATING AND 70% OF THE COOLING WILL BE ACCOMPLISHED WITH SOLAR ENERGY. THE AUXILIARY FUEL IS NATURAL GAS.
TECHNIQUE	THE COLLECTOR IS MADE OF ALUMINUM, WITH 16 UNITS EACH 3 FEET WIDE AND 16 FEET LONG, WITH A TUBE PATTERN WHICH WAS DESIGNED TO INSURE EQUAL FLOW AMONG THE COLLECTOR PANELS. TWO DOUBLE STRENGTH NON-TEMPERED COVER GLASSES ARE USED. TWO AND A HALF INCHES OF FIBERGLAS INSULATION ARE USED UNDER THE COLLECTOR.

THE FLUID CIRCULATED THROUGH THE COLLECTORS IS WATER PLUS ETHYLENE GLYCOL, AND A SHELL-AND-TUBE HEAT EXCHANGER IS USED TO TRANSFER THE COLLECTED HEAT TO PURE WATER WHICH IS STORED IN A 1,000 GALLON VERTICAL NON-PRESSURIZED TANK. WATER FROM THIS TANK SERVES A NUMBER OF PURPOSES. IT HELPS TO HEAT THE DOMESTIC HOT WATER THROUGH AN EXCHANGER WHICH IS CONNECTED IN SERIES WITH AN 80 GALLON PREHEAT TANK AND A 40 GALLON GAS FIRED HOT WATER HEATER. IT PROVIDES HEAT FOR THE HOUSE IN WINTER BY CIRCULATING THROUGH A COIL CONNECTED IN THE OUTLET OF THE AIR CONDITIONER. IN SUMMER, THE HOT WATER PROVIDES THE HEAT NEEDED TO ACTIVATE THE ARKLA LITHIUM BROMIDE WATER CHILLER. THE CHILLED WATER COOLS THE AIR WHICH IS DRAWN BACK THROUGH THE RETURN DUCTS FROM THE HOUSE AND THEN IS SUPPLIED THROUGH THE DELIVERY DUCTS.

THE AUXILIARY HOT WATER BOILER PROVIDES WATER FOR A SEPARATE COIL IN THE DELIVERY DUCT WHEN ADDITIONAL HEAT IS NEEDED DURING THE WINTER, AND IT ALSO PROVIDES HEAT IN SUMMER WHENEVER THE SUN-HEATED WATER TEMPERATURE FALLS TOWARDS 185 F, AT WHICH TEMPERATURE THE LITHIUM BROMIDE SYSTEM WOULD NO LONGER FUNCTION. |
MATERIAL	FIBERGLAS INSULATION, DOUBLE GLAZED ALUMINUM COLLECTOR, ETHYLENE GLYCOL, SHELL-AND-TUBE HEAT EXCHANGER, 1,000 GALLON NON-PRESSURIZED TANK, 80 GALLON PREHEAT TANK, 40 GALLON GAS FIRED HOT WATER HEATER, ARKLA LITHIUM BROMIDE WATER CHILLER, AUXILIARY HOT WATER BOILER
AVAILABILITY	ADDITIONAL INFORMATION MAY BE OBTAINED FROM RICHARD L. CROWTHER, AIA, ARCHITECTS GROUP, 2830 EAST THIRD AVE., DENVER, COLORADO 80206.
REFERENCE	ARIZONA STATE UNIVERSITY, COLLEGE OF ARCHITECTURE COLORADO STATE UNIVERSITY HOUSE AT FORT COLLINS, THE WASHINGTON, D.C., IN SOLAR-ORIENTED ARCHITECTURE, PUBLISHED BY THE AIA RESEARCH CORPORATION, JANUARY 1975, PP 131-134

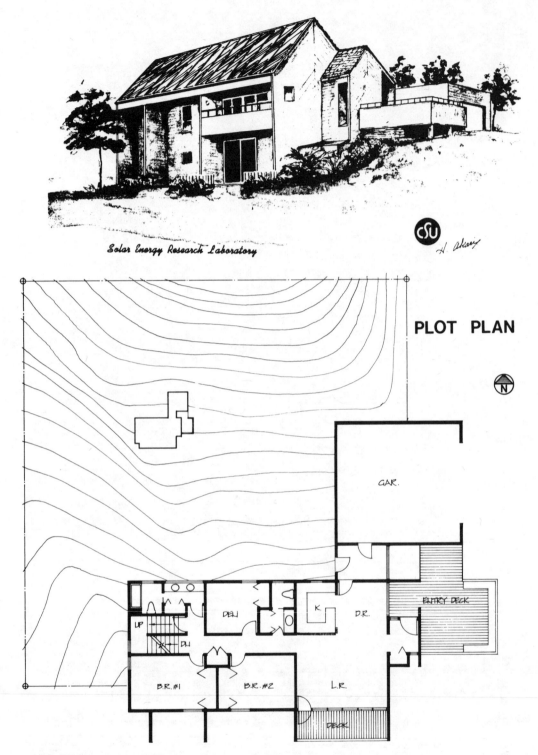

Solar Energy Research Laboratory

CSU
H. akary

PLOT PLAN

N

GAR.

ENTRY DECK

DEN

UP

K.

D.R.

B.R. #1

B.R. #2

L.R.

DECK

FLOOR PLAN 1/16" = 1'-0"

264

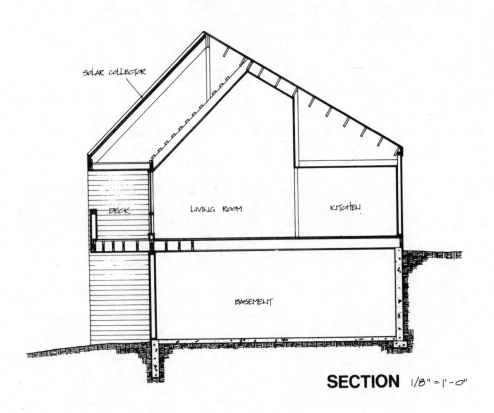

SECTION 1/8" = 1'-0"

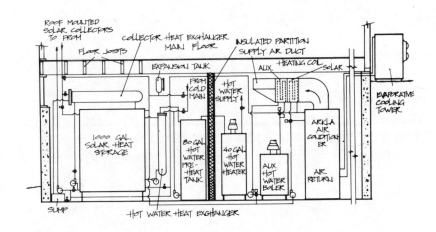

SYSTEM SCHEMATICS

TITLE	COMBS RESIDENCE
CODE	
KEYWORDS	SOLAR HEATING MIRROR SWIMMING POOL
AUTHOR	JAMES LAMBETH
DATE & STATUS	THE COMBS RESIDENCE WAS COMPLETED IN 1973.
LOCATION	SPRINGDALE, ARKANSAS
SCOPE	THIS HOUSE IS AN APPLICATION OF RESEARCH IN SOLAR RADIATION AS A WAY OF HEATING INTERIORS AND EXTERIOR SURFACES.
TECHNIQUE	THE SWIMMING POOL IS PLACED TO COLLECT THE MAXIMUM AMOUNT OF SOLAR GAIN IN THE WINTER AND TO DECREASE GAIN UNTIL MID-SUMMER. THIS IS DONE USING A MIRRORED SURFACE ABOVE THE POOL AREA. THE MIRROR IS ANGLED TO THE SOUTH AT 75 DEGREES. THE SIDE WALLS ARE AT 45 DEGREES TO THE SOUTH, ADMITTING WINTER SUNRISE AND SUNSET, AND BLOCKING OFF SUMMER SUNRISE AND SUNSET. THIS GIVES THE COMBS RESIDENCE A 27% GAIN AT MID-WINTER, HEATING THE WATER AND THE ENCLOSED PORTION OF THE DECK.
MATERIAL	MIRRORED SURFACE
AVAILABILITY	ADDITIONAL INFORMATION MAY BE OBTAINED FROM JAMES LAMBETH, AIA, 1591 CLARK ST., FAYETTEVILLE, ARKANSAS 72701.
REFERENCE	ARCHITECTURAL RECORD COMBS RESIDENCE, SPRINGDALE, ARKANSAS, BY J. LAMBETH ARCHITECTURAL RECORD, DECEMBER 1972, PP.102-103

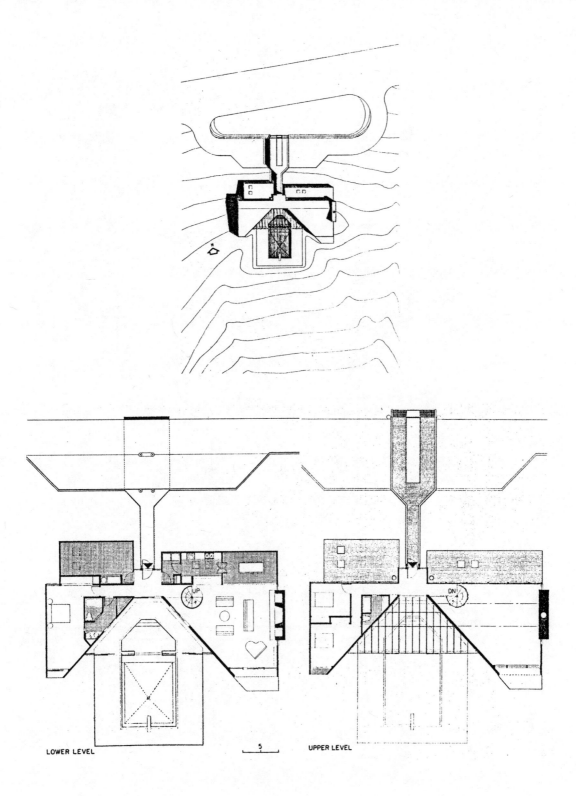

LOWER LEVEL   5   UPPER LEVEL

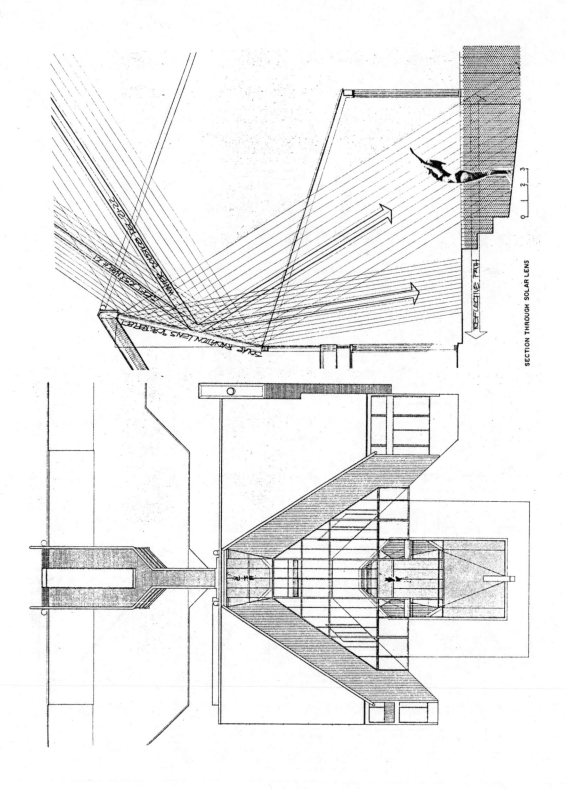

SECTION THROUGH SOLAR LENS

REFLECTIVE PATH

SOLAR RADIATION LENS TO REFLECTOR

WINTER SOLSTICE 12-22

FALL EQUINOX 9-22

0 1 2 3

TITLE	CONCORD NATIONAL BANK
CODE	
KEYWORDS	SOLAR COLLECTOR PANEL HEATING COOLING HOT WATER
AUTHOR	KENNETH F. PARRY ASSOCIATES; CA CROWLEY ENGINEERING CO.
DATE & STATUS	THE CONCORD NATIONAL BANK WAS BUILT IN 1976.
LOCATION	CONCORD, NEW HAMPSHIRE
SCOPE	THE CONCORD NATIONAL BANK IS A 1-STORY BUILDING, AND CONTAINS 2,000 SQ. FT. OF SPACE.
TECHNIQUE	THE STRUCTURE IS DESIGNED TO GET UP TO 65% OF ITS COOLING AND 50% OF ITS SPACE HEATING FROM THE SUN THROUGH 26 SOLAR COLLECTORS AND 4 HEAT DUMP PANELS DESIGNED BY THE DAYSTAR CORP. OF BURLINGTON, MASSACHUSETTS. THE SYSTEM FEEDS SOLAR HEATED WATER FOR DOMESTIC HOT WATER AND SPACE HEATING DEMANDS, BACKED UP BY AN ELECTRIC HEAT PUMP. IN THE AIR CONDITIONING MODE, THE COLLECTORS DIRECTLY INJECT HEATED WATER REQUIRED FOR AN ARKLA ABSORPTION CYCLE AIR CONDITIONING UNIT.
MATERIAL	SOLAR COLLECTOR, HEAT DUMP PANEL
AVAILABILITY	ADDITIONAL INFORMATION MAY BE OBTAINED FROM DAYSTAR CORP., 90 CAMBRIDGE STREET, BURLINGTON, MASSACHUSETTS 01803.
REFERENCE	DAYSTAR CORPORATION ALL-COPPER SOLAR COLLECTORS FOR THREE NEW ENGLAND BANKS BURLINGTON, MASSACHUSETTS, DAYSTAR CORP. PAMPHLET IFS-6, 1976, 1 P

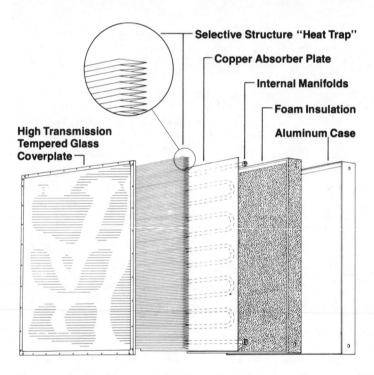

Selective Structure "Heat Trap"

Copper Absorber Plate

Internal Manifolds

Foam Insulation

Aluminum Case

High Transmission Tempered Glass Coverplate

271

TITLE	COOS BAY HOUSE

CODE

KEYWORDS	GRAVITY AIR SYSTEM SOLAR HEATED STORAGE TANK ELECTRIC RESISTANCE HEATERS COLLECTOR HOUSE
AUTHOR	HENRY MATHEW
DATE & STATUS	THIS HOUSE WAS BUILT IN 1967, AND HAS BEEN OWNER OCCUPIED CONTINUOUSLY SINCE THEN.
LOCATION	COOS BAY, OREGON
SCOPE	THE LIVING-ROOM/KITCHEN PORTION IS HEATED DIRECTLY BY THE GRAVITY AIR SYSTEM AROUND THE SOLAR HEATED STORAGE TANK. THE BEDROOMS AND BATHROOMS ARE SUPPLEMENTARILY HEATED AS NECESSARY BY ELECTRIC RESISTANCE HEATERS. THE HOUSE WAS OCCUPIED WITHOUT SOLAR HEAT FOR ONLY THE FIVE MONTH PERIOD OF AUGUST THROUGH DECEMBER 1967. THE COST OF ELECTRICITY USED FOR HEATING ONLY WAS $54 FOR THIS PERIOD. THE COST FOR THE SAME FIVE MONTH PERIOD, OVER A FOUR YEAR AVERAGE WITH SOLAR HEAT WAS $25.
TECHNIQUE	THE SOLAR HEAT COLLECTION SYSTEM CONSISTED OF 410 SQUARE FEET OF COLLECTOR FROM 1967 TO 1974. ON JANUARY 1, 1974, 273 SQUARE FEET OF COLLECTOR WERE ADDED FOR A TOTAL OF 683 SQUARE FEET. THE COLLECTORS ARE SINGLE-GLAZED, AND MADE OF GALVANIZED STEEL PIPES AND ALUMINUM SHEETS. THE HEAT TRANSFER FLUID IS WATER. ORDINARY HOUSEHOLD ALUMINUM FOIL IS USED ON THE ROOF TO REFLECT MORE SUNLIGHT TO THE COLLECTORS. THE STORAGE SYSTEM CONSISTS OF AN 8,000 GALLON UNINSULATED STORAGE TANK, WHICH IS PROTECTED FROM MOISTURE BY A SHEET OF PLASTIC FILM. THE STORAGE TANK IS UNUSUALLY LARGE DUE TO THE LARGE NUMBER OF SUNLESS DAYS OF THE NORTHWESTERN UNITED STATES. THE HEAT FROM THE TANK IS TRANSFERRED TO THE ROOMS BY GRAVITY AIR CIRCULATION AROUND THE TANK. THE AUXILIARY HEATING SYSTEM CONSISTS OF PORTABLE ELECTRIC RESISTANCE UNITS.
MATERIAL	8,000 GALLON UNINSULATED STORAGE TANK, SHEET OF PLASTIC FILM, WATER, A SINGLE GLAZED SOLAR COLLECTOR MADE OF GALVANIZED STEEL PIPES AND ALUMINUM SHEETS, PORTABLE ELECTRIC RESISTANCE UNITS
AVAILABILITY	ADDITIONAL INFORMATION MAY BE OBTAINED BY WRITING TO HENRY MATHEW, ROUTE 3, BOX 768, COOS BAY, OREGON 97420. THE COST FOR THIS INFORMATION IS $10.00.
REFERENCE	REYNOLDS, JOHN S. PART TWO: A SOLAR HOUSE AT COOS BAY, OREGON EUGENE, OREGON, IN SOLAR ENERGY PACIFIC NORTHWEST BUILDINGS, PUBLISHED BY THE CENTER FOR ENVIRONMENTAL RESEARCH, SCHOOL OF ARCHITECTURE AND ALLIED ARTS, UNIVERSITY OF OREGON, 1974, PP 25-31 ARIZONA STATE UNIVERSITY, COLLEGE OF ARCHITECTURE COOS BAY HOUSE WASHINGTON, D.C., IN SOLAR-ORIENTED ARCHITECTURE, PUBLISHED BY THE AIA RESEARCH CORPORATION, JANUARY 1975, P 39

TITLE	COPPER DEVELOPMENT ASSOCIATION DECADE 80 SOLAR HOUSE
CODE	CDA HOUSE, DECADE 80 SOLAR HOUSE
KEYWORDS	PROTOTYPE HOME COOLING HEATING SOLAR PANEL TRANSPORT STORAGE MEDIUM INSULATED TANK AIR CONDITIONING ABSORPTION UNIT SILICON CELLS ANGLE GREENHOUSE TINTED GLASS EXCHANGER HOUSE
AUTHOR	M. ARTHUR KOTCH
DATE & STATUS	THE CDA HOUSE IS COMPLETED AS OF 1976.
LOCATION	TUCSON, ARIZONA
SCOPE	THE DECADE 80 SOLAR HOUSE IS A PROTOTYPE HOME WHICH RECEIVES UP TO 75% OF ITS COOLING AND 100% OF ITS HEATING FROM THE SUN. PRACTICALLY EVERYTHING IN THIS INNOVATIVE HOUSE CAN BE RUN ON STORED-UP SOLAR ENERGY FROM THE CLIMATE CONTROL SYSTEM TO THE SECURITY SYSTEM, STEREO, CLOCKS, EVEN THE KITCHEN TELEVISION. THE CDA SOLAR HOME FEATURES NEW PRODUCTS SUCH AS A COMBINATION COPPER ROOF AND SOLAR ENERGY PANEL SYSTEM, A BRASS DOOR AND WINDOW SYSTEM, AND A COPPER SPRINKLER SYSTEM.

ALTHOUGH LIGHTING AND MAJOR APPLIANCES ARE POWERED BY CONVENTIONAL ELECTRICAL SOURCES, OVERALL ENERGY SAVINGS ARE NOW ESTIMATED AT 90% IN THE WINTER, AND 80% IN THE SUMMER OVER ORDINARY HOMES.

TECHNIQUE	THE SOLAR ENERGY PANEL SYSTEM CONSISTS OF 2' X 8' COPPER SHEETS, 0.016 INCH THICK, LAMINATED TO PLYWOOD COMBINED WITH UNIQUE, RECTANGUALR COPPER TUBES TO CARRY THE ENERGY SYSTEM'S TRANSPORT AND STORAGE MEDIUM - WATER. THE COPPER PANELS AND COPPER TUBE ARE BLACKENED TO BETTER ABSORB RADIANT HEAT, TRANSFERRING THIS HEAT TO WATER CIRCULATING THROUGHOUT THE HOUSE IN COPPER TUBE. ON CLOUDY DAYS WHEN LESS ENERGY IS AVAILABLE FROM THE SUN - ALTHOUGH THERE ALWAYS IS SOME - A BURIED INSULATED STORAGE TANK PROVIDES STAND-BY ENERGY IN THE FORM OF 3,000 GALLONS OF SOLAR HEATED WATER.

AIR CONDITIONING FOR THE CDA SOLAR HOUSE IS PROVIDED BY TWO STANDARD LITHIUM BROMIDE WATER ABSORPTION UNITS MODIFIED FOR HOT WATER FIRING.

IN ADDITION TO THE SOLAR COLLECTOR PANELS, THE HOUSE'S ENERGY SYSTEM ALSO INCORPORATES SILICON SOLAR CELLS ON THE ROOF TO CONVERT THE SUN'S ENERGY TO LOW VOLTAGE POWER FOR SELECTED ELECTRICAL SYSTEMS AND APPLIANCES. SOLAR CELLS ALSO PROVIDE STAND-BY POWER FOR THE HOME'S OVERALL SECURITY SYSTEM IN CASE OF ELECTRICAL FAILURE.

EXTERIOR AND INTERIOR DOOR AND WINDOW FRAMES ARE HELD TOGETHER WITH A SPECIAL THERMAL BREAK MATERIAL WITH AN INSULATING, FIRE-RETARDANT URETHANE FOAM CORE THAT PREVENTS HEAT LOSS AND GUARDS AGAINST CONDENSATION ON THE INTERIOR SURFACE FRAME.

SOLAR PANELS ON THE ROOF OF THE CONNECTING GUEST WING ARE USED NOT ONLY TO HEAT THE SWIMMING POOL IN THE WINTER, BUT ALSO TO COOL THE WATER IN THE SUMMER. DURING SUMMER NIGHTS, POOL WATER CIRCULATES THROUGH THE ROOF/COLLECTOR SYSTEM AND IS COOLED BY THE DESERT AIR FOR SWIMMING COMFORT DURING THE HOT DAYS. THE GUEST ROOF SLOPES AT APPROXIMATELY A 40 DEGREE ANGLE TO FAVOR SWIMMING POOL HEATING DURING TUCSON'S SPRING AND FALL. THE HOUSE ROOF SLOPES AT APPROXIMATELY A 27 DEGREE ANGLE TO FAVOR SUMMER COOLING CONDITIONS WHEN A RELATIVELY LARGE AMOUNT OF ENERGY IS REQUIRED TO KEEP THE HOUSE AIR CONDITIONED.

COPPER WAS SELECTED FOR THE SOLAR ENERGY COLLECTION SYSTEM BECAUSE IT CONDUCTS HEAT UP TO EIGHT TIMES BETTER THAN ANY OTHER MATERIAL FEASIBLE FOR SOLAR PANELS, AND HAS A HIGHER RESISTANCE TO CORROSION.

SPECIAL GLASS PERFORMS TWO OPPOSITE FUNCTIONS: IN THE SOLAR PANELS, IT ATTRACTS AND TRAPS SOLAR ENERGY; IN THE WINDOWS, IT REFLECTS AND EXCLUDES SOLAR HEAT. IN THE SOLAR PANELS, TWO PANELS OF CLEAR TEMPERED GLASS SEPARATED BY AN INSULATING AIR

SPACE TRANSMIT A HIGH LEVEL OF SOLAR ENERGY TO THE BLACK COATED
COPPER ABSORBER PLATE AND INSULATE AGAINST HEAT LOSS. FURTHER
SHORT-WAVE LENGTH RADIATION PASSING THROUGH THE GLASS IS
CONNECTED TO A LONGER WAVE LENGTH, WHICH THE GLASS THEN TRAPS
FOR GREATER EFFICIENCY - THE SAME PRINCIPLE AS IN GREENHOUSES.

WINDOWS ON THE POOL-SIDE OF THE HOUSE ARE GLAZED WITH SPECIAL
SOLAR BRONZE TINTED, DOUBLE-GLASS INSULATION UNITS THAT REDUCE
HEAT GAIN BY 40% FOR LOWER AIR-CONDITIONING LOADS AND IMPROVE
INDOOR VISUAL COMFORT. ON THE SUN SIDE OF THE HOUSE IS GLAZING
THAT IS HIGHLY REFLECTIVE, ALMOST MIRROR-LIKE, TINTED MORE IN
THE COPPER TONES, AND OF EVEN HIGHER INSULATING QUALITY.

THE TWO ARKLA LITHIUM BROMIDE WATER ABSORPTION UNITS TOGETHER
HAVE A TOTAL COOLING CAPACITY OF 72,000 BTU/HR OR SIX TONS.
THE SAME UNITS ALSO PROVIDE 120,000 BTU/HR - 60,000 BTU/HR
EACH OF HEATING DURING THE WINTER, AND INCLUDE TWO 1,200 CFM
FANS FOR CIRCULATING THE AIR. THE HEAT FROM THE WATER CIRCULATING
THROUGH THE SOLAR PANELS IS TRANSFERRED TO A SECOND CLOSED WATER
SYSTEM WITH A 3,000 GALLON INSULATED STORAGE TANK BURIED
UNDERGROUND.

A STANDARD COMMERCIAL, AUXILIARY WATER HEATER CAPABLE OF
PROVIDING UP TO 170,000 BTU/HR OF THERMAL OUTPUT IS AVAILABLE
ON A STAND-BY BASIS. THE HOUSE'S SOLAR SYSTEM ALSO INCORPORATES
A HEAT EXCHANGER AND AN AUXILIARY STAND-BY HEATER TO PROVIDE
HOT WATER FOR HOUSEHOLD USE.

MATERIAL
COPPER ROOF AND SOLAR ENERGY PANEL SYSTEM, BRASS DOOR AND
WINDOW SYSTEM, COPPER SPRINKLER SYSTEM, BLACKENED 2' X 8' COPPER
SHEETS .016 INCH THICK, BLACKENED RECTANGULAR COPPER TUBES,
WATER, INSULATED STORAGE TANK, LITHIUM BROMIDE WATER ABSORPTION
UNIT, SILICON SOLAR CELLS, URETHANE FOAM CORE, CLEAR TEMPERED
GLASS, REFLECTIVE GLASS, BRONZE TINTED DOUBLE GLASS INSULATION
UNITS, WATER HEATER, HEAT EXCHANGER

AVAILABILITY
ADDITIONAL INFORMATION MAY BE OBTAINED FROM ESTES JONES, COPPER
DEVELOPMENT ASSOCIATION INC., 405 LEXINGTON AVE., NEW YORK,
N.Y. 10017.

REFERENCE
COPPER DEVELOPMENT ASSOCIATION, INC.
DECADE 80 SOLAR HOUSE IN TUCSON NOW COMPLETED
NEW YORK, N.Y., COPPER DEVELOPMENT ASSOCIATION INC. NEWS,
1976, 11 P

PROFESSIONAL BUILDER
COPPER MAKES SOLAR ENERGY RIGHT AT HOME
PROFESSIONAL BUILDER, NOVEMBER 1975, PP 17-28

ARIZONA STATE UNIVERSITY, COLLEGE OF ARCHITECTURE
DECADE 80 SOLAR HOUSE
WASHINGTON, D.C., SOLAR-ORIENTED ARCHITECTURE, PUBLISHED BY
THE AIA RESEARCH CORPORATION, JANUARY 1975, P 75

Outdoor living side of "Decade 80 Solar House" has solar-heated private hot pool outside master bath (extreme left), solar-heated swimming pool, and (extreme right) guest apartment with private entrance. First totally integrated solar collector-roof system will provide 75% of the cooling and nearly 100% of house's heating needs.

Front of the "Decade 80 Solar House" is the "sun side" so that is where the solar-collector roof is located—the industry's first such totally integrated system.

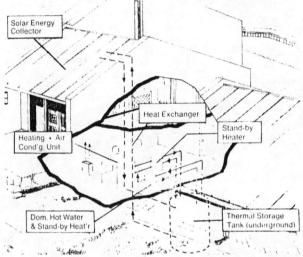

Solar Energy Collector

Heat Exchanger

Stand-by Heater

Heating • Air Cond'g. Unit

Dom. Hot Water & Stand-by Heat'r

Thermal Storage Tank (underground)

Schematic of the CDA "Decade 80 Solar House" shows mechanical room located under the den study containing the solar heating and air-conditioning unit plus its stand-by heater, a solar heater for domestic water and its stand-by heater. Buried under this for efficiency and economy is the thermal storage tank.

In integrated solar collector-roof system for CDA solar house, copper piping is rectangular rather than round for both aesthetic and functional reasons. A rectangular tube provides more intimate contact, thereby collecting more heat energy.

TITLE	CORONA DEL SOL HIGH SCHOOL
CODE	
KEYWORDS	SCHOOL SOLAR COLLECTOR BERM INSULATION LANDSCAPING WATER LOW COMPACT EXPOSURE
AUTHOR	MICHAEL GOODWIN OF MICHAEL AND KEMPER GOODWIN, LTD.
DATE & STATUS	THIS BUILDING WAS UNDER CONSTRUCTION IN LATE 1976.
LOCATION	TEMPE, ARIZONA
SCOPE	THE ARCHITECT PREDICTS A PAYBACK PERIOD OF ONLY 8 YEARS FOR THIS 264,000 SQ. FT. SCHOOL. IN ADDITION, OPERATING COSTS ARE EXPECTED TO BE 1 MILLION DOLLARS LESS THAN A MORE CONVENTIONAL DESIGN OVER THE NEXT 20 YEARS.
TECHNIQUE	SOLAR COLLECTORS ARE THE MAIN FEATURE IN CORONA DEL SOL'S SCHEME FOR ENERGY PRODUCTION. HEAT LOSS FROM THE WATER-TO-AIR SYSTEM IS MINIMIZED BY RUNNING THE MECHANICAL DUCTS DIRECTLY BELOW THE SOLAR COLLECTORS ON THE ROOF. EARTH BERMS ARE USED TO REDUCE ENERGY LOSS FROM THE LOW, COMPACT BUILDING. THE SCHOOL'S MAIN STRUCTURAL MASS IS SITUATED ON AN ANGLE OF MINIMUM EXPOSURE TO THE SUN. THE EXTENSIVE USE OF INSULATION IN THE INTERIOR, AND THE NATURAL INSULATION AFFORDED BY LANDSCAPING ON THE PERIMETER FURTHER HELPS TO CONSERVE ENERGY IN THE SCHOOL.
MATERIAL	SOLAR COLLECTOR, EARTH BERMS, INSULATION, LANDSCAPING
AVAILABILITY	ADDITIONAL INFORMATION MAY BE OBTAINED FROM MICHAEL AND KEMPER GOODWIN, LTD., 115 EAST 5TH STREET, TEMPE, ARIZONA.
REFERENCE	HARACK, THOMAS ENERGY SAVING PROJECTS ON STREAM COAST TO COAST BUILDING DESIGN AND CONSTRUCTION, NOVEMBER 1976, PP 43-46

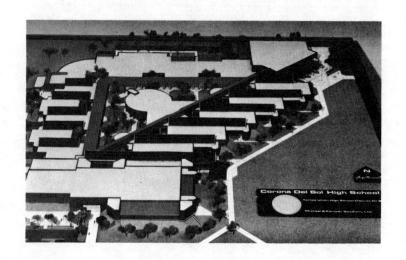

TITLE	COTE NORD PROTOTYPE HOUSE
CODE	
KEYWORDS	HOUSE COLLEGE INSULATION WINDOW AIR TYPE COLLECTOR TURBULENCE HEAT TRANSFER STORAGE GRAVITY CONVECTION
AUTHOR	B. MCCLOSKEY
DATE & STATUS	THE COTE NORD PROTOTYPE HOUSE WAS COMPLETED IN EARLY 1976.
LOCATION	LA MACAZA, QUEBEC, CANADA
SCOPE	THE BUILDING IS A ONE-STORY, 3-BEDROOM HOUSE WITH A CRAWL SPACE, BUT WITH NO BASEMENT OR ATTIC. THE FOUNDATIONS ARE OF CONCRETE BLOCKS. THE WALLS ARE OF HORIZONTAL, SQUARED-OFF LOGS.
TECHNIQUE	THE INSULATION IS THICK, OF PRESSED-MOSS PANELS AND FIBERGLAS IN THE WALLS. THERE ARE SMALL WINDOW AREAS ON THE EAST, NORTH AND WEST SIDES. THE BUILDING HAS A HUMUS COMPOSTING TOILET.

THE COLLECTOR IS A 500 SQ. FT., AIR TYPE COLLECTOR, 36' X 14', MOUNTED ON THE VERTICAL SOUTH WALL. THE HEART OF THE COLLECTOR IS A CORRUGATED SHEET OF ALUMINUM THAT HAS A NON-SELECTIVE BLACK COATING. IT IS DOUBLE GLAZED WITH GLASS SHEETS. BE- TWEEN THE GLAZING AND THE BLACK CORRUGATED SHEET, THERE IS A BLACK PAINTED SHEET OF EXPANDED METAL LATH, WHICH INCREASES THE TURBULENCE OF THE AIRFLOW AND ENLARGES THE HEAT-TRANSFER AREA. THE AIR IS DRIVEN DOWNWARD THROUGH THE COLLECTOR BY A 3,000 CFM FAN SITUATED IN THE CRAWL SPACE BENEATH THE SOUTH HALF OF THE HOUSE. THE AIR THEN PASSES INTO THE STORAGE SYSTEM.

THE STORAGE SYSTEM CONSISTS OF 1,150 CUBIC FEET OF STONES, WITH AN AVERAGE DIAMETER OF 4", IN AN INSULATED RECTANGULAR BIN IN THE CRAWL SPACE UNDER THE NORTH SIDE OF THE HOUSE. AIR FROM THE COLLECTOR IS DRIVEN NORTHWARD, THEN UPWARD, WITHIN THE BIN. IT THEN TRAVELS BACK TO THE TOP OF THE COLLECTOR VIA ROOMS OR VIA DUCTS INCORPORATED IN THE MAIN-STORY FLOOR. IF THE ELECTRIC SUPPLY FAILS, AND THE FAN IS NOT OPERATING, HEAT FROM THE BIN OF STONES WILL CONTINUE TO TRAVEL TO THE ROOMS BY GRAVITY CONVECTION.

MATERIAL	PRESSED-MOSS PANELS; FIBERGLAS; HUMUS COMPOSTING TOILET; CORRUGATED SHEET OF ALUMINUM WITH NON-SELECTIVE BLACK COATING; GLASS SHEETS; BLACK PAINTED SHEET OF EXPANDED METAL LATH; 3,000 CFM FAN; 1,150 CUBIC FEET OF 4" DIAMETER STONES; INSULATED RECTANGULAR BIN
AVAILABILITY	ADDITIONAL INFORMATION MAY BE OBTAINED FROM PROF. BRIAN MCCLOSKEY, SHELTER SYSTEMS GROUP, SCHOOL OF ARCHITECTURE, MCGILL UNIVERSITY, P.O.BOX 6070, STATION A, MONTREAL, P.Q., H3C 3GI, CANADA.
REFERENCE	SHURCLIFF, W.A. SOLAR HEATED BUILDINGS, A BRIEF SURVEY CAMBRIDGE, MASSACHUSETTS, W.A. SHURCLIFF, MARCH 1976, 212 P

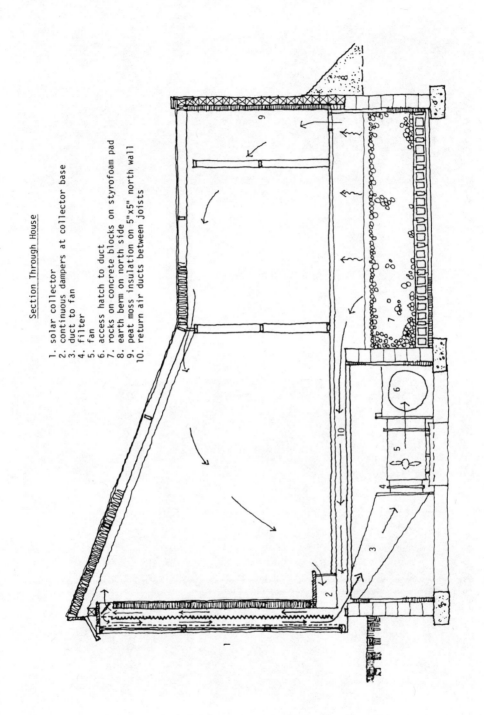

Section Through House

1. solar collector
2. continuous dampers at collector base
3. duct to fan
4. filter
5. fan
6. access hatch to duct
7. rocks on concrete blocks on styrofoam pad
8. earth berm on north side
9. peat moss insulation on 5"x5" north wall
10. return air ducts between joists

283

Basement Rock Storage

Hand Crank to Open and Close Dampers

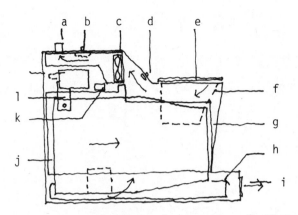

Humus Composting Toilet

a) vent (pressure chamber)
b) on/off switch
c) fan
d) rotation switch
e) toilet seat
f) funnel

g) drain
h) mould box
i) mould box container
j) heater
k) timer
l) motor support

(m) rotation
      motor

TITLE	COTTRELL-VISALIA SOLAR HOUSE
CODE	
KEYWORDS	PARABOLIC TROUGH CONCENTRATING COLLECTORS REFLECTOR HOUSE INSULATED
AUTHOR	R. COTTRELL AND R. SARTAIN
DATE & STATUS	THE COTTRELL-VISALIA HOUSE WAS BUILT IN 1974.
LOCATION	VISALIA, CALIFORNIA
SCOPE	THE COTTRELL-VISALIA HOUSE IS A MODERN THREE BEDROOM HOUSE. THE HOUSE CONCISTS OF TWO STORIES, AND A ROOF WHICH FACES DUE SOUTH AND IS TILTED 15 DEGREES FROM THE HORIZONTAL. THE HOUSE DOES NOT HAVE A BASEMENT, ATTIC OR CARPORT. IT IS ESTIMATED THAT 80% OF THE HEAT WILL BE GENERATED BY THE SUN.
TECHNIQUE	THE PRINCIPLE METHOD OF SOLAR COLLECTION IS DONE BY 400 SQUARE FEET PARABOLIC TROUGH CONCENTRATING COLLECTORS. REFLECTORS IN THE CONCENTRATING COLLECTORS ARE ROTATED SLOWLY TO KEEP THE SUN'S RAYS IN FOCUS. ONCE THE HEAT IS COLLECTED, IT IS STORED IN AN 8,000 GALLON CONCRETE BLOCK TANK, WHICH IS LOCATED UNDERNEATH THE MASTER BEDROOM. THE TANK IS WATER PROOFED BY POOL PLASTER, AND IS INSULATED WITH 2 INCH STYROFOAM. THE HEAT IS DISTRIBUTED BY A CONVENTIONAL FAN-COIL SYSTEM AND DUCTS. SINCE IT IS ESTIMATED THAT 80% OF THE BUILDING'S HEATING NEEDS WILL BE MET BY THE SUN, A FRANKLIN STOVE HAS BEEN PROVIDED AS A MEANS FOR PROVIDING ADDITIONAL HEAT. THE COOLING SYSTEM CONSISTS OF 60 F WATER TAKEN FROM A WELL ON THE 8 ACRE SITE. THE COOL WATER IS CIRCULATED THROUGH THE FAN COILS, THEN USED TO IRRIGATE A WALNUT GROVE WHICH IS ON THE SITE.
MATERIAL	PARABOLIC TROUGH CONCENTRATING COLLECTORS, CONCRETE BLOCK STORAGE TANK, STYROFOAM INSULATION, FRANKLIN STOVE, WELL WATER
AVAILABILITY	REFER TO THE FOLLOWING REFERENCE FOR ADDITIONAL INFORMATION.
REFERENCE	ARIZONA STATE UNIVERSITY, COLLEGE OF ARCHITECTURE COTTRELL-VISALIA SOLAR HOUSE WASHINGTON, D.C., SOLAR-ORIENTED ARCHITECTURE, PUBLISHED BY THE AIA RESEARCH CORPORATION, JANUARY 1975, P 50

TITLE	CRAB SCHEME
CODE	
KEYWORDS	SOLAR HOUSING INSULATION FLAT PLATE INSULATED BACK UP SOLID STATE CONTROLLER SPHERE ZONE DISTRIBUTION RECOVERY INSOLATION
AUTHOR	JOINT VENTURE, INC., ARCHITECTS, ENVIRONMENTALISTS, VISIONARIES
DATE & STATUS	THE CRAB HAS BEEN DESIGNED, BUT NOT BUILT, AS OF 1975.
LOCATION	DENVER REGIONAL CORRIDOR
SCOPE	THE UNIQUE CLIMATE OF DENVER OFFERS AN EXCELLENT SITE FOR THE UTILIZATION OF SOLAR BUILDING HEATING SINCE THE WINTER SKIES ARE CLEAR AND TEMPERATURES NOT EXTREME. THE ALTITUDE IS HIGH WITH CONSEQUENT HIGH LEVELS OF INSOLATION OF 70%, AND SNOWFALL IS NOT PERSISTENT.

FOR EACH 100 DWELLING UNITS OF CONSTRUCTION THERE ARE 45 TWO-BEDROOM UNITS, 45 THREE-BEDROOM UNITS, 10 HANDICAPPED HOUSING UNITS, 10 NEIGHBORHOOD CENTERS, AND 1 COMMUNITY CENTER. EACH TWO-BEDROOM UNIT OCCUPIES 1,090 SQUARE FEET FOR A TOTAL OF 49,050 SQUARE FEET. EACH THREE-BEDROOM UNIT OCCUPIES 1,350 SQUARE FEET FOR A TOTAL OF 60,750 SQUARE FEET. EACH HANDICAPPED UNIT OCCUPIES 1,200 SQUARE FEET FOR A TOTAL OF 12,000. EACH NEIGHBORHOOD CENTER OCCUPIES 530 SQUARE FEET FOR A TOTAL OF 5,300 SQUARE FEET. EACH COMMUNITY CENTER OCCUPIES 4,420 SQUARE FEET. THE TOTAL OF THE ENTIRE COMPLEX IS 131,520 SQUARE FEET. THE 1,000 SQUARE FEET OF COLLECTOR AREA PROVIDES 75 PERCENT OF THE HEAT LOAD.

THE 1,000 SQUARE FEET OF COLLECTOR AREA PROVIDES 75 PERCENT OF THE HEAT LOAD. WITH AN INSTALLED AREA OF 1,000 SQUARE FEET, THE TOTAL EXTRA SOLAR COST IS ESTIMATED TO BE FROM $10,400 TO $13,400.

TECHNIQUE	THE REQUIREMENTS FOR COOLING IN THE METROPOLITAN DENVER AREA ARE SMALL AND CAN EASILY BE MET BY CONTROLLED AND NATURAL VENTILATION AND BUILDING ORIENTATION.

SINCE THE SUN IS ALWAYS SOUTH OF DENVER IN WINTER AND LOW IN THE SKY, THE COLLECTOR IS NORMALLY TILTED UP FROM THE HORIZONTAL TO FACE SOUTH AT AN ANGLE OF ABOUT 15 DEGREES PLUS THE LOCAL LATITUDE. IN THE PRESENT DESIGNS, AN ANGLE OF 58 DEGREES CAN BE CONSTRUCTED BY AN 8/5 SLOPE. IN ADDITION, THE COLLECTOR MAY FACE EAST OR WEST 20 DEGREES FROM SOUTH IF LOCAL OBSTRUCTIONS EXIST WHICH BLOCK THE AFTERNOON OR MORNING SUN, RESPECTIVELY. THE COLLECTOR TYPICALLY REPRESENTS 60-80 PERCENT OF THE TOTAL SOLAR SYSTEM COST.

SIX INCHES OF FIBERGLAS INSULATION WERE USED TO MINIMIZE HEAT LOSS. IN ORDER TO ACCOMODATE THIS INSULATION IT WAS NECESSARY TO USE 2X6 WALL FRAMING STUDS. THIS INCREASE IN INSULATION PERMITTED A REDUCTION IN THE AMOUNT OF COLLECTOR AREA.

OTHER METHODS WHICH HELP CONSERVE ENERGY INCLUDE A ZONED HEAT DISTRIBUTION SYSTEM. THIS REQUIRES PLACING ADDITIONAL THERMOSTATS AND CONTROLS RELATING TASK AND FUNCTION TO THE AMOUNT OF HEAT DELIVERED TO VARIOUS PARTS OF THE HOUSE. FOR EXAMPLE, ROOMS USED ONLY DURING THE EVENING OR ONLY FOR SPECIAL PURPOSES ARE NOT HEATED WHEN ANOTHER PART OF THE HOUSE CALLS FOR HEAT. RECOVERY OF ENERGY FROM HOT WATER FLOWING OUT DRAINS IS ALSO CONSIDERED.

BECAUSE OF THE LARGE AMOUNT OF COLLECTION SURFACE (1,000 SQUARE FEET) REQUIRED FOR THE FLAT PLATE MECHANICAL SYSTEMS, THE DESIGN BEGAN BY ASSUMING A LARGE DOMINANT PLANE FACING SOUTH WITH THE HOUSING AS AN OUTGROWTH BLOB ATTACHED TO THE NORTH. NEXT, THE HOUSING UNITS WERE ORIENTED ON AN EAST/WEST AXIS, ALLOWING FOR NATURAL FLOW-THROUGH VENTILATION AND EXTERIOR EXPOSURE. BY BERMING THE NORTH SIDE, PROTECTING THE NORTH AIRLOCK ENTRY FROM PREVAILING NORTHWEST WINDS AND INTRODUCING THE 45 DEGREE ANGLE PASSIVE NORTH SIDE ROOF FORM, A KIND OF 'BACK DOOR' RESULTED. FURTHER, BY RECESSING AND OVERHANGING

THE SOUTH SIDE UNDER THE COLLECTION PLATE AND SUGGESTING A
MORE RECEPTIVE AND WARM SENSE OF ENTRY, A 'FRONT DOOR' RESULTED.
FENESTRATION AND FORM FRAGMENTATION ON THE NORTH 45 DEGREE
ANGLE ROOF WERE HELD TO A MINIMUM TO REDUCE HEAT LOSS.

THE COLLECTOR IS PLACED BETWEEN STRUCTURAL MEMBERS IN THE SOUTH
ROOF OF THE BUILDING. IT FORMS AN INTEGRAL PART OF THE ENTIRE
SOUTH ROOF, THEREBY SAVING ABOUT 50 CENTS PER SQUARE FOOT ON
THE ROOFING MATERIALS. DURING THE WINTER SEASON, THE COLLECTOR
OPERATES AT A TEMPERATURE OF ABOUT 150 F WITH WATER FLOW RATE
IN THE VICINITY OF 10 POUNDS PER HOUR PER SQUARE FOOT OF
COLLECTOR AREA.

THE ENERGY REMOVED FROM THE COLLECTOR IS CONDUCTED TO THE
STORAGE TANK BY A HOT WATER/ANTIFREEZE SOLUTION. SINCE PVC
PLASTIC PIPING HAS A MAXIMUM USE TEMPERATURE OF 150 F, ABOVE
WHICH IT DISTORTS, IT CANNOT BE USED IN SOLAR SYSTEMS. ORDINARY
COPPER PIPE IS EMPLOYED THROUGHOUT. THE STORAGE TANK IS PLACED
NEAR THE COLLECTOR TO MINIMIZE PIPING COSTS. A CIRCULATING
PUMP WILL MAINTAIN FLOW RATES IN THE
VICINITY OF 24 GPM TO THE 1,000 SQUARE FOOT COLLECTOR. ELECTRI-
CALLY OPERATED, MOTORIZED VALVES ARE REQUIRED TO CONTROL THE
FLOW RATES TO AND FROM THE COLLECTOR.

STORAGE CONSISTS OF A 2,700 GALLON CONCRETE TANK PLACED BENEATH
THE BUILDING. IT WILL BE INSULATED WITH TWO INCHES OF RIGID
STYROFOAM ON THE EARTH SIDES TO MINIMIZE HEAT LOSS. THE TANK
SHOULD BE LINED WITH PLASTIC FILM MATERIAL CAPABLE OF WITH-
STANDING TEMPERATURES UP TO 200 F FOR 20 YEARS OR MORE WITHOUT
DETERIORATION.

THE DISTRIBUTION SYSTEM CONSISTS OF A PUMP AND PIPES WHICH
DELIVER HEATED WATER TO EACH DWELLING UNIT IN THE BUILDING. THER-
MOSTATS IN EACH DWELLING UNIT CONTROL THE OPERATION OF A SUITABLY
SIZED FAN COIL LOCAL DISTRIBUTOR SYSTEM. SINCE STORAGE TEMPERA-
TURES AVERAGE AROUND 110 F, FAN COIL UNITS ARE THE ONLY
FEASIBLE CONVENTIONAL HEAT TRANSFER DEVICE.

CONTROLS ARE PROVIDED IN CONSULTATION WITH ELECTRONICS MANU-
FACTURERS WHO HAVE DEVELOPED SPECIAL INTEGRATED CIRCUIT DESIGNS
FOR APPLICATION TO A FLAT PLATE WATER-COOLED SYSTEM.

MATERIAL   COPPER COLLECTOR ABSORBER PLATES, TEDLAR BONDED ACRYLIC FORTIFIED
                FIBERGLAS PANELS, LOW-IRON GLASS, SOLID STATE CONTROLLER,
                SELECTIVE BLACK ABSORPTION SURFACE, CONCRETE STORAGE TANK,
                COVERED WITH 2 INCHES OF STYROFOAM AND LINED WITH A PLASTIC FILM

AVAILABILITY   ADDITIONAL INFORMATION ON THE CRAB SCHEME FOR MULTI-FAMILY
                ENVIRONMENTALISTS, VISIONARIES, 1406 PEARL ST., BOULDER,
                COLORADO 80302.

REFERENCE   JOINT VENTURE, INC.
                HERE COMES THE SUN - 1981
                BOULDER, COLORADO, MARCH 1975, JOINT VENTURE & FRIENDS, PP 7-59

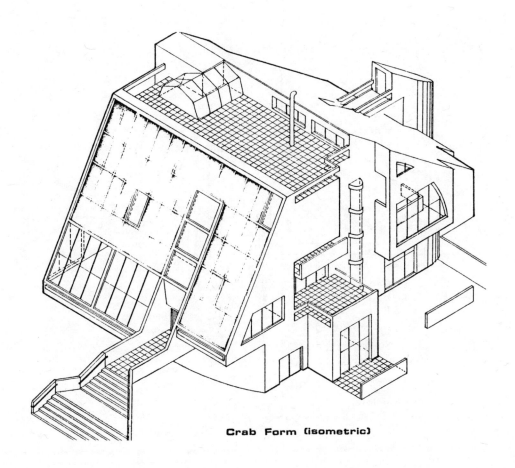

Crab Form (isometric)

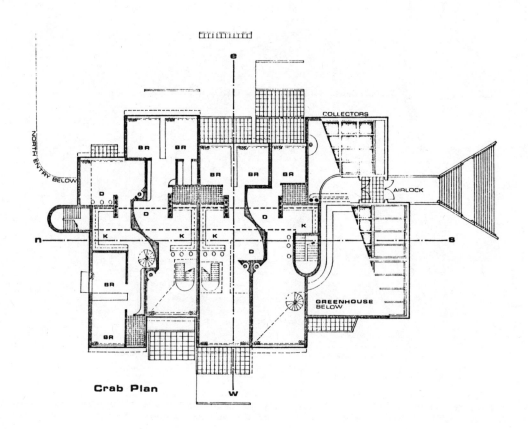

Crab Plan

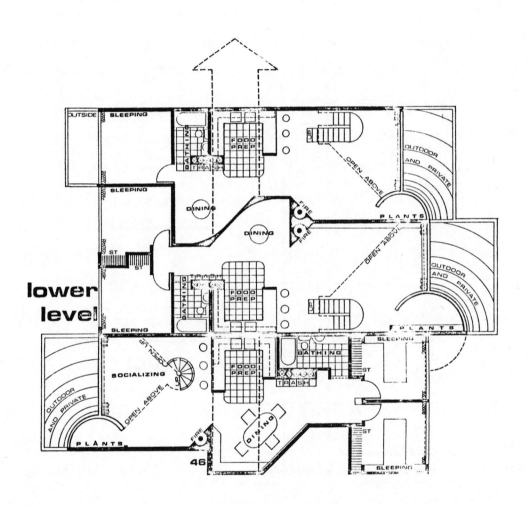

lower
level

46

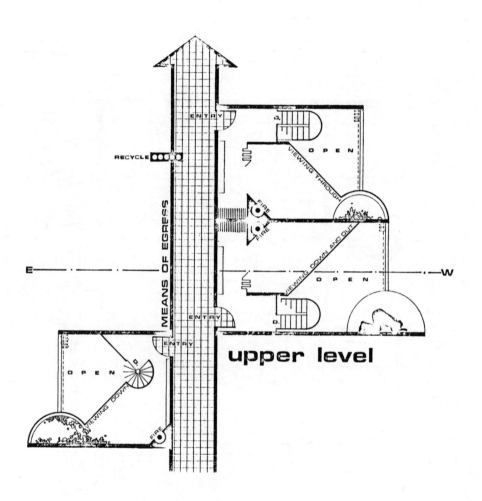

upper level

294

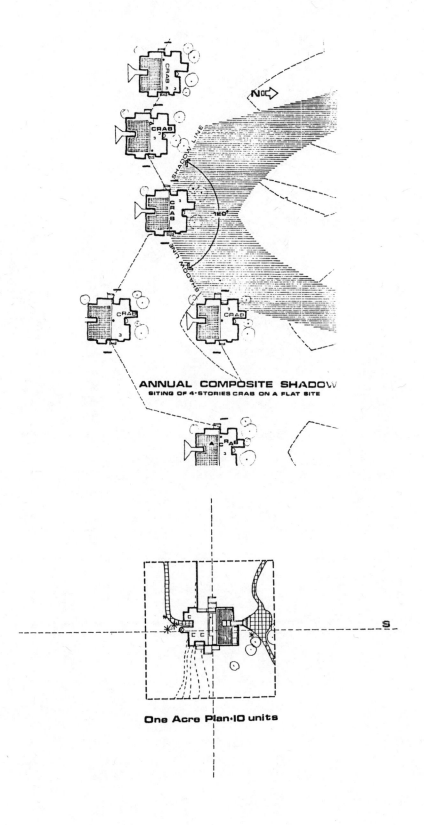

ANNUAL COMPOSITE SHADOW
SITING OF 4-STORIES CRAB ON A FLAT SITE

One Acre Plan·10 units

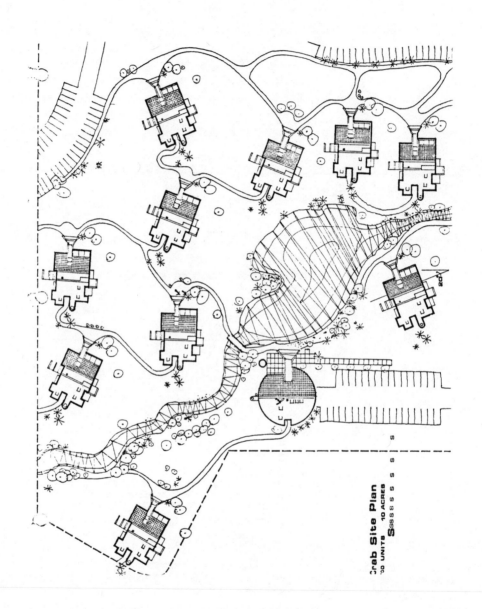

Crab Site Plan
00 UNITS   10 ACRES
SSSSSS S S S

296

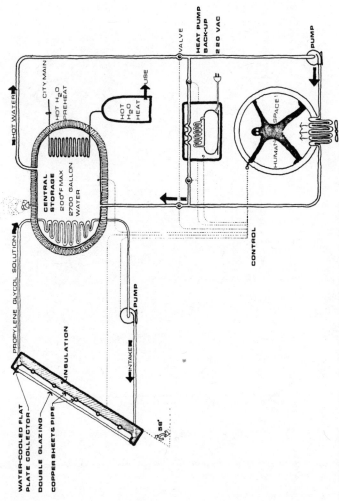

Water-Cooled Flat Plate System

WATER-COOLED FLAT PLATE COLLECTOR

DOUBLE GLAZING

COPPER SHEET & PIPE

INSULATION

PROPYLENE GLYCOL SOLUTION

INTAKE

PUMP

56°

CENTRAL STORAGE
200°F MAX
2700 GALLON WATER

HOT WATER

CITY MAIN

HOT H_2O PREHEAT

HOT H_2O HEAT

USE

VALVE

HEAT PUMP BACK-UP

220 VAC

PUMP

HUMAN SPACE

CONTROL

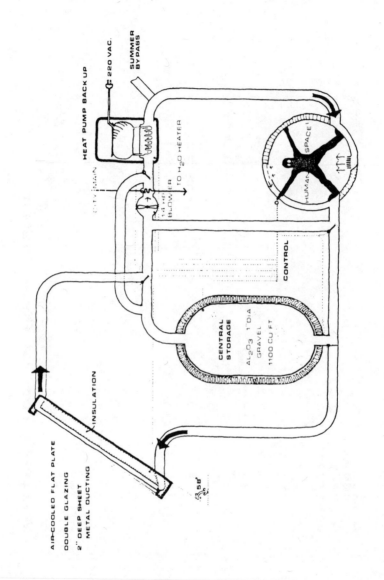

Air-Cooled Flat Plate System

AIR-COOLED FLAT PLATE

DOUBLE GLAZING

2" DEEP SHEET
METAL DUCTING

INSULATION

CENTRAL
STORAGE

Al₂O₃ 1" DIA
GRAVEL
1100 CU FT

CONTROL

HUMAN SPACE

1/4 HP
BLOWER

TO H₂O HEATER

CITY MAIN

HEAT PUMP BACK UP

220 VAC.

SUMMER
BY PASS

TITLE	CRAIN HOUSE
CODE	
KEYWORDS	SALT BOX HOUSE AIR TYPE COLLECTOR STORAGE MASONRY BIN
AUTHOR	H.A. WADE
DATE & STATUS	THE CRAIN HOUSE WAS COMPLETED IN THE FALL OF 1975.
LOCATION	FLAGSTAFF, ARIZONA
SCOPE	THE CRAIN HOUSE IS A 2-STORY, WOOD-FRAME, SALT-BOX TYPE HOUSE. IT HAS 3 BEDROOMS, 2 BATHROOMS, ATTIC, BUT NO BASEMENT. THE TOTAL FLOOR AREA, INCLUDING A 2-CAR GARAGE, IS 2,300 SQ. FT. THE LIVING ROOM AND FAMILY ROOM HAVE CATHEDRAL CEILINGS. HEAVY INSULATION IS USED IN THE FOUNDATION WALLS TO MINIMIZE HEAT LOSS. 75% OF THE HEATING IS PROVIDED BY SOLAR ENERGY.
TECHNIQUE	THE SOLAR COLLECTOR AREA IS 750 SQ. FT. IT IS AN AIR-TYPE COLLECTOR MOUNTED ON THE ROOF, SLOPING 45 DEGREES FROM THE HORIZONTAL. THE HEART OF THE COLLECTOR CONSISTS OF A SHEET OF RIBBED, NON-GALVANIZED STEEL THAT HAS AN ON-SITE-APPLIED, NON-SELECTIVE-BLACK COATING ON THE UPPER SURFACE. THE GLAZING IS SINGLE SHEET OF DOUBLE-STRENGTH GLASS. AIR FLOWS INTO THE SPACE BEHIND THE BLACK METAL SHEET. THE SPACE IS NOMINALLY 4" THICK, BUT CONTAINS MANY BAFFLES THAT PROJECT 3" INTO SPACE; BAFFLES INCREASE TURBULENCE AND HEAT TRANSFER. AIR FLOWS UPWARD IN THIS SPACE. IT IS DRIVEN BY A 3/4 HP BLOWER SITUATED IN A CLOSET AT THE EAST END OF THE LOFT. BEHIND THE AIRSPACE IS A LAYER OF FOIL-FACED FIBERGLAS BATTS.
	THE STORAGE SYSTEM CONSISTS OF 60 TONS OF 3-TO-6" DIAMETER ROUNDED GRANITE STONES IN A TALL, SLENDER MASONRY BIN FORMING THE WEST HALF OF THE NORTH WALL OF THE MAIN PORTION OF THE HOUSE. THE BIN IS STRENGTHENED BY THE INCLUSION OF MANY VERTICAL AND HORIZONTAL 1/2" DIAMETER STEEL RODS. AIR FROM THE COLLECTOR ENTERS THE BIN AT THE BOTTOM, VIA 8 8" STEEL PIPES. AIR PASSES UPWARD THROUGH THE BIN TO THE PLENUM AT THE TOP, AND THENCE IT RETURNS TO HE BOTTOM OF THE COLLECTOR--OR PASSES INTO ROOMS VIA FLOOR-LEVEL PERIMETER REGISTERS. RETURN AIR LEAVES ROOMS VIA REGISTERS AT CEILING HEIGHT. AUTOMATICALLY CONTROLLED DAMPERS ARE PROVIDED.
	A HEATILATOR-TYPE FIREPLACE WITH TWO MANUALLY SWITCHED FANS FOR SENDING WARM AIR TO THE MAIN ROOMS IS PROVIDED FOR AUXILIARY HEATING. COOLING IN SUMMER IS ACCOMPLISHED AT NIGHT WHEN COOL OUTDOOR AIR IS CIRCULATED THROUGH THE BIN OF STONES, COOLING IT. DURING HOT DAYS, ROOM AIR IS CIRCULATED THROUGH THE BIN AND COOLED.
MATERIAL	6" FIBERGLAS INSULATION; INSULATED FOUNDATION WALLS; SHEET OF RIBBED, NON-GALVANIZED STEEL; NON-SELECTIVE-BLACK COATING; SINGLE SHEET OF DOUBLE-STRENGTH GLASS; 3/4 HP BLOWER; FOIL-FACED FIBERGLAS BATTS; HEATILATOR-TYPE FIREPLACE
AVAILABILITY	ADDITIONAL INFORMATION MAY BE OBTAINED FROM THE CONTINENTAL COUNTRY CLUB, 5455 COUNTRY CLUB DRIVE, FLAGSTAFF, ARIZONA.
REFERENCE	SHURCLIFF, W.A. SOLAR HEATED BUILDINGS, A BRIEF SURVEY CAMBRIDGE, MASSACHUSETTS, W.A. SHURCLIFF, MARCH 1976, 212 P

TITLE	CROFT SOLAR HOUSE
CODE	
KEYWORDS	HOUSE CLERESTORY INSULATION DRAPES PASSIVE SOLAR RADIATION COLLECTING WINDOW STORAGE WIDE EAVES
AUTHOR	P. OKHUYSEN, J.L. FERGUSON, W.L. CROFT
DATE & STATUS	THE CROFT HOUSE WAS COMPLETED IN 1974.
LOCATION	LONGVIEW, MISSISSIPPI
SCOPE	THE BUILDING IS A 1-1/2-STORY, 3-BEDROOM, 2,600 SQ. FT. HOUSE, EMPLOYING POST-AND-BEAM CONSTRUCTION WITH BRICK VENEER. THE HOUSE INCLUDES 3 BATHROOMS, A KITCHEN-DINING AREA, DEN, FOYER, WORKROOM, MEZZANINE, AND A BRIDGE OVER THE DEN. THERE IS ALSO A GARAGE.
TECHNIQUE	A CLERESTORY RUNS THE LENGTH OF THE HOUSE. THE INSULATION IS TO R-30. THE WINDOWS ARE DOUBLE GLAZED WITH GOOD AIR-TIGHTNESS. THE SLOPING SOUTH ROOF IS OF 0.032" ALUMINUM. HEAVY DRAPES COVER THE WINDOWS AT NIGHT.
	THE SOLAR HEATING SYSTEM IS OF THE PASSIVE TYPE. THE TOTAL WINDOW AREA IS 450 SQ. FT. THE LARGEST SOLAR-RADIATION-COLLECTING WINDOW AREA IS ON THE SOUTH FACE OF THE CLERESTORY, AND IS ABOUT 52' X 5'. THE RADIATION ENTERING HERE PENETRATES DEEP INTO THE HOUSE. THE SOUTH FACE OF THE FIRST STORY INCLUDES TWO LARGE AREAS OF WINDOWS, ADMITTING RADIATION TO THE SOUTH PORTION OF THE HOUSE. THE SOUTH-SLOPING ROOF ADJACENT TO THE CLERESTORY, HAVING A REFLECTIVE ALUMINUM SURFACE, REFLECTS MUCH ADDITIONAL RADIATION THROUGH THE CLERESTORY WINDOWS. THE STORAGE IS PROVIDED BY THE MASSIVE COMPONENTS OF THE HOUSE ITSELF.
	COOLING IN SUMMER IS ACCOMPLISHED BY WIDE EAVES, WHICH EXCLUDE RADIATION FROM ALL OF THE SOUTH WINDOWS IN SUMMER. THE ALUMINUM ROOFS REFLECT INCIDENT RADIATION TOWARD THE SKY.
MATERIAL	DOUBLE GLAZED WINDOWS, 0.032" ALUMINUM, DRAPES
AVAILABILITY	ADDITIONAL INFORMATION MAY BE OBTAINED FROM PABLO OKHUYSEN, 214 LAMPKIN ROAD, STARKVILLE, MISSISSIPPI 39759.
REFERENCE	OKHUYSEN, PABLO THE MISSISSIPPI ARCHITECT 5, 4, WINTER 1974, P 15
	SHURCLIFF, W.A. SOLAR HEATED BUILDINGS, A BRIEF SURVEY CAMBRIDGE, MASSACHUSETTS, W.A. SHURCLIFF, MARCH 1976, 212 P

TITLE	CROWTER SOLAR GROUP HOME
CODE	
KEYWORDS	AIR HOUSE HEAT HOT WATER COIL COLLECTOR SOLAR BLOWER MOTOR INSULATION PEBBLE STORAGE FORCED BACKUP
AUTHOR	SOLARON CORP.
DATE & STATUS	THE HOME WAS COMPLETED IN 1975.
LOCATION	MOUNTAINS WEST OF DENVER, COLORADO
SCOPE	THIS HOUSE UTILIZES THE AIR COLLECTION SYSTEM FOR ITS DOMESTIC HEATING PURPOSES. IT DEMONSTRATES THE FLEXIBILITY AND SAFETY OF AIR COLLECTORS OVER COLLECTORS WHICH USE WATER AS A HEAT TRANSFER MEDIUM.
TECHNIQUE	THE SYSTEM CONSISTS OF AIR HEATING COLLECTORS THAT CONTAIN A FLAT ABSORBER/HEAT EXCHANGER AND INTERNAL MANIFOLDING. AUTOMATIC DAMPERS, BLOWERS AND A MOTOR DIRECT THE FLOW OF AIR THROUGH THE SYSTEM. A HOT WATER COIL CAN ALSO BE ATTACHED TO HEAT WATER FOR DOMESTIC USE. A PROPERLY INSULATED WOODEN OR CONCRETE BOX HOLDS DRY PEBBLES, WHICH ACT AS A HEAT STORAGE MEDIUM. A CONVENTIONAL FORCED AIR HEATING SYSTEM ACTS AS A BACKUP FOR THIS SYSTEM.
MATERIAL	AIR HEATING COLLECTORS, DAMPERS, BLOWERS, MOTOR, HOT WATER COIL, INSULATED STORAGE CONTAINER, PEBBLES, FORCED AIR HEATER AS BACKUP
AVAILABILITY	ADDITIONAL INFORMATION MAY BE OBTAINED FROM SOLARON CORP., 4850 OLIVE STREET, COMMERCE CITY, COLORADO 80022.
REFERENCE	SOLARON CORP. SOLARON CORP. SOLAR ENERGY SYSTEMS: THE SOLARON AIR HEATING SYSTEM COMMERCE CITY, COLORADO, SOLARON CORP. PAMPHLET, 102-1, SEPTEMBER 1975, 3 P

302

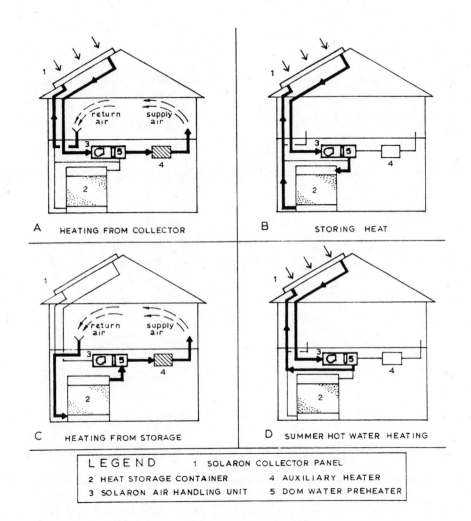

A HEATING FROM COLLECTOR

B STORING HEAT

C HEATING FROM STORAGE

D SUMMER HOT WATER HEATING

LEGEND	1 SOLARON COLLECTOR PANEL
2 HEAT STORAGE CONTAINER	4 AUXILIARY HEATER
3 SOLARON AIR HANDLING UNIT	5 DOM WATER PREHEATER

TITLE	CROWTHER CONDOMINIUM
CODE	
KEYWORDS	SOLAR FLAT PLATE AIR COLLECTOR SYSTEM SPACE HEATING HEAT EXCHANGER NATURAL VENTILATE CONVECTION CONDOMINIUM
AUTHOR	RICHARD L. CROWTHER & ASSOCIATES
LOCATION	DENVER, COLORADO
DATE & STATUS	THE CROWTHER CONDOMINIUM HAS NOT YET BEEN BUILT.
SCOPE	THE CONDOMINIUM USES FLAT PLATE AIR COLLECTOR SYSTEMS FOR WATER AND SPACE HEATING.
TECHNIQUE	THE COLLECTED HOT AIR WILL BE CIRCULATED THROUGH A HEAT EXCHANGER TO HOT WATER COILS IN AIR HANDLING UNITS. THE PROJECTS WILL ALSO INCLUDE SUCH ENERGY CONSERVING FEATURES AS WINDOWS DESIGNED TO RESIST SOLAR HEAT GAINS IN SUMMER, BUT USE DIRECT SOLAR RADIATION TO REDUCE WINTER HEATING LOADS: FEW OR SMALL WINDOWS WILL BE USED ON THE NORTH, EAST AND WEST SIDES OF THE BUILDING TO REDUCE WINTER HEAT LOSSES. THE BUILDINGS WILL NATURALLY VENTILATE THEMSELVES THROUGH CONVECTUON.
MATERIAL	FLAT PLATE AIR COLLECTOR, HEAT EXCHANGER, HOT WATER COILS, AIR HANDLING UNIT.
AVAILABILITY	ADDITIONAL INFORMATION MAY BE OBTAINED FROM RICHARD L. CROWTHER ARCHITECTS GROUP, CROWTHER/SOLAR GROUP, 2830 EAST THIRD AVE., DENVER, COLORADO 80206.
REFERENCE	VILLECCO, MARGUERITE SUN POWER ARCHITECTURE PLUS, SEPTEMBER/OCTOBER, 1974, P 92

TITLE	CROWTHER SOLAR OFFICE BUILDING
CODE	
KEYWORDS	OFFICE BUILDING INSULATION FLAT PLATE AIR TYPE COLLECTOR STORAGE SYSTEM
AUTHOR	R.L. CROWTHER
DATE & STATUS	CONSTRUCTION OF THE CROWTHER SOLAR OFFICE BUILDING HAS NOT YET BEGUN.
LOCATION	DENVER, COLORADO
SCOPE	THE BUILDING IS A THREE-LEVEL MODERN OFFICE BUILDING, 69' X 43', WITH ONE LEVEL BELOW GRADE, OPENING TO A BELOW-GRADE COURTYARD. THE TOTAL FLOOR AREA IS 7,000 SQ. FT.
TECHNIQUE	THE INSULATION IS EXCELLENT. THE FIRST FLOOR VIEW-WINDOW AREA IS A 180 SQ. FT. TOTAL OF DOUBLE-GLAZED, REGRESSED GLASS. THE COLLECTOR AREA IS 600 SQ. FT. THE COLLECTOR IS A FLAT-PLATE, AIR-TYPE COLLECTOR, SLOPING 53 DEGREES FROM THE HORIZONTAL. IT IS DOUBLE-GLAZED WITH TEMPERED GALSS. THE HORIZONTAL AREA AT THE SOUTH EDGE OF THE COLLECTOR SERVES TO DIRECT MORE RADIATION TOWARD THE COLLECTOR. THE STORAGE SYSTEM CONTAINS 4,000 GALLONS OF WATER. THE WATER IS IN TWO INSULATED, CONCRETE TANKS IN THE BASEMENT.
MATERIAL	FLAT-PLATE, AIR-TYPE COLLECTOR; 4,000 GALLONS OF WATER; TWO INSULATED, CONCRETE TANKS
AVAILABILITY	ADDITIONAL INFORMATION MAY BE OBTAINED FROM R.L. CROWTHER, AIA, ARCHITECTS GROUP, 2830 EAST THIRD AVE., DENVER, COLORADO 80206.
REFERENCE	SHURCLIFF, W.A. SOLAR HEATED BUILDINGS, A BRIEF SURVEY CAMBRIDGE, MASSACHUSETTS, W.A. SHURCLIFF, MARCH 1976, 212 P

TITLE	CSIRO SOLAR BUILDING
CODE	
KEYWORDS	BUILDING AIR TYPE COLLECTOR STORAGE
AUTHOR	D.J. CLOSE, R.V. DUNKLE AND K.A. ROBESON
DATE & STATUS	THE BUILDING WAS RETROFITTED FOR SOLAR ENERGY IN 1964. THE SYSTEM WAS STILL IN USE IN 1975.
LOCATION	HIGHETT, AUSTRALIA
SCOPE	THE BUILDING FLOOR PLAN IS 180' X 100'. THE CENTRAL PORTION, CONSISTING OF ONE VERY TALL STORY, IS FLANKED ON THE NORTH AND SOUTH BY TWO-STORY BAYS.
TECHNIQUE	THE BUILDING AIMS 10 DEGREES EAST OF NORTH. THERE IS VERY LITTLE INSULATION. THERE IS REFLECTIVE INSULATION ON SOME WALLS. THE SOLAR HEATING WAS APPLIED JUST TO A SMALL PORTION OF THE SOUTH (SHADED) SIDE; IT WAS APPLIED TO A TWO-STORY PORTION, 30' X 30', WITH A FLOOR AREA OF 900 SQ. FT. PER STORY, I.E. 1,800 SQ. FT. IN ALL.
	THE COLLECTOR CONSISTS OF 600 SQ. FT. GROSS AREA OF AIR-TYPE COLLECTOR. IT IS MOUNTED ON THE ROOF, WHICH SLOPES 7 DEGREES FROM THE HORIZONTAL. THE COLLECTOR INCLUDES FOUR PARALLEL EAST-WEST ARRAYS, EACH 2-1/2' X 60', AND EACH SLOPING 60 DEGREES FROM THE HORIZONTAL. THE ARRAYS ARE 6' APART ON CENTERS TO MINIMIZE SHADING OF ONE ARRAY BY THE NEXT. THE ABSORBING SHEET CONSISTS OF COPPER CLAD STEEL FORMED TO PROVIDE 60 DEGREES V'S 1" HIGH. EACH SHEET HAS A SELECTIVE BLACK COATING CONSISTING OF OXIDIZED COPPER.
	THE SINGLE GLAZING IS OF 24-OZ. GLASS. BEHIND THE ABSORBING SHEET IS AN AIRSPACE (DUCT) WITH A CROSS SECTION OF 1-1/2 SQ. FT. THIS DUCT IS INSULATED WITH 1" OF INSULWOOL AND ALUMINUM FOIL. AN AIRSTREAM PASSES EAST ALONG ONE ARRAY, AND WEST ALONG ANOTHER. TWO AIRSTREAMS ARE USED IN ALL. THE TOTAL AIR-FLOW RATE IS 200 TO 900 CFM. DAMPERS ARE AVAILABLE FOR ADJUSTING THE FLOW-RATE TO MAINTAIN THE DESIRED TEMPERATURE RISE. THE LENGTHS OF EACH DUCT TO OR FROM THE STORAGE SYSTEM IS 100'. THE DUCTS HAVE 1" TO 2" OF INSULATION.
	THE STORAGE SYSTEM CONSISTS OF 1,100 CUBIC FEET OF 3/4" DIAMETER STONES, I.E. ABOUT 50 TONS OF STONES, IN A SET OF THREE 8' IN DIAMETER STEEL TANKS, INSULATED WITH 1-1/2" POLYURETHANE FOAM, ADJACENT TO THE BUILDING. THE ORIGINAL STORAGE BINS ARE CURRENTLY BEING REPLACED BY A SINGLE BRICK BIN.
	DURING THE SUMMER, THE STORAGE SYSTEM IS COOLED AT NIGHT BY A 2,500 CFM FLOW OF OUTDOOR AIR. WATER IS SPRAYED INTO THE AIR, JUST BEFORE IT ENTERS THE STORAGE SYSTEM, TO MAKE IT EVEN COLDER. DURING THE DAY, AIR IS BLOWN FROM THE STORAGE SYSTEM TO THE ROOMS AND, EN ROUTE TO THE ROOMS, RECEIVES A FURTHER SPRAY TO PRODUCE ADDITIONAL COOLING.
MATERIAL	COPPER CLAD STEEL; SELECTIVE BLACK COATING OF OXIDIZED COPPER; 24-OZ. GLASS; 1" OF INSULWOOL AND ALUMINUM FOIL; 50 TONS OF 3/4" DIAMETER STONES
AVAILABILITY	ADDITIONAL INFORMATION MAY BE OBTAINED FROM THE DIVISION OF MECHANICAL ENGINEERING, COMMONWEALTH SCIENTIFIC AND INDUSTRIAL RESEARCH ORGANIZATION (CSIRO), P.O.BOX 26, HIGHETT, VIC. 3190, MELBOURNE, VICTORIA, AUSTRALIA.
REFERENCE	SHURCLIFF, W.A. SOLAR HEATED BUILDINGS, A BRIEF SURVEY CAMBRIDGE, MASSACHUSETTS, W.A. SHURCLIFF, MARCH 1976, 212 P

TITLE	CULP HOUSE
CODE	
KEYWORDS	HOUSE WINDOWS INSULATED COLLECTOR TRICKLING WATER TYPE ORIEN-TATION STORAGE GRAVITY FLOW
AUTHOR	SUNSTRUCTURES, INC.
DATE & STATUS	THE CULP HOUSE IS SCHEDULED FOR COMPLETION IN 1976.
LOCATION	POMEROY, OHIO
SCOPE	THE BUILDING IS A TWO-STORY, 1,500 SQ. FT., WOODEN A-FRAME HOUSE WITH A BASEMENT.
TECHNIQUE	THERE ARE NO WINDOWS ON THE NORTH. THE WINDOWS ON THE EAST, SOUTH AND WEST ARE DOUBLE GLAZED. THE HOUSE IS INSULATED THROUGH-OUT WITH 6" OF FIBERGLAS BATT.

THE COLLECTOR IS A 750 SQ. FT., TRICKLING-WATER TYPE COLLECTOR, MOUNTED ON THE ROOF, SLOPING 55 DEGREES FROM THE HORIZONTAL. THE HOUSE FACES STRAIGHT SOUTH. THE GROSS DIMENSIONS OF THE COLLECTOR ARE 40' X 20', WITH INTERRUPTIONS BY THE SOUTH WINDOWS. THE DIMENSIONS OF THE GLASS COVERS ARE 76" X 34". ALONG THE UPPER 76" OF THE COLLECTOR, THE GLAZING IS DOUBLE. EACH GLASS SHEET IS 1/8" THICK, AND THE AIRSPACE BETWEEN THE SHEETS IS 3/8". THE LOWER PART OF THE COLLECTOR IS SINGLE GLAZED WITH 3/16" OF TEMPERED GLASS. THE BLACK COATING ON THE CORRUGATED ALUMINUM SHEET IS NON-SELECTIVE. THE COLLECTOR IS AN INTEGRAL PART OF THE ROOF, AND WAS CONSTRUCTED ON THE SITE, ON THE ROOF. IT IS BACKED BY 6" OF FIBERGLAS. THE LIQUID IS WATER, WITH NO ANTIFREEZE OR INHIBITOR. THE FLOWRATE IS 15 GPM, MAINTAINED BY A 1/3 HP CENTRIFUGAL PUMP. THE COLLECTOR WATER FLOWS DIRECTLY INTO THE STORAGE SYSTEM. THERE IS NO HEAT EXCHANGER.

THE STORAGE SYSTEM CONTAINS 3,600 GALLONS OF WATER IN A RECTANGU-LAR, 10' X 8' X 6' HIGH TANK OF CONCRETE BLOCKS, WATERPROOFED ON THE INSIDE WITH BITUMINOUS ASPHALT AND POLYETHYLENE SHEET. THE TANK INSULATION IS 3" OF POLYURETHANE FOAM. THE TANK IS SITU-ATED IN THE BASEMENT.

AUXILIARY HEAT IS PROVIDED BY A STONE MASONRY FIREPLACE NEAR THE CENTER OF THE MAIN FLOOR. THE PLAN-VIEW OUTSIDE DIMENSIONS OF THE STRUCTURE ARE 8' X 4'. THE FIREPLACE OPENING IS 4' WIDE X 3' HIGH. SPECIAL HEAT-SAVING, GRAVITY-FLOW, AIR-DUCTS ARE ON EITHER SIDE OF THE FIREPLACE.

MATERIAL	6" FIBERGLAS BATT; DOUBLE GLAZING; 1/8" THICK GLASS SHEET; 3/16" TEMPERED GLASS; NON-SELECTIVE BLACK COATING; CORRUGATED ALUMINUM SHEET; 6" OF FIBERGLAS; 1/3 HP CENTRIFUGAL PUMP; 3,600 GALLONS OF WATER; RECTANGULAR, 10' X 8' X 6' HIGH TANK OF CONCRETE BLOCKS; BITUMINOUS ASPHALT; POLYETHYLENE SHEET; 3" POLYURETHANE FOAM
AVAILABILITY	ADDITIONAL INFORMATION MAY BE OBTAINED FROM SUNSTRUCTURES, INC., 225 E. LIBERTY STREET, ANN ARBOR, MICHIGAN 48104.
REFERENCE	SHURCLIFF, W.A. SOLAR HEATED BUILDINGS, A BRIEF SURVEY CAMBRIDGE, MASSACHUSETTS, W.A. SHURCLIFF, MARCH 1976, 212 P

TITLE	CUSTOM LEATHER BOUTIQUE BUILDING
CODE	
KEYWORDS	LEATHER BOUTIQUE WINDOWS COLLECTOR ORIENTATION STORAGE HEAT PUMPS
AUTHOR	J. DEVRIES OF SOL-R-TECH
DATE & STATUS	THE CUSTOM LEATEHR BOUTIQUE BUILDING WAS COMPLETED IN 1974.
LOCATION	WHITE RIVER JUNCTION, VERMONT

SCOPE

THE BUILDING IS A 1-1/2-STORY, U-SHAPED, 2,100 SQ. FT., WOOD-FRAME BUILDING WITH NO BASEMENT, AND WITH A CONCRETE SLAB FIRST-STORY FLOOR. 60% OF THE BUILDING IS SOLAR HEATED, NOT COUNTING THE ENERGY FROM THE HEAT-PUMP MOTORS. IF SUCH ENERGY IS INCLUDED, THE FIGURE IS ABOUT 80%. THE PERCENT OF THE WINTER'S HEAT-NEED, SUPPLIED BY AN AUXILIARY SOURCE, IS 20%.

PERFORMANCE HAS GENERALLY BEEN GOOD, BUT MINOR PROBLEMS WITH THE COLLECTORS, INCLUDING EXCESSIVE THERMAL CAPACITANCE ASSOCIATED WITH THE COLLECTOR BACKING AND INADEQUATE INSULATION ON TOP OF THE STORAGE TANK HURT THE PERFORMANCE IN THE WINTER OF 1974-1975. STEPS TO OVERCOME SOME OF THE DIFFICULTIES HAVE BEEN TAKEN, AND FURTHER STEPS HAVE BEEN PLANNED. THE SYSTEM IS NOW DEEMED CONSIDERABLY MORE COST-EFFECTIVE THAN ELECTRICAL HEATING, AND APPROXIMATELY EQUAL TO OIL-HEATING IN WINTER COMBINED WITH CONVENTIONAL AIR CONDITIONING IN SUMMER.

TECHNIQUE

THE THICKNESS OF THE FIBERGLAS INSULATION IS 4" IN THE WALL, AND 6" IN THE ROOF. ALL THE WINDOWS ARE DOUBLE GLAZED. THERE ARE NO WINDOWS ON THE VERTICAL SOUTH, EAST, OR WEST WALLS. THE BUILDING AIMS EXACTLY SOUTH. THE STREET AND DOORS ARE AT THE NORTHERN EXPOSURE.

THE COLLECTOR IS 910 SQ. FT. NET. IT IS A WATER-TYPE COLLECTOR, MOUNTED ON THE SOUTH ROOF, SLOPING 45 DEGREES FROM THE HORIZONTAL. THERE ARE 38 SOL-R-TECH 1974-TYPE PANELS, EACH 8' X 3'. THE HEART OF THE PANEL IS AN OLIN BRASS CO. ALUMINUM ROLL-BOND PLATE, OF TYPE 1100 ALUMINUM, WITH INTEGRAL EXPANDED PASSAGES FOR A COOLANT. THE BLACK COATING IS NON-SELECTIVE. THE COOLANT IS WATER WITH 500 PPM OF CHROMATE INHIBITOR CALLED BARTLETT CHEMICAL CO. INHIBITOR C. NO ANTIFREEZE IS USED; THE LIQUID IS DRAINED INTO THE STORAGE TANK BEFORE FREEZE-UP CAN OCCUR. THE GLAZING IS DOUBLE, OF 0.04" KALWALL SUN-LITE (FIBERGLAS AND POLYESTER). BEHIND THE COLLECTOR, IN THE ROOF, THERE IS 6" OF FIBERGLAS. A DEKOLABS, INC., DIFFERENTIAL CONTROLLER IS USED.

THE STORAGE SYSTEM CONTAINS 3,200 GALLONS OF WATER IN A RECTANGULAR, CONCRETE-BLOCK TANK, WITH INSIDE DIMENSIONS OF 17' X 7' X 5' HIGH, SITUATED IMMEDIATELY BELOW THE CONCRETE SLAB, NEAR THE CENTER OF THE BUILDING. THE TANK IS WATERPROOFED ON THE INTERIOR WITH EPOXY. 3" OF STYROFOAM INSULATION WAS APPLIED TO THE TOP OF THE TANK IN 1975.

THERE ARE THREE YORK TRITON DW-30H, WATER-TO-AIR, 2-1/2 TON CAPACITY, HEAT PUMPS. WHEN THE STORAGE TANK TEMPERATURE EXCEEDS 45 DEGREES F, THE HEAT PUMPS EXTRACT ENERGY FROM THE TANK OF WATER AND DELIVER ENERGY, AT 120 DEGREES F, TO THE COILS IN THE DUCTS OF THE FORCED-HOT-AIR SYSTEM. THE TYPICAL COP OF THE HEAT PUMPS IS 3. IF THE TANK TEMPERATURE EXCEEDS 90 DEGREES F, THE HEAT PUMPS REMAIN OFF, AND THE TANK WATER IS CIRCULATED DIRECTLY TO THE COILS IN THE DUCTS. IF THE TANK TEMPERATURE IS BELOW 45 DEGREES F, AUXILIARY HEAT IS INVOKED. AUXILIARY HEAT IS PROVIDED BY ELECTRIC HEATERS IN THE DUCTS.

COOLING IN SUMMER IS ACCOMPLISHED BY THE HEAT PUMPS. THEY DELIVER COLD WATER TO THE COILS IN THE DUCTS, AND DISSIPATE HEAT TO THE OUTDOORS VIA AN OUTDOOR COOLING TOWER. THE COLLECTOR CONTINUES TO COLLECT ENERGY AND DELIVER IT TO THE STORAGE TANK, TO HEAT THE DOMESTIC HOT WATER.

MATERIAL	DOUBLE GLAZED WINDOWS; SOL-R-TECH 1974-TYPE PANELS; OLIN BRASS CO. ALUMINUM ROLL-BOND PLATE OF TYPE 1100 ALUMINUM; NON-SELECTIVE BLACK COATING; WATER WITH 500 PPM CHROMATE INHIBITOR (BARTLETT CHEM. CO. INHIBITOR C); 0.04" KALWALL SUN-LITE (FIBERGLAS AND POLYESTER); 6" OF FIBERGLAS; DEKOLABS, INC. DIFFERENTIAL CONTROLLER; 3,200 GALLONS OF WATER; RECTANGULAR, CONCRETE-BLOCK TANK; CONCRETE SLAB; EPOXY; 3" OF STYROFOAM INSULATION; THREE YORK TRITON DW-30H, WATER-TO-AIR, 2-1/2 TON CAPACITY, HEAT PUMPS
AVAILABILITY	ADDITIONAL INFORMATION MAY BE OBTAINED FROM SOL-R-TECH, MILL ROAD, HARTFORD, VERMONT 05047.
REFERENCE	SHURCLIFF, W.A. SOLAR HEATED BUILDINGS, A BRIEF SURVEY CAMBRIDGE, MASSACHUSETTS, W.A. SHURCLIFF, MARCH 1976, 212 P

311

TITLE	DALLAS WORLD TRADE CENTER
CODE	DWTC
KEYWORDS	TRADECENTER THERMOSTAT ELECTRICAL POWER HEATING COOLING SWITCH LIGHT AIR CONDITIONING PLENUM OFFICE COMMERCIAL
AUTHOR	BERAN & SHELMIRE, ARCHITECTS
LOCATION	DALLAS, TEXAS
DATE & STATUS	CONSTRUCTION OF THE DALLAS WORLD TRADE CENTER WAS COMPLETED IN JULY, 1974.
SCOPE	THE DALLAS WORLD TRADE CENTER CONTAINS 4.3 MILLION SQUARE FEET IN 3,500 SHOWROOMS. EACH ROOM HAS ITS OWN THERMOSTAT SO THAT THE HEAT CAN BE REGULATED INDEPENDENTLY OF ANY OTHER SHOWROOM. THE OWNER IS FURNISHING ELECTRICAL POWER, HEATING AND COOLING. IN ORDER TO SAVE ENERGY, THERE ARE DEVICES BEING INSTALLED WHICH AUTOMATICALLY SHUT OFF ALL THE POWER IN A SHOWROOM THAT IS NOT BEING USED.
	THE BUILDING WILL BE CONSTRUCTED IN THREE PHASES. THE FIRST PHASE INCLUDES A 52 X 388 SEVEN STORY STRUCTURE CONTAINING 1.4 MILLION SQUARE FEET AND TWO FOUR STORY PARKING GARAGES. FIVE MORE FLOORS WILL PROBABLY BE ADDED WITHIN FIVE YEARS. THE THIRD CONSTRUCTION PHASE WILL BRING THE BUILDING UP TO ITS ULTIMATE HEIGHT OF 20 STORIES AT A TOTAL COST NOW ESTIMATED AT $80 MILLION.
TECHNIQUE	THE AUTOMATIC HEAT CONTROL SYSTEM IS CONTROLLED BY A DOOR SWITCH. WHEN THE DOOR IS CLOSED, LIGHTS AND THE AIR CONDITIONING EQUIP-MENT DO NOT OPERATE. WHEN THE DOOR IS OPEN, A POWERED MIXING UNIT ABOVE THE SHOWROOM CEILING IS ACTIVATED, FURNISHING COOLING OR HEATING AS REQUIRED BY THE ROOM'S THERMOSTAT. IF HEATED AIR IS REQUIRED, IT WILL BE PROVIDED BY A SMALL ELECTRIC HEATING COIL CONTAINED WITHIN THE MIXING BOX.
	A SINGLE DUCT LEADING FROM A PLENUM TO EACH SHOWROOM FEEDS THE COLD AIR, SUPPLIED AT 50 F, TO THE FAN POWERED MIXING UNIT SERVING THE SPACE. THE UNIT IS DESIGNED TO HANDLE THE QUANTITY OF AIR REQUIRED TO COOL A SHOWROOM TO 72 F IF IT IS USING THE MAXIMUM ELECTRICAL LOAD OF 12 WATTS PER SQUARE FOOT..
MATERIAL	THERMOSTAT, DOOR SWITCH, POWERED MIXING UNIT, ELECTRIC HEATING COIL, PLENUM
AVAILABILITY	ADDITIONAL INFORMATION ON THE DALLAS WORLD TRADE CENTER MAY BE OBTAINED FROM BERAN & SHELMIRE, 1900 EMPIRE LIFE BUILDING, DALLAS, TEXAS 75201.
REFERENCE	ENGINEERING NEWS RECORD BIG TRADE MART'S HVAC SYSTEM KEYED TO FLUCTUATIONS IN DEMAND ENGINEERING NEWS RECORD, MARCH 14, 1974

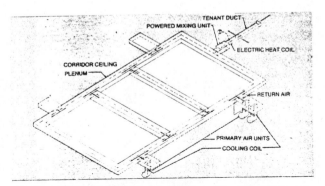

Powered mixing units serving each showroom draw cold air from corridor ceiling supply plenum that eliminated traditional ductwork.

TITLE	DAM ARCOLOGY
CODE	
KEYWORDS	DAM SOLAR ENERGIZED CITY GREENHOUSE CHIMNEY APSE EFFECT ARCOLOGY
AUTHOR	PAOLO SOLERI
DATE & STATUS	THE DESIGN FOR THE DAM ARCOLOGY IS COMPLETE, BUT NOT BUILT, AS OF 1976.
LOCATION	MIDDLE LATITUDE CLIMATES AND COLD STEPPE CLIMATES SUCH AS IRAN AND AFGHANISTAN
SCOPE	THE DAM ARCOLOGY IS DESIGNED TO HOUSE 13,000 PEOPLE. IT TYPIFIES THE DOUBLE USE OF THE APSE FORM FOR THE PURPOSE OF DAMMING A RIVER AND COLLECTING SOLAR ENERGY.
TECHNIQUE	THE DAM ARCOLOGY IS BASED ON SEVERAL EFFECTS OF THE SUN, THE GREENHOUSE, CHIMNEY, AND APSE EFFECTS. AN ARCOLOGY IS SITUATED DIRECTLY ABOVE A LARGE TERRACED GREENHOUSE WHICH SENDS UP WARM, MOIST AIR THROUGH THE CHIMNEY EFFECT INTO THE CITY WHERE IT MAY BE USED OR STORED. MEANWHILE, APSE-SHAPED ELEMENTS OF THE CITY CATCH SUNLIGHT IN THE WINTER, PROVIDING NATURAL HEAT, BUT IN THE SUMMER THEY SHIELD THE INHABITANTS FROM HOT RAYS, PROVIDING NATURAL COOLING.
MATERIAL	TRANSPARENT SHIELD
AVAILABILITY	ADDITIONAL INFORMATION MAY BE OBTAINED FROM PAOLO SOLERI, COSANTI FOUNDATION, 6433 DOUBLETREE ROAD, SCOTTSDALE, ARIZONA 85253.
REFERENCE	ARCOSANTI NEWSLETTER TWO SUNS EXHIBITION OPENS ARCOSANTI NEWSLETTER, SUMMER 1976, PP 1, 4-5

315

TITLE	DASBURG HOUSE
CODE	
KEYWORDS	HOUSE WINDOW AREA PASSIVE SOLAR ADOBE STORAGE ORIENTATION
AUTHOR	W. LUMPKINS
DATE & STATUS	THE DASBURG HOUSE WAS COMPLETED IN 1975.
LOCATION	SANTA FE, NEW MEXICO
SCOPE	THE BUILDING IS A ONE-STORY, ONE-BEDROOM, 1,200 (INTERIOR) SQ.FT. HOUSE, ABOUT 65' X 18'. THERE IS NO ATTIC, BASEMENT, OR GARAGE.
TECHNIQUE	THE BUILDING FACES 7 DEGREES EAST OF SOUTH. THE WALL INCLUDES TEN 4' X 6' WINDOWS; THE WEST WALL INCLUDES ONE SUCH WINDOW; THE EAST WALL INCLUDES TWO 3' X 3' WINDOWS. ALL OF THESE WINDOWS ARE THERMOPANE. THE NORTH WALL WINDOW AREA IS VERY SMALL. THE WINDOWS ARE COVERED AT NIGHT WITH MAGNETICALLY AFFIXED, 1" THICK SHEET OF POLYSTYRENE FOAM. THE NORTH WALL, AND THE SOUTH WALL REGIONS BETWEEN THE WINDOWS, ARE OF 2' THICK ADOBE. ALL THE WALLS ARE INSULATED ON THE OUTSIDE WITH 2" OF POLYURETHANE FOAM, PROTECTED BY A LAYER OF STUCCO CEMENT. THE FLOOR CONSISTS OF A SINGLE LAYER OF 2-1/4" THICK INDOOR-TYPE BRICKS LAYED ON SEVERAL INCHES OF SAND. THERE ARE NO RUGS ON THE FLOOR. THERE IS 3" OF POLYURETHANE BETWEEN THE DECKING AND THE 5 LAYERS OF ROOFING MATERIAL.
	THERE IS NO FORMAL COLLECTOR. THE LARGE SOUTH, EAST, AND WEST WINDOWS, TOTAL AREA 280 SQ. FT., ADMIT MUCH RADIATION, WHICH STRIKES THE WALLS AND THE FLOOR. THE LIGHT-COLORED SAND HAS BEEN LAID ON THE GROUND ADJACENT TO THE SOUTH SIDE OF THE BUILDING TO REFLECT RADIATION TOWARD THE WINDOWS.
	THERE IS NO FORMAL STORAGE SYSTEM. MUCH ENERGY IS STORED BY THE ADOBE WALLS AND THE FLOOR.
MATERIAL	THERMOPANE; 1" THICK SHEET OF POLYSTYRENE FOAM; 2' THICK ADOBE; 2" POLYURETHANE FOAM; STUCCO CEMENT; SINGLE LAYER OF 2-1/4" THICK INDOOR-TYPE BRICKS; SEVERAL INCHES OF SAND; 3" POLYURETHANE
AVAILABILITY	ADDITIONAL INFORMATION MAY BE OBTAINED FROM A.V. DASBURG, RT. 3, BOX 124M, SANTA FE, NEW MEXICO 87501.
REFERENCE	SHURCLIFF, W.A. SOLAR HEATED BUILDINGS, A BRIEF SURVEY CAMBRIDGE, MASSACHUSETTS, W.A. SHURCLIFF, MARCH 1976, 212 P

TITLE	DAVIS HOUSE
CODE	
KEYWORDS	SOLAR AIR COLLECTOR STORAGE GRAVITY SKYLID HOUSE
AUTHOR	STEVE BAER
LOCATION	ALBUQUERQUE, NEW MEXICO
DATE & STATUS	THE DAVIS HOUSE WAS BUILT IN 1972.
SCOPE	THE DAVIS HOUSE, WHICH IS LARGELY OWNER BUILT, OCCUPIES 1,200 SQUARE FEET AND HAS 2 FLOORS. THE WARM INTERIOR HAS A HOMEY SKI LODGE CHARACTER-OPEN KITCHEN, BALCONY, BEDROOMS, WALLS OF BOOKS AND A RUGGED FIRE PLACE.
TECHNIQUE	THE DAVIS HOUSE OPERATES ON SOLAR AIR COLLECTORS AND ROCK BED STORAGE FORMS THE FLOOR OF THE SOUTH-FACING PORCH. THE SOLAR AIR HEATING SYSTEM WORKS BY MEANS OF GRAVITY FLOW, WITHOUT THE BENEFIT OF FANS. AS A BUILDING, IT WORKS ABSOLUTELY SILENTLY. THE HEAT FLOWS GENTLY UPWARD. THE ONLY CONTROLS ARE STRINGS AND DAMPERS TO OPEN THE DUCTS.
MATERIALS	SOLAR AIR COLLECTORS, ROCK BED STORAGE, STRINGS, DAMPERS, DUCTS, SKYLID
AVAILABILITY	ADDITIONAL INFORMATION MAY BE OBTAINED FROM ZOMEWORKS, P.O.BOX 712, ALBUQUERQUE, NEW MEXICO 87103.
REFERENCE	COOK, JEFFREY VARIED AND EARLY SOLAR ENERGY APPLICATIONS OF NORTHERN NEW MEXICO, THE AIA JOURNAL, AUGUST 1974, P 40
	ARIZONA STATE UNIVERSITY, COLLEGE OF ARCHITECTURE DAVIS HOUSE WASHINGTON, D.C., SOLAR-ORIENTED ARCHITECTURE, PUBLISHED BY THE AIA RESEARCH CORPORATION, JANUARY 1975, PP 42, 111-114

In the Davis house, the passive solar air collector and its thermal storage are under the south-facing porch.

The controls for the solar-heated Davis house consist of strings to open and close dampers; there are no fans or thermostats

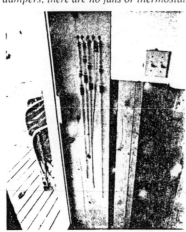

The patented "skylid" is an unpowered automatic solar heat trap. Louvers are tipped by the weight of freon that condenses in one of the connected sealed cannisters.

TITLE   DAY CARE CENTER

CODE

KEYWORDS   RADIANT HEATING LOUVER DECIDUOUS TREE DAY CARE CENTER WIND
               CATCHER SKYLIGHT ORIENTATION SITING UNDERGROUND

AUTHOR   WILLIAM F. ROGERS AND KIM WANG

DATE & STATUS THE DAY CARE CENTER WAS DESIGNED IN 1974, BUT HAS NOT BEEN BUILT.

LOCATION   THE DAY CARE CENTER IS TO BE BUILT IN NORTHERN NEW YORK STATE,
               ON A VALLEY LIKE SITE

SCOPE   THE VALLEY LIKE SITE HAS EXISTING DYNAMIC ENVIRONMENTAL
               PATTERNS THAT ARE ACCENTED TO REDUCE, AVOID AND SOLVE MANY OF
               THE ENVIRONMENTAL PROBLEMS NORMALLY HANDLED BY THE MECHANICAL
               SYSTEM, THE INTENT IS FOR THE MECHANICAL SYSTEM TO BE ONLY
               A BACK UP, THEREBY REDUCING ENERGY USE TO A MINIMUM.

TECHNIQUE   THE STRUCTURE, WHICH CONTAINS A MULTIPURPOSE SPACE, CLASSROOMS
               AND ADMINISTRATION AREAS, IS SITED ON A SOUTHERN SLOPE WITH A
               12-DEGREE SOUTHEAST ORIENTATION TO ALLOW FOR MAXIMUM HEAT GAIN
               IN WINTER, WITH THE SOUTHERN AND WESTERN SIDES OF THE BUILDING
               FLEXIBLY OPENED TO THE SUN'S RAYS. HEAT LOSS IS DIMINISHED BY
               THE BURIAL UNDERGROUND OF THE BUILDINGS NORTHERN SIDE. BOTH
               STRUCTURE AND EVERGREEN TREES PROTECT THE ENTRANCE PARKING
               AND PEDESTRIAN AREAS FROM NORTHWEST WIND AND SNOW. IN THE
               SUMMER, THE ROWS OF TREES CAPTURE MAXIMUM WIND, AND TREES ON
               THE WEST GIVE SUN PROTECTION. THE BUILDING'S EXTERIOR SOUTHERN
               LOUVERS AND PROTECTED NORTHERN SKYLIGHTS GIVE MAXIMUM SUN
               PROTECTION IN THE SUMMER. THE WEST WALL AND THE NORTHERN WIND
               CATCHERS CAPTURE THE WIND AND CIRCULATE IT THROUGH THE MULTI-
               PURPOSE AREA, PULLING AIR FROM CLASSROOMS AND OFFICES.

               ON A WINTER'S SUNNY DAY, THE EXTERIOR LOUVERS ARE ADJUSTED FOR
               MAXIMUM SUN EXPOSURE. THE PROTECTED SKYLIGHTS AND THE
               ADJUSTABLE LOUVERS PROVIDE MAXIMUM LIGHT. THE MULTIPURPOSE
               AREA, ACTING AS A GREEN HOUSE, ABSORBS THE SUN'S HEAT.
               BOTH HEAT AND AIR ARE CIRCULATED TO INTERIOR SPACES. WARMTH
               FROM THE MECHANICALLY HEATED CLASSROOM AND ADMINISTRATION
               AREAS IS CIRCULATED IN AND OUT.

               A MISTY WINTER DAY WILL HAVE MINIMUM HEAT LOSS BECAUSE THE
               EXTERIOR LOUVERS CLOSE OFF OUTSIDE COLD WHILE THE RADIATION
               PIPING RECREATES THE GREEN HOUSE EFFECT.

               THE MECHANICAL SYSTEM FEEDS DIRECTLY TO THE CLASSROOM AND
               ADMINISTRATION AREAS AND TO THE TOP OF THE MULTIPURPOSE ROOM.
               AIDED BY THE FLOOR RADIANT HEATING SYSTEM, RETURN AIR IS PULLED
               BACK TO THE MULTIPURPOSE ROOM. THE FRESH AIR SYSTEM IS
               INTEGRATED INTO THE SUPPLY PATTERN.

MATERIAL   EARTH, TREES, LOUVERS, SKYLIGHTS, WIND-CATCHERS, GREEN HOUSE

AVAILABILITY   ADDITIONAL INFORMATION MAY BE OBTAINED BY WRITING TO WILLIAM F.
               ROGERS, 100 FAIRVIEW SQUARE 2K, ITHACA, NEW YORK 14850.

REFERENCE   ROGERS, WILLIAM F. / WANG, KIM
               DAY CARE CENTER, A
               AIA JOURNAL, AUGUST 1974, P 45

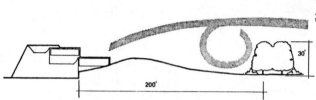

Summer:

— Maximum wind pulled down over deciduous trees-horizontal distance: tree height = 7:1

30'

200'

Winter:

— Minimum wind enters building since hill-crest lifts wind up

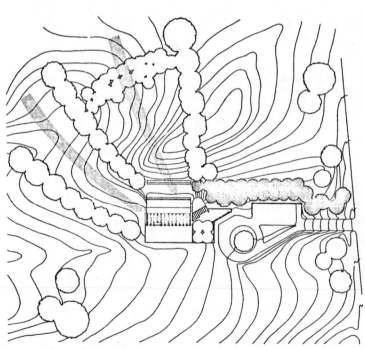

Southern slope location with 12° SE orientation allows:

Summer:

—Maximum wind captured by rows of tree.

—Minimum heat provided by trees to the west.

Winter:

—Maximum heat gain with southern and western sides flexibly opened to sun rays.

—Minimum heat loss with northern side buried underground.

— Maximum protection from the north-west wind and snow provided for entrance parking and pedestrian areas by the building and evergreen trees.

ENVIRONMENTAL STRATEGY: SITE

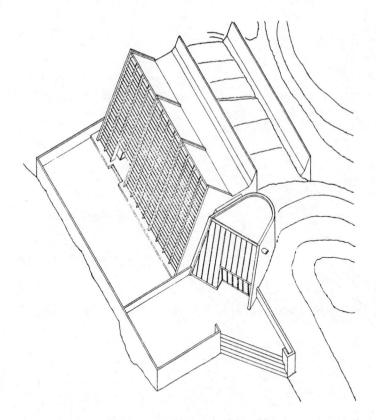

SOUTHEAST AXONOMETRIC

Summer

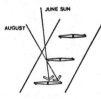

Winter Sunny

Winter Misty

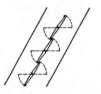

PERIOD	LOUVRE ANGLE	WINDOW EXPOSURE	VIEW
June	0°	0%	100%
July	0	0	100
Aug	0	0	100
Sept	−14	0	55
Oct	30	81	34
Nov	30	100	34
Dec	27	100	40
Jan	30	100	34
Feb	37	100	25
Mar	30	62	34
Apr	30	39	34
May	37	20	25

ENVIRONMENTAL STRATEGY:
LOUVRES

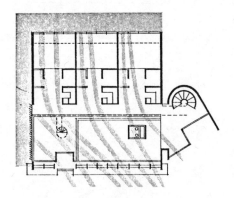

Summer:

— Maximum ventilation: west wall and northern windcatchers grab wind
— Maximum ventilation: wind going through multipurpose area pulls air from classrooms and administration area

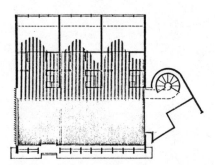

Winter:

— Maximum natural ventilation: multipurpose area absorbs suns heat and ventilation to interior spaces

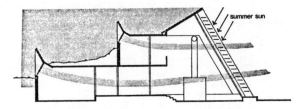

summer sun

Summer

— Maximum sun protection from exterior southern louvres and protected northern skylights.
— Maximum ventilation as northern windcatchers and southern openness pull ventilation through open plan.

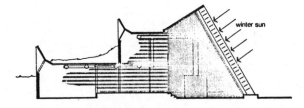

winter sun

Winter – sunny

— Maximum heat gain with exterior louvres adjusted to maximum sun exposure.
— Maximum heat circulation as multipurpose area acts as heat generating greenhouse, circulating air to and from mechanically heated class and administration area.

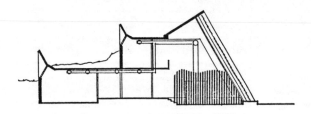

Winter – misty

— Minimum heat loss since exterior louvres close off outside cold.   Radiation piping recreates greenhouse effect.

ENVIRONMENTAL STRATEGY: BUILDING

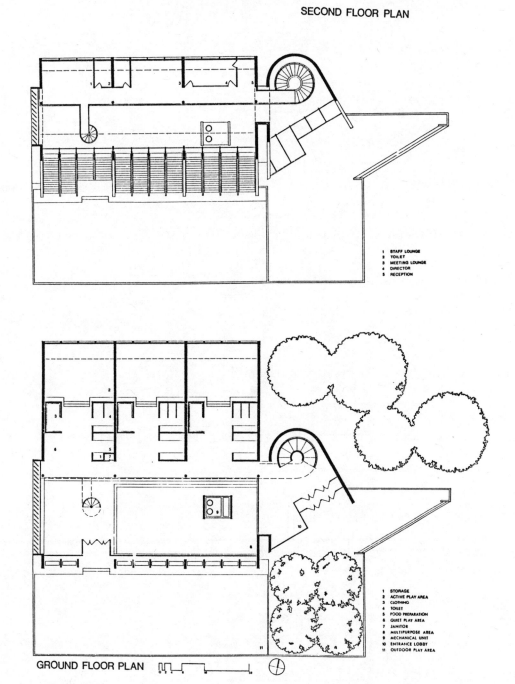

SECOND FLOOR PLAN

1 STAFF LOUNGE
2 TOILET
3 MEETING LOUNGE
4 DIRECTOR
5 RECEPTION

GROUND FLOOR PLAN

1 STORAGE
2 ACTIVE PLAY AREA
3 CLOTHING
4 TOILET
5 FOOD PREPARATION
6 QUIET PLAY AREA
7 JANITOR
8 MULTIPURPOSE AREA
9 MECHANICAL UNIT
10 ENTRANCE LOBBY
11 OUTDOOR PLAY AREA

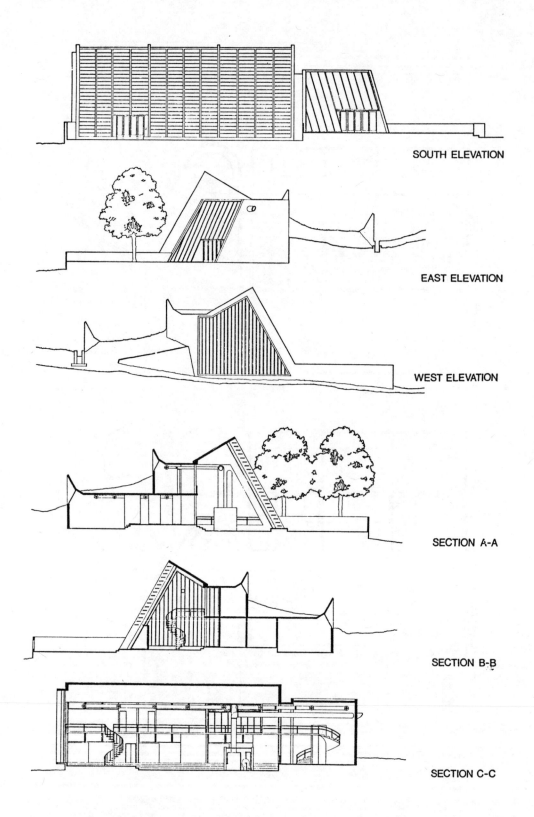

SOUTH ELEVATION

EAST ELEVATION

WEST ELEVATION

SECTION A-A

SECTION B-B

SECTION C-C

324

TITLE	DELAP HOUSE
CODE	
KEYWORDS	HOUSE ORIENTATION INSULATION NORTH WALL SUN PASSIVE SOLAR COLLECTOR REFLECTIVE GLASS HEAT SHADING STORAGE CONVECTION AUXILIARY PUMP UNDERGROUND LIGHT OVERHANGS NATURAL VENTILATION AIRCONDITIONING
AUTHOR	JAMES LAMBETH / JERRY WALL
DATE & STATUS	THE DELAP HOUSE WAS COMPLETED JANUARY 1976.
LOCATION	FAYETTEVILLE, ARKANSAS
SCOPE	THE PROJECT BUDGET FOR THIS THREE-BEDROOM, TWO-BATH HOUSE WAS $40,000 PLUS LAND COSTS. BUILDING ORIENTATION AND INSULATION WERE USED AS THE KEY FACTORS IN THE DESIGN. THE DESIGN CONSISTS OF A FAN-SHAPED PLAN WITH THE NORTH WALL AS THE SHORTEST END.

TECHNIQUE

A PORCH EXTENDING FROM THIS WALL DEFLECTS COLD NORTHERLY WINDS TO THE SIDES OF THE HOUSE. THE WINDOWLESS EAST AND WEST WALLS ARE ANGLED OUT TO ACCEPT THE WINTER SUN AND BLOCK OUT DIFFICULT SUMMER SUN.

THE KEY ELEMENT OF THE DESIGN IS A TOWERING WALL OF WINDOWS TO THE SOUTH. THESE WINDOWS FORM AN 860 SQ. FT. PASSIVE SOLAR COLLECTOR, WHICH SUPPLIES THE HOUSE WITH 60-75% OF ALL HEATING NEEDS. EACH SQUARE METER OF CLEAR INSULATED GLASS EXPOSED TO THE SOUTH HEATS UP 10 CUBIC METERS OF THE HOUSE'S INTERIOR.

REFLECTIVE EXTENSIONS OF THE EAST AND WEST WALLS RADIATE HEAT BACK TO THE COLLECTOR. THESE EXTENSIONS CAN BE SHUT DURING WINTER SUNLESS DAYS TO PROVIDE ADDITIONAL INSULATION. ON UN- SEASONABLY WARM DAYS IN THE SPRING AND THE FALL WHEN THE SUN IS NOT YET HIGH IN THE SKY, THE WALL EXTENSIONS PROVIDE SHADING FOR THE INTERIOR.

THE SOLAR SYSTEM USES THE 12 IN. THICK SOUTH THERMAL WALL AS A COLLECTOR AND PARTIAL STORAGE AREA. THE SUN'S RAYS ABSORBED INTO THE BLACKENED SURFACE OF THE WALL CAUSE AN INCREASE IN THE WALL TEMPERATURE. THIS TEMPERATURE DIFFERENTIAL CAUSES HEAT TO BE CONDUCTED THROUGH THE WALL AND THEN TRANSFERRED TO THE INTERIOR OF THE HOUSE BY CONVECTION. THE HEAT STORAGE CAPACITY OF THE MASONRY WALL PERMITS CONTINUED CONVECTION LONG AFTER THE SUN'S RAYS ARE GONE. THE AIR SPACE BETWEEN THE WINDOW WALL AND MASONRY WALL ALSO HEATS UP BY CONVECTION. THE CIRCULATION PROCESS CAN BE AIDED BY A THERMO-SWITCHED FAN IN THE RETURN AIR DUCT.

THE HEATED AIR IS DRAWN INTO RETURN AIR DUCTS WHICH LEAD TO A ROCK BED HEAT STORAGE AREA LOCATED IN THE BELOW-GROUND CRAWL SPACE. WHEN NEEDED, THIS HEAT IS DISTRIBUTED THROUGHOUT THE HOUSE BY A DUCT SYSTEM. AN AUXILIARY HEAT PUMP PROVIDES HEAT WHEN TEMPERATURES OF THE AIR IN THE ROCK BED STORAGE ARE INADE- QUATE FOR HEATING.

THE INTERIOR CONCRETE WALLS PLUS A STONE FIREPLACE AND THE UNDERGROUND STONE STORAGE PROVIDE 488 CU. FT. OF THERMAL STORAGE. THIS SUPPLIES ENOUGH ENERGY TO PROVIDE HEAT THROUGH FOUR SUN- LESS DAYS.

THE HOUSE, WHICH IS FULL OF LIGHT AND SPACE, IS ORIENTED TO CATCH THE LOWER WINTER SUNRAYS AND TO REPEL THE SUMMER RAYS WITH FOUR-FOOT OVERHANGS. NORTH AND SOUTH OPENINGS PROVIDE NATURAL VENTILATION, AND CONVENTIONAL AIR-CONDITIONING IS NEEDED FOR ONLY TWO MONTHS OF THE YEAR.

MATERIAL

CLEAR INSULATED GLASS, REFLECTIVE EXTENSIONS, 12 IN. THICK SOUTH THERMAL WALL, BLACKENED SURFACE, THERMO-SWITCHED FAN, RETURN AIR DUCT, HEATED AIR, ROCK BED, HEAT PUMP, STONE FIREPLACE, OVERHANGS

AVAILABILITY

FURTHER INFORMATION MAY BE OBTAINED FROM JAMES LAMBETH, AIA, 1591 CLARK ST., FAYETTEVILLE, ARKANSAS 72701.

REFERENCE

AMERICAN INSTITUTE OF ARCHITECTS
CASE STUDY NO. 7: DELAP HOUSE
WASHINGTON, D.C., IN ENERGY NOTEBOOK, PUBLISHED BY THE AMERICAN INSTITUTE OF ARCHITECTS, 1975, PP CS-31 TO CS-33

AIR CIRCULATION DIAGRAM (FIGURE 6 a)

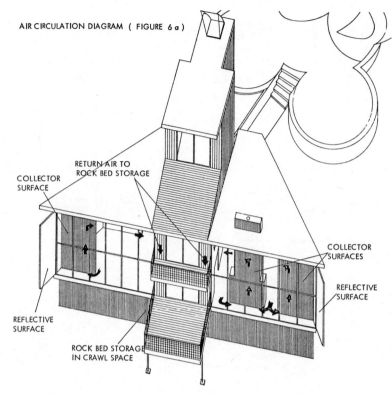

COLLECTOR
SURFACE

RETURN AIR TO
ROCK BED STORAGE

COLLECTOR
SURFACES

REFLECTIVE
SURFACE

REFLECTIVE
SURFACE

ROCK BED STORAGE
IN CRAWL SPACE

SOUTH

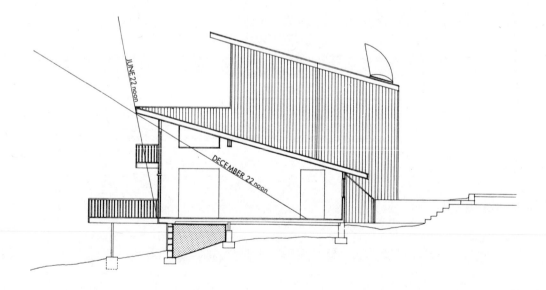

JUNE 22 noon

DECEMBER 22 noon

TITLE	DENVER COMMUNITY COLLEGE
CODE	
KEYWORDS	SCHOOL SOLAR COLLECTOR INSULATION PANEL STORAGE COMMUNITY COLLEGE
AUTHOR	BRIDGERS AND PAXTON CONSULTING ENGINEERS INC., ALBUQUERQUE, NEW MEXICO
DATE & STATUS	THE DENVER COMMUNITY COLLEGE WILL BE BUILT IN MAY, 1977.
LOCATION	WESTMINSTER, COLORADO
SCOPE	THE CLASSROOM BUILDING PLANNED FOR DENVER COMMUNITY COLLEGE WILL ACCOMODATE 3,517 FULL TIME EQUIVALENT STUDENTS. THE SOLAR COLLECTOR SYSTEM, PLUS ADDITIONAL MATERIAL TO BETTER INSULATE THE STRUCTURE, WILL COST $736,780. THE TOTAL COST OF THE BUILDING WILL BE ABOUT TWELVE MILLION DOLLARS.
	IN ORDER TO MAKE THE SYSTEM WORK, PART OF THE $736,780 WILL BE USED FOR THE FOLLOWING: DOUBLE WINDOWS, EXTRA WALL AND ROOF INSULATION, A CONCRETE ROOF STRUCTURE FOR THE SOLAR PANELS, HEAT RECOVERY UNITS FOR EXHAUST AIR, STORAGE TANK AND TEMPERATURE CONTROLS. THE HEAT PUMPS COULD OPERATE AS AN AIR-CONDITIONING UNIT IN THE SUMMER MONTHS. THE ROOF PANELS, ALTHOUGH NOT YET SELECTED, WILL BE FIXED AT A 53 DEGREE 10 MINUTE ANGLE FACING SOUTH.
	THE BUILDING WILL BE 1,000 FEET LONG AND WILL STRETCH ACROSS THE BROW OF A HILL DROPPING 25 FEET FROM THE SOUTHEAST TO THE NORTHEAST CORNER. THE STRUCTURAL CONCRETE COLUMNS, BEAMS AND ROOF TRUSSES CAN BE CAST AT THE SITE OR PRECAST.
TECHNIQUE	THE BUILDING WILL CONTAIN A 200,000 GALLON UNDERGROUND STORAGE RESERVOIR AS A SOURCE TO PUMP WATER THROUGH THE BUILDING. IF THE RESERVOIR IS SUFFICIENTLY HOT, THE BUILDING WILL BE HEATED BY DIRECT CIRCULATION. IF NOT, THE HEAT WILL BE EXTRACTED BY TWO HEAT PUMPS, AND THEN SENT THROUGH THE BUILDING. THERE WILL ALSO BE A SMALL BOILER UNIT TO ACT AS BACKUP, ALTHOUGH THE SYSTEM SHOULD MAINTAIN 100 PERCENT OF THE BUILDING HEAT LOAD.
MATERIAL	SOLAR COLLECTOR PANELS, INSULATION, DOUBLE WINDOWS, HEAT PUMPS, UNDERGROUND WATER STORAGE RESERVOIR, BOILER, HEAT RECOVERY UNITS
AVAILABILITY	FURTHER INFORMATION MAY BE OBTAINED BY WRITING TO ROBERT KULA, DIRECTOR OF PLANNING FOR THE COMMUNITY COLLEGE OF DENVER, 1009 GRANT ST., DENVER, COLORADO 80203.
REFERENCE	CHASE, DENNIS PROSPECTS GETTING BRIGHTER FOR SOLAR ENERGY COLLEGE & UNIVERSITY BUSINESS, AUGUST 1974, PP 36-38
	VILLECCO, MARGUERITE SUN POWER ARCHITECTURE PLUS, SEPTEMBER/OCTOBER 1974, PP 85-99

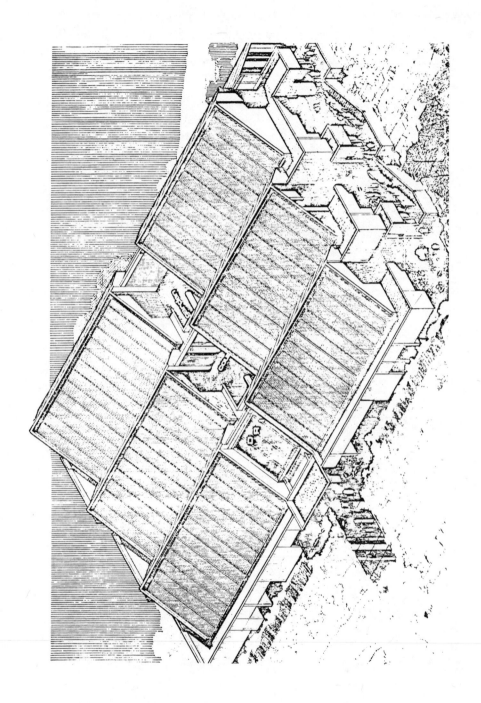

TITLE	DENVER HOUSE
CODE	
KEYWORDS	OVERLAPPED PLATE SOLAR COLLECTOR REFLECTIVE DRAPERY SHOJI SCREENS HOUSE
AUTHOR	GEORGE LOF
DATE & STATUS	THE DENVER HOUSE WAS CONSTRUCTED IN 1956.
LOCATION	DENVER, COLORADO
SCOPE	THE DENVER HOUSE IS DESIGNED SO THAT IT IS CONVENIENT TO ADD A SOLAR HEATING SYSTEM TO IT AS AN APPLIANCE OR A PIECE OF EQUIPMENT RATHER THAN INCORPORATING IT AS AN INTEGRAL PART OF THE STRUCTURE. THE HOUSE WAS THUS DESIGNED WITH A FLAT ROOF ON WHICH WERE PLACED TWO BANKS OF SLOPING (45 DEGREE ANGLE) SOLAR COLLECTORS, EACH 6 FEET HIGH AND 50 FEET LONG FOR A TOTAL COLLECTOR AREA OF 600 SQUARE FEET.

THE HOUSE HAS A HEAT LOSS RATE BETWEEN 20,000 AND 25,000 BTU PER DEGREE DAY. DURING THE WINTER OF 1959-60 THIS SYSTEM PROVIDED 26 PERCENT OF THE HEATING LOAD PLUS A PORTION OF THE HEAT NEEDED FOR DOMESTIC HOT WATER. |
| TECHNIQUE | THE ONE STORY, 2,100 SQUARE FOOT HOME HAS MANY FEATURES FOR COLLECTING HEAT AND KEEPING IT INSIDE THE HOUSE. AMONG THEM, SOUTH-FACING WINDOWS, REFLECTIVE-LINED DRAPERIES, AND SHOJI SCREENS ON THE WEST DESIGNED TO ACT AS ONE-WAY MIRRORS THAT CAN BE REVERSED TO REFLECT HEAT OUTWARD OR TO RETAIN HEAT INSIDE.

THE COLLECTORS ARE BASED ON THE OVERLAPPED PLATE PRINCIPLE. THE HEAT IS STORED IN TWO COLUMNS OF 1.5 TO 2.0 INCH GRAVEL. EACH COLUMN IS 3 FEET IN DIAMETER AND 18 FEET HIGH FOR A TOTAL OF ABOUT 12 TONS OF ROCK. |
MATERIAL	REFLECTIVE-LINED DRAPERIES, ROCK, GRAVEL, SHOJI SCREENS, SOLAR COLLECTORS
AVAILABILITY	FURTHER INFORMATION MAY BE OBTAINED FROM BRUCE ANDERSON, TOTAL ENVIRONMENTAL ACTION, BOX 47, HARRISVILLE, N.H. 03450.
REFERENCE	ANDERSON, BRUCE DENVER HOUSE HARRISVILLE, N.H., IN SOLAR ENERGY AND SHELTER DESIGN, PUBLISHED BY TOTAL ENVIRONMENTAL ACTION, JANUARY 1973, PP 133-134

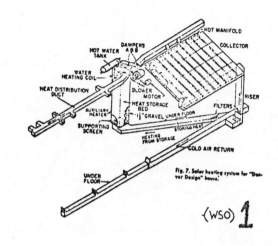

HOT MANIFOLD
COLLECTOR
DAMPERS A & B
HOT WATER TANK
WATER HEATING COIL
HEAT DISTRIBUTION DUCT
BLOWER MOTOR
HEAT STORAGE BED
RISER
AUXILIARY HEATER
1½" GRAVEL UNDER FLOOR
FILTERS
SUPPORTING SCREEN
STORING HEAT
HEATING FROM STORAGE
COLD AIR RETURN

Fig. 7. Solar heating system for "Denver Design" house.

UNDER FLOOR

(WSO) 1

TITLE   DICKINSON HOUSE

CODE

KEYWORDS   SOLAR PANEL FIBERGLAS CONCRETE TANK HEAT STORAGE HOUSE COLLECTOR

AUTHOR   SPENCER DICKINSON AND KALWALL CORP.

DATE & STATUS   THE DICKINSON HOUSE WAS COMPLETED IN 1975.

LOCATION   JAMESTOWN, RHODE ISLAND

SCOPE   THE DICKINSON RESIDENCE IS ONE OF A SERIES OF HOMES BUILT BY
                 SPENCER DICKINSON TO ILLUSTRATE THE PRACTICALITY AND ECONOMY
                 IN SOLAR HOMES. HERE AN ATTEMPT HAS BEEN MADE TO USE STANDARD
                 COMMERCIAL COMPONENTS WHICH ARE RELIABLE AND AVAILABLE.

TECHNIQUE   THIS HOME USES A FIBERGLAS PANEL COLLECTOR COVER ASSEMBLY.
                 HEAT IS STORED IN A CONCRETE WATER TANK UNDER THE HOUSE.

MATERIAL   FIBERGLAS COLLECTOR COVER ASSEMBLY, CONCRETE WATER TANK

AVAILABILITY   ADDITIONAL INFORMATION MAY BE OBTAINED FROM SOLAR HOMES, INC.,
                 2 NARRAGANSETT AVENUE, JAMESTOWN, RHODE ISLAND 02835.

REFERENCE   SOLAR NEWS
                 DICKINSON RESIDENCE
                 SOLAR NEWS, EDITION NO. 2, 1975, P 4

TITLE	DIMETRODON CONDOMINIUM
CODE	
KEYWORDS	CONDOMINIUM WINDMILL WATER TYPE COLLECTOR TRICKLING STORAGE
AUTHOR	W. MACLAY, R. TRAVERS, J. SANFORD
DATE & STATUS	BY DECEMBER 1976, ALL FOUNDATIONS WERE COMPLETE. THE #6 AND #7 UNITS WERE NEARLY COMPLETE, #8 AND #10 WERE UNDER CONSTRUCTION, AND THE COMMUNAL BUILDING AND THE COURTYARD WERE COMPLETE. THE COLLECTOR AND THE AUTOMATIC CONTROL SYSTEM WERE 20% COMPLETE, AND THE STORAGE SYSTEM WAS 100% COMPLETE. ONE OF THE WIND ELECTRIC GENERATING PLANTS WAS BEING USED AS A SUPPLEMENTARY WATER HEATER. AS MORE ARE ADDED IN, THEY WILL UTILIZE THE SYNCHRONOUS INVERTER ON HAND. THE PROJECT WAS BEGUN IN 1971; THE SOLAR COLLECTOR HAS BEEN IN OPERATION SINCE 1975.
LOCATION	WARREN, VERMONT
SCOPE	THE CONDOMINIUM IS ON THE SLOPE OF PRICKLY MOUNTAIN. IT INCLUDES 10 CONTIGUOUS DWELLINGS (#1, #2, ...#10), EACH 16' X 28', AND A 24' X 28' COMMUNAL BUILDING AROUND A CENTRAL COURTYARD. EACH DWELLING UNIT HAS A FULL BASEMENT, EXCEPT FOR #1, #2, #9 AND #10. THERE IS A WINDMILL ON THE TOWER IN THE COURTYARD WHICH PRODUCES ELECTRICITY.
TECHNIQUE	THE COLLECTOR IS 2,700 SQ. FT. GROSS, 2,500 SQ. FT. NET. IT IS A WATER-TYPE COLLECTOR, MOUNTED ON A 60 DEGREE ANGLE FROM THE HORIZONTAL. THERE IS A SLOPING SURFACE OF THE SOUTH SET OF THE STRUCTURES (#4, #5, THE COMMUNAL BUILDING, #6, AND #7). THE COLLECTOR IS 88' LONG, AND 31' WIDE. IT IS OF THE MODIFIED THOMASON TRICKLING WATER TYPE, AND EMPLOYS CORRUGATED GALVANIZED IRON SHEETS. THERE IS DOUBLE GLAZING, CONSISTING OF TWO SHEETS OF KALWALL CORP. SUN-LITE FIBERGLAS-REINFORCED POLYESTER. THE SHEETS ARE HELD 1-1/2" APART BY U-SHAPED SPACER BARS. THE ASSEMBLY IS MOUNTED WITH AN INNER SHEET 1" FROM THE CORRUGATED METAL, THE SPACING BEING ESTABLISHED BY BLOCKS OF RUBBER CUT FROM AUTOMOBILE TIRES. THE COLLECTOR AIR-SPACE IS VENTED IN SUMMER. THE CIRCULATION OF WATER TO THE COLLECTOR IS CONTROLLED AUTOMATICALLY.

THE STORAGE SYSTEM CONSISTS OF 12,000 GALLONS OF WATER IN THREE STEEL STORAGE TANKS, 4,000 GALLONS EACH, SURROUNDED BY 200 TONS OF FIST-SIZE STONES IN THE BASEMENT OF #6, AND #7.

AUXILIARY HEAT IS FROM AN AEROGENERATOR, A CENTRAL WOOD FURNACE, AND GAS HEATERS. |
MATERIAL	CORRUGATED GALVANIZED IRON SHEETS; TWO SHEETS OF KALWALL CORP. SUN-LITE FIBERGLAS-REINFORCED POLYESTER; BLOCKS OF RUBBER CUT FROM AUTOMOBILE TIRES; 12,000 GALLONS OF WATER; THREE STEEL STORAGE TANKS, 4,000 GALLONS EACH; 200 TONS OF FIST-SIZE STONES; AEROGENERATOR; CENTRAL WOOD FURNACE; GAS HEATERS
AVAILABILITY	ADDITIONAL INFORMATION MAY BE OBTAINED FROM R. TRAVERS, DIMETRODON CORP., RT. 1, BOX 160, WARREN, VERMONT 05674.
REFERENCE	SHURCLIFF, W.A. SOLAR HEATED BUILDINGS, A BRIEF SURVEY CAMBRIDGE, MASSACHUSETTS, W.A. SHURCLIFF, MARCH 1976, 212 P

333½

TITLE	DISCOVERY 76
CODE	
KEYWORDS	SOLAR HEATING COOLING TRACKING CONCENTRATOR COLLECTOR HOUSE
AUTHOR	TODD HAMILTON, DR. GERALD LOWERY, DR. TOM LAWLEY, DR. ERNEST BUCKLEY
DATE & STATUS	DISCOVERY 76 WAS COMPLETED FEBRUARY 1976, AND IS CURRENTLY BEING TESTED.
LOCATION	ARLINGTON, TEXAS
SCOPE	DISCOVERY 76 IS A PROJECT FOR SOLAR ENERGY RESEARCH FUNDED LARGELY BY THE TEXAS ELECTRIC SERVICE COMPANY AND NEARLY SEVENTY OTHER LOCAL AND NATIONAL FIRMS. THE HOUSE WILL ALLOW VARIOUS COMBINATIONS OF HEATING AND COOLING MODES, EXAMINE ALTERNATE COLLECTORS, STORAGE METHODS, AND OTHER COMPONENTS FOR A COMPARATIVE ANALYSIS. THE TEST PERIOD IS FIVE YEARS IN SCOPE RUNNING FROM 1975-1980. DURING THIS PERIOD, INTERIM REPORTS WILL BE MADE PUBLIC AS WILL THE FINAL MAJOR ANALYSIS.
	DURING THE TEST PERIOD, THE THREE-BEDROOM HOUSE WILL BE OCCUPIED BY EITHER A YOUNG FACULTY MEMBER AND HIS FAMILY, OR GRADUATE STUDENTS. THE RESIDENTIAL CONDITIONS ARE INTENDED TO TYPLIFY THE MIDDLE INCOME SINGLE FAMILY RESIDENCE, WHICH IS DOMINANT IN THE SOUTHWEST.
TECHNIQUE	DISCOVERY 76 USES 42 ROOF MOUNTED NORTHRUP COLLECTORS WHICH TRACK THE SUN ON ITS DAILY PATH FROM EAST TO WEST.
MATERIAL	NORTHRUP CO. COLLECTORS, ARKLA/SERVEL ABSORPTION A/C UNIT, LENNOX ELECTRIC COIL FURNACE, LENNOX WATER-TO-WATER HEAT PUMP, 3-1200 GALLON FIBERGLAS STORAGE TANKS, RHEEM CO. 60 GALLON HOT WATER DOMESTIC TANK
AVAILABILITY	ADDITIONAL INFORMATION MAY BE OBTAINED FROM TODD HAMILTON, ASSISTANT PROFESSOR OF ARCHITECTURE, UNIVERSITY OF TEXAS, DEPARTMENT OF ARCHITECTURE, ARLINGTON, TEXAS.
REFERENCE	ARCHITECTURAL RECORD SEEKING ALTERNATIVE FUELS, UTILITIES SPONSOR SOLAR ENERGY APPLICATIONS ARCHITECTURAL RECORD, MID-OCTOBER 1975, P 24

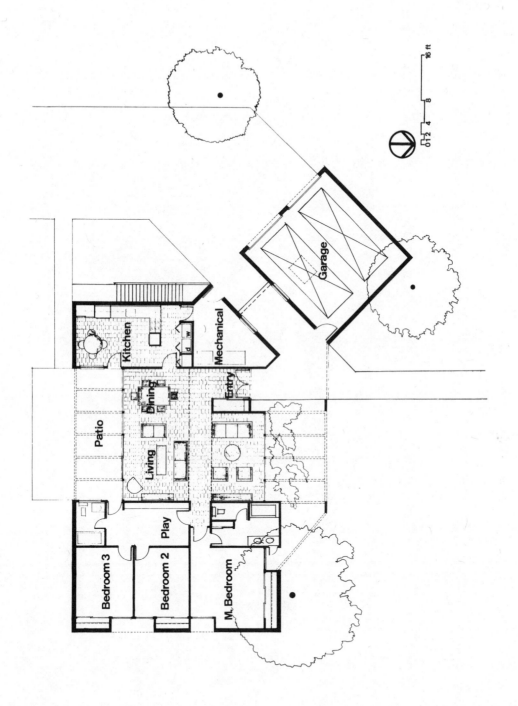

Garage

Kitchen

d w

Mechanical

Dining

Entry

Patio

Living

Play

Bedroom 3

Bedroom 2

M. Bedroom

0 1 2 4   8   16 ft

338

TITLE	DIY-SOL, INC. HOUSE
CODE	
KEYWORDS	HOUSE AIR TYPE SOLAR COLLECTOR ROCK STORAGE CONTROL
AUTHOR	DIY-SOL, INC.
DATE & STATUS	THE DIY-SOL, INC. HOUSE WAS RETROFITTED FOR SOLAR HEATING IN 1974.
LOCATION	MARLBORO, MASSACHUSETTS
SCOPE	THE BUILDING IS AN EXISTING 1-1/2-STORY, SPLIT-LEVEL, WOOD-FRAME HOUSE WITH 4 BEDROOMS, 1-1/2 BATHS, A SHALLOW ATTIC, AND A HALF-BASEMENT.
TECHNIQUE	THE THICKNESS OF THE INSULATION IS 8" IN THE ATTIC, AND 1" IN THE WALLS.
	THE COLLECTOR CONTAINS 200 SQ. FT. GROSS, 190 SQ. FT. NET. IT IS AN AIR TYPE COLLECTOR, WHICH WAS ASSEMBLED IN PLACE ON THE VERTICAL WALL FACING 26 DEGREES WEST OF SOUTH. IT CONSISTS OF 7 VERTICAL PANELS, EACH 2' WIDE, 4" THICK, AND 12' TO 16' HIGH. THE BLACK COATING IS OF THE SPECIAL, SELECTIVE TYPE. THE GLAZING IS DOUBLE; THE INNER LAYER IS OF 1-MIL TEDLAR, AND THE OUTER LAYER IS EITHER 4-MIL OF TEDLAR, MOUNTED WITH THE USE OF HEAT TO MAKE THE MATERIAL TAUT, OR 40-MIL OF KALWALL SUN-LITE. THE AIR IS DRIVEN BY A 185-WATT CENTRIFUGAL BLOWER, AND TRAVLES ON BOTH SIDES, FRONT AND BACK, OF THE ABSORBING LAYER. IT PICKS UP ENERGY AND CARRIES IT EITHER TO THE ROOMS OR TO THE HEAT EXCHANGER SERVING THE STORAGE SYSTEM. THE HEAT EXCHANGER EMPLOYS THE CORE OF AN AUTOMOBILE RADIATOR.
	THE STORAGE SYSTEM CONTAINS 750 GALLONS OF WATER IN A CONCRETE SEPTIC TANK, WATERPROOFED WITH SEARS EPOXY PAINT. THE CIRCULATION OF WATER FROM THE TANK TO THE HEAT EXCHANGER IS BY A VERY SMALL, 1/100 HP CENTRIFUGAL PUMP. THE CONTROL SYSTEM, INCLUDING THE BLOWER, WATER PUMP, AND DAMPERS, EMPLOYS A SOLID STATE ELECTRONIC SYSTEM AND THERMISTOR TEMPERATURE SENSORS. THERE ARE FOUR MODES OF OPERATION.
MATERIAL	SELECTIVE BLACK COATING; 1-MIL TEDLAR; 4-MIL TEDLAR; 40-MIL KALWALL SUN-LITE; 185-WATT CENTRIFUGAL BLOWER; HEAT EXCHANGER; AUTOMOBILE RADIATOR; 750 GALLONS OF WATER; CONCRETE SEPTIC TANK; SEARS EPOXY PAINT; 1/100 HP CENTRIFUGAL PUMP; CONTROL SYSTEM; SOLID STATE ELECTRONIC SYSTEM; THERMISTOR TEMPERATURE SENSORS
AVAILABILITY	ADDITIONAL INFORMATION MAY BE OBTAINED FROM F. RAPP, 29 HIGHGATE ROAD, MARLBORO, MASSACHUSETTS 01752.
REFERENCE	SHURCLIFF, W.A. SOLAR HEATED BUILDINGS, A BRIEF SURVEY CAMBRIDGE, MASSACHUSETTS, W.A. SHURCLIFF, MARCH 1976, 212 P

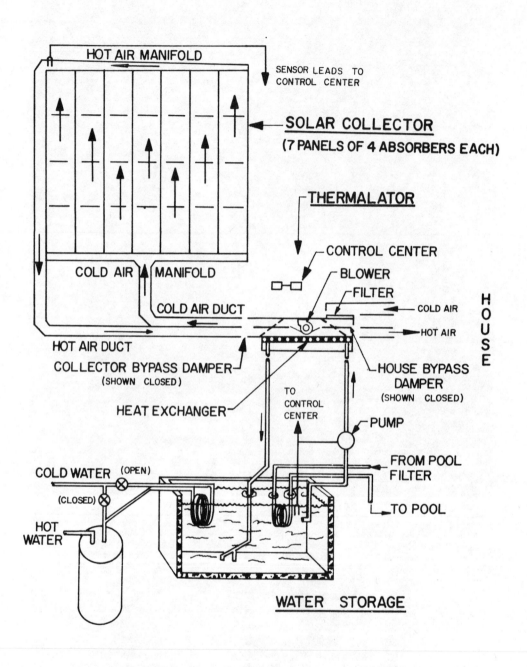

DIY—SOL HEATING SYSTEM: WATER STORAGE

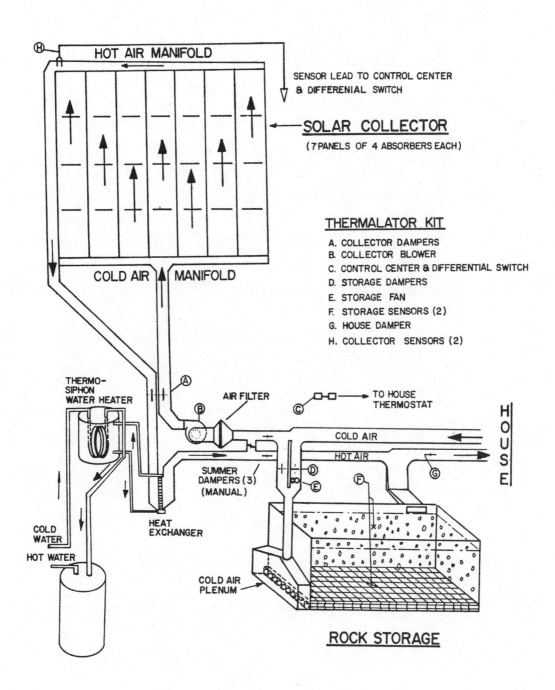

DIY-SOL HEATING SYSTEM: ROCK STORAGE

341

TITLE	EAMES HOUSE
CODE	
KEYWORDS	HOUSE AIR TYPE COLLECTOR SOLAR
AUTHOR	D. WATSON
DATE & STATUS	THE EAMES HOUSE IS COMPLETE AS OF DECEMBER 1976.
LOCATION	GROTON, CONNECTICUT
SCOPE	THE BUILDING IS A TWO-STORY, 2-BEDROOM, 1,800 SQ. FT. (44' X 24') HOUSE WITH A BASEMENT AND AN ATTIC.

TECHNIQUE

THE TYPICAL THICKNESS OF THE FIBERGLAS INSULATION IN THE WALLS AND THE CEILINGS IS 5-1/2". THE WINDOWS ARE DOUBLE GLAZED. THE LIVING ROOM IS OF A CLERESTORY TYPE. THE BUILDING AIMS SOUTHWEST. THE SHUTTERS COVER THE WINDOWS AT NIGHT.

THE MAIN COLLECTOR HAS 335 SQ. FT. NET. THE COLLECTOR IS OF THE AIR TYPE. IT IS MOUNTED ON THE ROOF, AND SLOPES 60 DEGREES FROM THE HORIZONTAL. THE COLLECTOR INCLUDES 18 SUNWORKS AIR-TYPE PANELS, EACH 7' X 3'. THE HEART OF THE PANEL IS A COPPER SHEET WITH SELECTIVE BLACK COATING. THE SINGLE GLAZING IS 3/16" TEMPERED GLASS. THE AIR IS CIRCULATED AT 120 CFM IN THE SPACE BEHIND THE BLACK SHEET. THE BLOWER POWER IS 1/3 HP.

THE SUPPLEMENTARY COLLECTOR IS A GREENHOUSE, WITH PLAN DIMENSIONS OF 14' X 11'. IT HAS A LARGE GLAZED AREA OF ABOUT 14' X 11', SLOPING 60 DEGREES FROM THE HORIZONTAL, AND IS GLAZED WITH INSULATED GLASS.

THE STORAGE SYSTEM OCCUPIES 1,250 CUBIC FEET OF ABOUT 70 TONS OF 2-TO-4" DIAMETER STONES IN A 26' X 12' X 6' HIGH BIN BENEATH THE FLOOR. THE BIN OF STONES RECEIVES HOT AIR FROM THE MAIN COLLECTOR, AND ALSO FROM A SPECIAL FIREPLACE. THERE IS A SMALL, SEPARATE BIN-OF-STONES STORAGE SYSTEM BENEATH THE GREENHOUSE.

THE HEATILATOR FIREPLACE CAN SUPPLY HEAT TO THE BIN OF STONES. THERE IS A WOOD-BURNING STOVE IN THE KITCHEN-LIVING AREA. THIS AREA CAN BE ISOLATED FROM THE REST OF THE HOUSE TO PERMIT MAINTAINING ADEQUATE TEMPERATURE HERE, EVEN IN THE EVENT OF AN ELECTRIC POWER CUT-OFF, WHICH WOULD MAKE THE MAIN BLOWER AND GREENHOUSE FAN INOPERATIVE.

THE SPECIAL, SMALL, ADJUSTABLE-TILT COLLECTOR IS SITUATED ABOVE THE GREENHOUSE. THIS DEVICE HAS FIVE SUNWORKS DOMESTIC-HOT-WATER PANELS. THE PANEL TILTS ARE ADJUSTED MANUALLY, FROM SEASON TO SEASON.

MATERIAL	18 SUNWORKS AIR TYPE PANELS; COPPER SHEET WITH SELECTIVE BLACK COATING; 3/16" TEMPERED GLASS; INSULATED GLASS; 70 TONS OF 2-TO-4" DIAMETER STONES; 26' X 12' X 6' HIGH BIN; HEATILATOR FIREPLACE; SUNWORKS DOMESTIC-HOT-WATER PANELS
AVAILABILITY	ADDITIONAL INFORMATION MAY BE OBTAINED FROM DONALD WATSON, BOX 401, GUILFORD, CONNECTICUT 06437.
REFERENCE	WATSON, DONALD DESIGNING AND BUILDING A SOLAR HOUSE CHARLOTTE, VERMONT, GARDEN WAY PUBLISHING, JANUARY 1977, 240 P

343

TITLE	ECCLI HOUSE
CODE	
KEYWORDS	HOUSE LABORATORY EARTH BERMS GREENHOUSE TRICKLING WATER COLLECTOR PASSIVE INSULATED
AUTHOR	E. ECCLI
DATE & STATUS	THE ECCLI HOUSE WAS COMPLETED IN 1974.
LOCATION	NEW PALTZ, NEW YORK
SCOPE	THE BUILDING IS A ONE-STORY HOUSE-LABORATORY, 15' X 18'. IT HAS A BASEMENT, BUT NO ATTIC.
TECHNIQUE	THERE ARE EARTH BERMS ON THE NORTH, EAST AND WEST SIDES. THE FIBERGLAS INSULATION IN THE WALLS IS 6". THERE ARE TWO ROOMS: AN INSTRUMENT ROOM AT THE WEST SIDE WITH A COLLECTOR, AND A GREENHOUSE ROOM AT THE EAST SIDE WITH LARGE DOUBLE-GLAZED SOUTH-FACING WINDOWS. AN ENTRANCE FOYER, WHICH IS EMPLOYED TO REDUCE THE LOSS OF WARM AIR WHEN PEOPLE ENTER, IS AT THE NORTHEAST CORNER. THE SOUTH PORTON OF THE ROOF IS 23 DEGREES FROM THE HORIZONTAL; THE SOUTH WALL IS 60 DEGREES FROM THE HORIZONTAL.
	THE MAIN COLLECTOR, ON THE WEST PORTION OF THE SLOPING SOUTH WALL, HAS 80 SQ. FT. OF CORRUGATED ALUMINUM. WATER TRICKLES DOWN THE VALLEYS OF THESE SHEETS, AND IS COLLECTED AND DELIVERED TO THE STORAGE SYSTEM. THE GLAZING CONSISTS OF TWO SHEETS OF 3/16" GLASS WITH AN 1/2" AIRSPACE BETWEEN. ADDITIONAL, PASSIVE COLLECTION IS VIA 160 SQ. FT. OF DOUBLE GLAZED WINDOWS ON THE EAST PORTION OF THE SLOPING SOUTH WALL AND THE SOUTH PORTION OF THE ROOF. ABOVE THESE LATTER WINDOWS, THERE IS A COVER OF INSULATING MATERIAL, THAT IS LOWERED (CLOSED) AT NIGHT TO PREVENT THE ESCAPE OF HEAT. THE OTHER WINDOWS ARE ALSO INSULATED AT NIGHT.
	THE MAIN STORAGE EMPLOYS 450 GALLONS OF WATER IN TWO 275 GALLON STEEL TANKS IN THE BASEMENT. A FAN BLOWS COLD ROOM AIR PAST THESE TANKS TO WARM THIS AIR. ADDITIONAL STORAGE IS PROVIDED BY THE BRICK AND SAND OF THE GREENHOUSE FLOOR.
MATERIAL	GREENHOUSE; DOUBLE GLAZED WINDOWS; CORRUGATED ALUMINUM; 3/16" GLASS; 450 GALLONS OF WATER; TWO 275 GALLON STEEL TANKS; FAN; BRICK; SAND
AVAILABILITY	REFER TO THE FOLLOWING REFERENCE FOR FURTHER INFORMATION.
REFERENCE	SHURCLIFF, W.A. SOLAR HEATED BUILDINGS, A BRIEF SURVEY CAMBRIDGE, MASSACHUSETTS, W.A. SHURCLIFF, MARCH 1976, 212 P

TITLE	ECOLOGICAL RESEARCH CENTER
CODE	ECO-CENTER
KEYWORDS	ECOTECTURE ECOLOGICAL RESEARCH SOLAR COLLECTOR ISOTHERM VEGETATION GREENHOUSE STORAGE FORM INSULATION OVERHANG NATURAL VENTILATION
AUTHOR	HECTOR OSSA
DATE & STATUS	THE ECO-CENTER WAS DESIGNED DURING THE SPRING OF 1975.
LOCATION	DESERT AREA, SOUTHWESTERN U.S.A.
SCOPE	THE ECO-CENTER DESIGN IS AN AWARD WINNING ENTRY IN THE INTER-ROYAL 1975 ECOLOGICAL RESEARCH CENTER DESIGN COMPETITION. THE ECO-CENTER IS A UNIVERSITY, WHICH IS ORIENTED TO THE CONCERNS OF ENVIRONMENT AND ECOLOGY. THE DESIGN MAKES USE OF ARCHITECTURAL FORMS RESPONSIVE TO THE CLIMATOLOGICAL CONDITIONS EXISTING ON THE SITE; SOLAR COLLECTION TECHNOLOGIES, THE ISOTHERM PRINCIPLES, AND ENERGY CONSERVATION MEASURES SUITABLE FOR ADAPTION TO THE NATURAL TERRAIN.
TECHNIQUE	THE ECO-CENTER IS COMPOSED OF SEVERAL BUILDINGS, WHICH EMPLOY VARIOUS MEANS OF ENERGY CONSERVATION AND ECOLOGICAL MEASURES. THE CENTER INCLUDES AN ALGAE POND FOR GROWING ALGAE TO BE USED FOR PROTEIN SUPPLEMENT; A GREENHOUSE WHICH IS USED TO GROW VEGE-TABLES AND ALSO TO RETAIN HEAT IN WINTER; A BUILDING WHICH CONTAINS WATER STORAGE ON ITS ROOF AND ABSORBS COOL NIGHT AIR, WHICH COOLS ON SUMMER DAYS. THERE IS ALSO EXTENSIVE USE OF SOLAR COLLECTORS FOR HEATING THROUGHOUT THE ENTIRE CENTER.
	THE BUILDING FORMS WERE ALSO CHOSEN TO CONSERVE ENERGY. IN ONE BUILDING, TOP SOIL IS USED ON THE ROOF AS INSULATION AND ALSO TO SUPPORT PLANT LIFE. OVERHANGS ARE USED TO KEEP OUT THE SUMMER SUN AND ALLOW IN THE WINTER SUN. A TRUSSED ROOF SYSTEM, OPEN AT TWO ENDS, ALLOWS FOR EXTENSIVE NATURAL VENTILATION. IN ANOTHER BUILDING, AIR, WHICH IS GROUND TEMPERATURE, CIRCULATES THROUGH THE BUILDING YEAR ROUND. IN ADDITION, TREES AND SHRUBBERY ARE USED TO KEEP SUMMER SUN OFF THE BUILDINGS.
MATERIAL	SOLAR COLLECTORS, ALGAE, WATER, TOP SOIL, TRUSSED ROOF SYSTEM
AVAILABILITY	ADDITIONAL INFORMATION MAY BE OBTAINED FROM HECTOR OSSA, 1081 E. 27TH ST., BROOKLYN, NEW YORK.
REFERENCE	OSSA, HECTOR ECOLOGICAL RESEARCH CENTER BROOKLYN, N.Y., STUDY PROJECT PRESENTATION AT PRATT INSTITUTE, 1975, 7 P

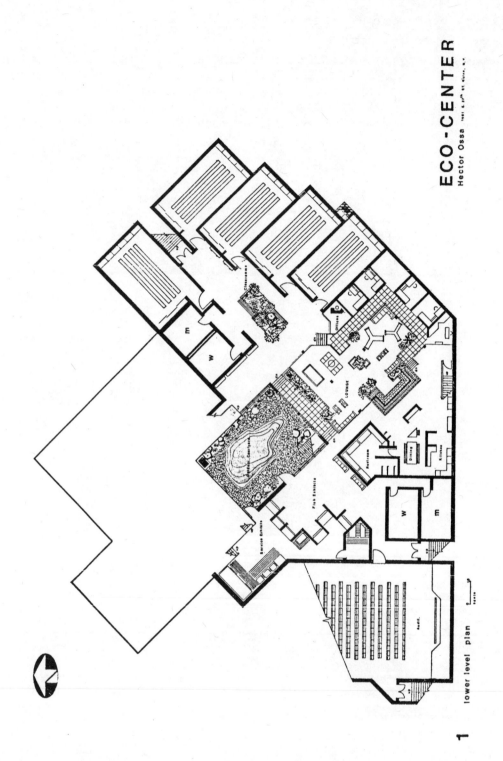

ECO-CENTER
Hector Ossa

1 lower level plan

346

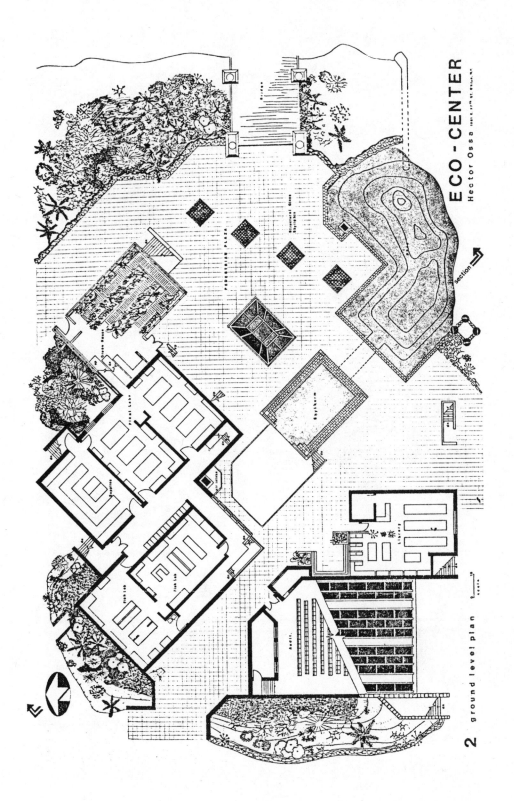

ECO - CENTER
Hector Ossa

section

2 ground level plan

347

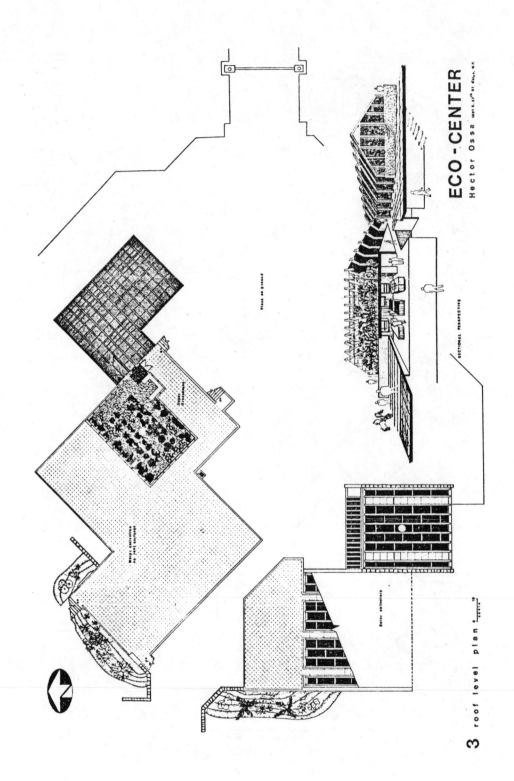

ECO - CENTER

Hector Ossa West.27th st.&6ay, n.y.

SECTIONAL PERSPECTIVE

3 roof level plan

348

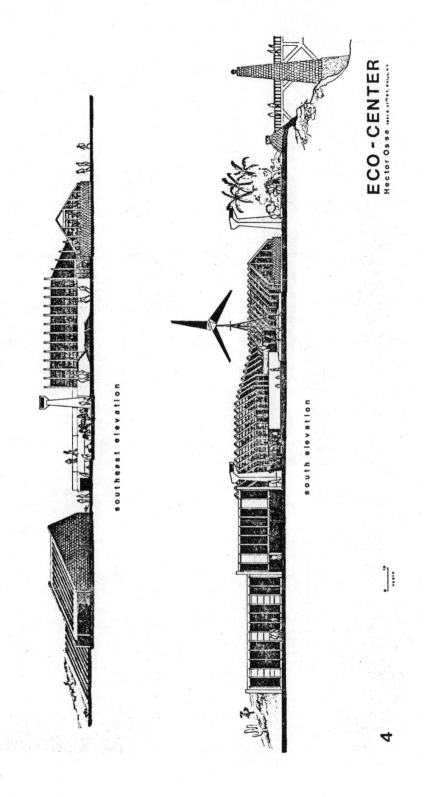

southeast elevation

south elevation

ECO - CENTER
Hector Ossa

4

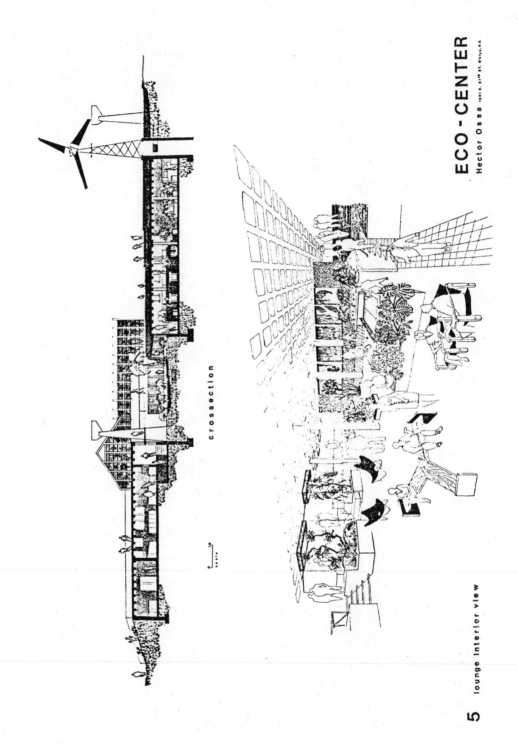

ECO-CENTER

Hector Ossa 1961, 3rd St. Spokane

crossection

5 lounge interior view

DESIGN APPROACH

ENVIRONMENTAL PARAMETERS

BUILDING WITH RENEW-
ABLE RESOURCES, BUT
WHICH ALSO BECOME AN
INHERENT ELEMENT OF
FORM WHICH ARE
MAINTAINABLE

BUILDING WITH MATERIALS
WHICH WOULD OTHERWISE
BE DISCARDED AT THE END OF
THE BUILDING LIFE

UTILIZING POLLUTION FREE
WIND ENERGY & USING
MINIMUM RESOURCES
PUMPING WATER, FINE DROP
IN EFFORT FOR MACHINING
& CLEANING

RECYCLING WATER AND
SEPARATING OUT POLLU-
TION - DRINKING & COOK-
ING MANAGEABLE AND
REGENERATE, & TOILET
PLUMBING

RELATE ALL COMPONENTS
THROUGHOUT
IT BY TERMS OF MOD-
ULAR COORDINATION

DISTRIBUTING SOLAR
ENERGY FOR HEATING
OF FACILITY IN TIME,
INCREASING, & HEATING
OF WATER

REFLECT INCREASING
TRANSMISSION
WITH AN ACCESS LOAD
& HEATING,
SELECTION, DAYLIGHT
AS ENTHUSIASTIC

ENERGY FLOW within these parameters

SYMBOLS

○ SOURCE
○ STORAGE
◇ SELF-
    MAINTAINING
    WORKGATE
△→□ HEAT SINK
     COLLECTOR

DOUGH
ENERGY
MODEL
from
New Alchemy
Institute

*Dr. Okum
T. Okum
ENVIRONMENT
POWER & SOCIETY

ELECT. MOTOR VALVE
PUMP
△ Collector
    (relay)
△ Public
   House

WIND

SUN
THERMAL

SOLAR

TRAIN

1 CIRCULATION OF PIPES AND
  WATER FREQUENT BOUND
  WORKGATES
2 WATER SOURCE BOUND
  & EARTH TREATMENT
3 POTABLE WATER, PLANTS
  POTABLE CEILING FROM
  TERRESTRIAL PLANTS TO
  TILAPIA (FISH)
4

8 ELEVATED TEMPERATURE
  CUPOLA FOR TILAPIA
  GREENHOUSE HEALTH
  FISH IN GREENHOUSE
9 RETURN HEAT WATER
  FISH PIT AREAS CURE
  PONDS
10 GREENHOUSE
11 INTERIOR PLANTS,
    VEGETABLE & OTHER
    GARDENING

RELATING BIOCLIMATIC & SOCIAL FUNCTION for the public

CLIMATOLOGICAL BUILDING FORM

ALGAE POND FOR
PROTEIN SUPPLEMENT

INTERIOR DARK, CONTROLLED
MICROCLIMATE & LIFESTYLE

STORM WATER UTILIZED
& STORED IN FACILITY

GREENHOUSES DOUBLING AS
BIO-SPHERES FOR RESTORATION
OF HEAT IN WINTER

WIDE APPLICATION OF SOLAR
COLLECTORS FOR DISTRIBUTING

CONTINUOUS TOPSOIL
STRATA OVER ROOFS

METHODS OF SUN
PROTECTION

EVAPORATIVE
COOLING

VENTILATION FACILITATED
BY ROOF STRUCTURE

... AND FOR COOLING
BY NIGHT SKY RADIA-
TION IN BOTTLES

UTILIZATION OF GROUND
TEMPERATURE YEAR ROUND

AIR COOLED BY WATER
AND VEGETATION

VENTILATION BY AIR
FLOW THROUGH COURT

SYNTHESIS

INTEGRATION OF THE LEARNING EXPERIENCE WITH
THE LIFE SUPPORT SYSTEMS, SYMBIOSIS OF AS
MANY NATURAL, ECOLOGICAL & ENERGY PROCESSES
AS POSSIBLE WITH HUMAN USES & LIFESTYLE

ECO - CENTER
Hector Ossa

6

TITLE	ECOLOGY HOUSE
CODE	
KEYWORDS	UNDERGROUND INSULATION ATRIUM HOUSE
AUTHOR	JOHN E. BARNARD, JR.
DATE & STATUS	THE ECOLOGY HOUSE WAS BUILT IN THE WINTER OF 1972. IT WAS TESTED OUT WITH A COUPLE LIVING IN IT FOR ONE YEAR.
LOCATION	MARSTON MILLS, MASSACHUSETTS
SCOPE	THE ECOLOGY HOUSE IS DESIGNED TO BE AN UNDERGROUND HOUSE THAT WOULD HAVE PRIVACY, BE DUST AND POLLEN FREE, EASY AND INEXPENSIVE TO BUILD, HEAT AND MAINTAIN, COMFORTABLE, ATTRACTIVE AND MARKETABLE. GOING UNDERGROUND IS A SOLUTION TO SUCH PROBLEMS AS CONSERVATION OF LAND IN DENSE POPULATION AREAS, OF ENERGY AND OF WOOD PRODUCTS. IT WAS SITED TO CONFORM TO LOCAL ZONING REGULATIONS. IT HAS A BEDROOM, BATH, KITCHEN AND A LIVING-DINING ROOM. THE HOUSE COSTS ONLY 60 PERCENT OF A CONVENTIONAL DWELLING TO HEAT, AND THE COST TO BUILD IS ABOUT 25 PERCENT LESS THAN THE USUAL FRAME CONSTRUCTION. THE ECOLOGY HOUSE COSTS ABOUT $24,000, JUST OVER $20 PER SQUARE FOOT OF LIVING SPACE, AND UTILITY ROOM.
TECHNIQUE	EXCEPT FOR THE BATHROOM, ALL THE ROOMS HAVE ONE WALL OF GLASS PANELS OPENING ON TO A 300 SQUARE FOOT ATRIUM. THE HOUSE IS ENTERED BY WALKING DOWN STAIRS FROM GROUND LEVEL INTO THE ATRIUM, WHICH IS A BRICK-PAVED OPENING THAT SERVES AS A GARDEN, A PLACE TO SUN-BATHE OR TO DINE. IT MAY BE COVERED WITH A PLASTIC BUBBLE IN THE COLD SEASON. THE ATRIUM FACES SOUTH SO THAT EACH ROOM CAN BE SUNNY, EVEN IN WINTER. THE ONLY WOOD IN THE ECOLOGY HOUSE IS A PLYWOOD SHIELD TO COVER A FOUNTAIN PUMP IN THE ATRIUM. THE WALLS ARE OF POURED CONCRETE REINFORCED WITH STEEL RODS. STYROFOAM INSULATES THE OUTSIDE OF THE HOUSE AND IS APPLIED OVER THE ROOF MADE OF 8 INCH PRECAST PANELS. THE EARTH COVER IS AT LEAST 12 INCHES. FOR WATERPROOFING, THERE IS A THREE-PLY PITCH AND ASBESTOS FELT MEMBRANE. AS TO THE MECHANICAL SYSTEM, THERE ARE AIR INDUCTION AND AIR EXHAUST EQUIPMENT, AN ELECTRONIC AIR FILTER, DEHUMIDIFYING DEVICES AND AIR CONDITIONING. THE SEWAGE GOES INTO A SUMP, IS PUMPED INTO A SEPTIC TANK, AND THEN GOES TO A LEACHING FIELD.
MATERIAL	EARTH, POURED REINFORCED CONCRETE WALL, PRECAST CONCRETE ROOF PANELS, STYROFOAM INSULATION
AVAILABILITY	ADDITIONAL INFORMATION ABOUT THE ECOLOGY HOUSE MAY BE OBTAINED BY WRITING TO JOHN BARNARD, JR., 60 MAIN ST., OSTERVILLE, MASSACHUSETTS 02655.
REFERENCE	AIA JOURNAL SAVING BY GOING UNDERGROUND AIA JOURNAL, FEB 1974, PP 48-49

354

TITLE	EDDY HOUSE
CODE	
KEYWORDS	AIR SOLAR COLLECTORS THERMOSIPHON PASSIVE SKYLID WOODBURNING STOVE GREENHOUSE
AUTHOR	DESIGNED BY TRAVIS PRICE, SOLAR COLLECTORS BY SUNWORKS, SKYLIDS BY ZOMEWORKS
DATE & STATUS	THE ADDITION TO THE EDDY HOUSE WAS COMPLETED IN 1976.
LOCATION	RHODE ISLAND
SCOPE	THE ADDITION TO THE EDDY HOUSE IS SOLAR HEATED AND WILL BE USED IN WINTER WITH 3 ROOMS FROM THE ALREADY EXISTING HOUSE. IN THE EXISTING HOUSE, ELECTRIC BILLS WERE RUNNING FROM $300 TO $400 A MONTH. THIS COST SHOULD BE ELIMINATED BY THE SOLAR HEATED ADDITION.
TECHNIQUE	THE ADDITION IS EXTREMELY WELL INSULATED, WITH THREE INCHES OF SPRAYED POLYURETHANE ON THE FOUNDATION, 3-1/2 INCHES OF FIBER-GLAS BATT PLUS ONE INCH OF STYROFOAM ON ALL EXTERIOR WALLS, AND 6-1/2 INCH BATT IN ALL ROOF JOISTS. ALL WINDOWS ARE DOUBLE PANE.
	THE ADDITION'S STRIKING SOUTH-FACING FACADE SUPPORTS 18 THREE-BY-SEVEN FOOT SUNWORKS AIR-TYPE COLLECTORS, WHICH STORE HEAT IN A 1,000 CU. FT. ROCK BIN.
	LIQUID-TYPE SUNWORKS COLLECTORS PROVIDE DOMESTIC HOT WATER. THE COLLECTORS ARE MOUNTED ON A SMALL, FREESTANDING STRUCTURE NEXT TO THE ADDITION, AND OPERATE ON THE THERMOSIPHON PRINCIPLE.
	THERE IS YET A THIRD SOLAR HEATING "SYSTEM" AT WORK VIA THE PASSIVE APPROACH: DIRECT HEATING OF THE HOUSE THROUGH SOUTH-FACING WINDOWS. THE PASSIVE SYSTEM MAKES UP THE MISSING 20 TO 30% OF HEAT NEEDED AND ELIMINATES THE NEED FOR A FOSSIL-FUEL BACK-UP SYSTEM.
	A SMALL WOODBURNING JOTUL STOVE PROVIDES THE ONLY BACK-UP FOR THE ADDITION. A GREENHOUSE, LOCATED UNDER THE AIR-TYPE COLLECTORS, IS INSULATED AT NIGHT BY SKYLID PANELS.
	THE SKYLIDS FIBERGLAS-FILLED ALUMINUM SANDWICHES OPERATE AUTOMATICALLY VIA AN INGENIOUS ARRANGEMENT OF TWO FREON CANNISTERS. DURING THE DAY, THE SUN'S HEAT CAUSES FREON IN ONE CANNISTER TO EXPAND INTO A CANNISTER ON THE PANEL'S OPPOSITE SIDE. THE SHIFT IN WEIGHT PIVOTS THE PANEL OPEN; AT NIGHT THE PROCESS REVERSES TO CLOSE IT.
MATERIAL	POLYURETHANE, FIBERGLAS BATT, STYROFOAM, DOUBLE PANE WINDOWS, AIR-TYPE COLLECTORS, ROCK BIN, LIQUID-TYPE COLLECTORS, WOOD-BURNING JOTUL STOVE, GREENHOUSE, SKYLID PANELS, FIBERGLAS, ALUMINUM, FREON
AVAILABILITY	REFER TO THE FOLLOWING REFERENCE FOR ADDITIONAL INFORMATION.
REFERENCE	STEPLER, RICHARD SOLAR ARCHITECTURE - IT'S MORE THAN PUTTING COLLECTORS ON THE ROOF POPULAR SCIENCE, JULY 1976, PP 48-52, 96-97

AIR-TYPE SPACE-HEATING SYSTEM

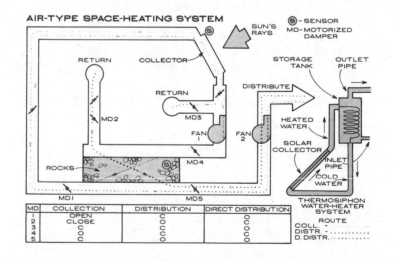

SUN'S RAYS

Ⓢ – SENSOR
MD– MOTORIZED DAMPER

RETURN

COLLECTOR

RETURN

MD2

MD3

FAN 1

FAN 2

DISTRIBUTE

STORAGE TANK

OUTLET PIPE

HEATED WATER

SOLAR COLLECTOR

INLET PIPE

COLD WATER

ROCKS

Ⓢ

MD4

MD1

MD5

THERMOSIPHON WATER-HEATER SYSTEM

MD	COLLECTION	DISTRIBUTION	DIRECT DISTRIBUTION
1	OPEN	C	C
2	CLOSE	C	C
3	C	C	C
4	C	C	C
5	C	O	O

ROUTE
COLL. –
DISTR. – . . . .
D. DISTR. . . . .

356

TITLE   EDMONDSON HOME

CODE

KEYWORDS   SOLAR COLLECTOR PANEL HOT WATER HEAT HOME

AUTHOR   THE SYSTEM USED IN THIS HOUSE WAS MANUFACTURED BY WILLIAM B.
                 EDMONDSON, INVENTOR. SOLARSAN IS A PATENTED TRADEMARK OF
                 SOLAR WATER HEATERS.

DATE & STATUS   THE SOLARSAN SYSTEM WAS INSTALLED INTO THE EDMONDSON HOME IN
                 SEPTEMBER, 1974.

LOCATION   SAN DIEGO, CALIFORNIA

SCOPE   THIS WATER HEATING SYSTEM WAS INSTALLED TO PROVIDE FOR DOMESTIC
                 HOT WATER NEEDS OF THIS SMALL HOME. THE HEATERS CAN BE INSTALLED
                 IN ANY HOME AND CAN BE EXPECTED TO FURNISH AT LEAST 95% OF
                 HOT WATER NEEDS AS IT DID IN THIS HOUSE.

TECHNIQUE   THE SOLARSAN HEATING SYSTEM BASICALLY CONSISTS OF FLAT PLATE
                 ROOF MOUNTED SOLAR COLLECTORS, WHICH CAN MOST EASILY BE USED
                 TO HEAT WATER FOR DOMESTIC USE, BUT CAN ALSO BE ADAPTED TO
                 SATISFY MANY HEATING NEEDS.

MATERIAL   SOLAR COLLECTOR PANELS

AVAILABILITY   ADDITIONAL INFORMATION MAY BE OBTAINED FROM SOLAR ENERGY DIGEST,
                 P.O.BOX 17776, 7401 SALERNO STREET, SAN DIEGO, CALIFORNIA 92117.

REFERENCE   SOLAR ENERGY DIGEST
                 SOLARSAN SOLAR WATER HEATERS
                 SOLAR ENERGY DIGEST, 1976, 2 P

TITLE	ELDERLY APARTMENTS
CODE	
KEYWORDS	ELDERLY HOUSING APARTMENT HOT WATER SOLAR COLLECTOR ANTIFREEZE FLUID HIGHRISE ELECTRIC HEAT EXCHANGER
AUTHOR	SAMUEL, MATTHEW W. AND ROGER W. STERN OF B-L ASSOCIATES; LOUIS MAZZINI OF NEW ENGLAND SOLAR SYSTEMS OF BROOKLINE, MASS.
DATE & STATUS	THE SOLAR SYSTEM WAS PUT INTO OPERATION IN OCTOBER 1976.
LOCATION	1550 BEACON STREET, BROOKLINE, MASSACHUSETTS
SCOPE	THE PROJECT CONSISTS OF A 16-STORY, 230-APARTMENT BUILDING FOR THE ELDERLY. THE BUILDING WAS AN ALL-ELECTRIC BUILDING. THE SOLAR CONCEPT WAS EXPLORED TO COMBAT RISING ELECTRIC ENERGY COSTS. THE SOLAR DESIGN, ENGINEERING AND INSTALLATION WERE FINANCED BY THE MASS. HOUSING FINANCE AGENCY (MHFA) FOR $96,000. IT IS EXPECTED THAT THE COST WILL BE PAID BACK IN ABOUT 9 YEARS, AND THAT THE SOLAR SYSTEM WILL REDUCE HOT WATER ELECTRIC USAGE BY AS MUCH AS 60%.
TECHNIQUE	THE SOLAR HEATING SYSTEM EMPLOYS 88 SOLAR COLLECTORS, MANU-FACTURED BY DAYSTAR CORP. OF BURLINGTON, MASSACHUSETTS. IT PROVIDES HOT WATER VIA A CONVENTIONAL HEAT EXCHANGER. THE COLLECTORS, STORAGE TANK AND CONTROLS ARE INSTALLED ON TOP OF THE BUILDING.
	BY USING A FLUID CALLED SOLARGARD, A NONTOXIC ANTIFREEZE DEVELOPED BY DAYSTAR, THE SOLAR SYSTEM WILL BE OPERATIONAL 365 DAYS A YEAR. THE SUN'S ENERGY WILL HEAT UP THE FLUID AS IT PASSES THROUGH THE COLLECTORS. THE FLUID IS THEN PIPED TO A HEAT EXCHANGER WHERE THE FLUID HEATS THE BUILDING'S INCOMING WATER BEFORE THE WATER GOES INTO THE EXISTING 2,000 GALLON ELECTRIC WATER HEATER. THE ELECTRIC SYSTEM IS USED, ACCORDING TO THE EXPERTS, ONLY WHEN THE SOLAR SYSTEM IS PROVIDING WATER AT LESS THAN 140 DEGREE TEMPERATURE, AND ONLY TO MAKE UP THE DIFFERENCE.
	IT WAS SAID THAT IN THREE TEST RUNS THE SYSTEM HAS MET OR EXCEEDED THE 140 DEGREE WATER TEMPERATURE REQUIREMENT EACH TIME. SINCE THE ELECTRIC HOT WATER HEATING EQUIPMENT WILL ONLY COME ON INFREQUENTLY, THE DEMAND TIME FOR ELECTRICITY WILL BE SUBSTANTIALLY REDUCED.
MATERIAL	SOLAR COLLECTOR, HEAT EXCHANGER, STORAGE TANK, CONTROL, SOLARGARD FLUID
AVAILABILITY	ADDITIONAL INFORMATION MAY BE OBTAINED FROM DAYSTAR CORP., 90 CAMBRIDGE STREET, BURLINGTON, MASSACHUSETTS 01803.
REFERENCE	BOSTON SUNDAY GLOBE ONE SOLAR SYSTEM HEATING WATER FOR 230 APARTMENTS BOSTON SUNDAY GLOBE, OCTOBER 17, 1976, 1 P

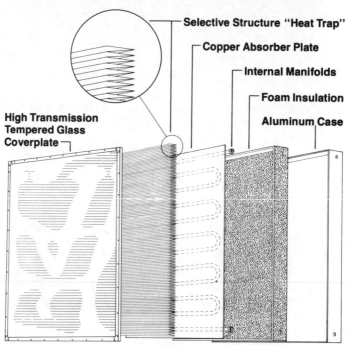

Selective Structure "Heat Trap"

Copper Absorber Plate

Internal Manifolds

Foam Insulation

Aluminum Case

High Transmission Tempered Glass Coverplate

TITLE	ELECTRA III HOME
CODE	
KEYWORDS	HEATING SWIMMING POOL OVERHANG DOOR WINDOW POSITIONING BREEZES HIGH ROOF HIP ATTIC VENTILATION SOLAR HEAT PUMP HOME HOUSE
AUTHOR	CORAL RIDGE PROPERTIES, INC.; AND WESTINGHOUSE ELECTRIC CORP.
DATE & STATUS	THE ELECTRA III HOME WAS OPENED IN NOVEMBER 1974.
LOCATION	CORAL SPRINGS, FLORIDA
SCOPE	THE 3,200 SQUARE FOOT HOME SAVES MORE THAN 50 PERCENT IN ENERGY CONSUMPTION OVER A SIMILAR HOME IN THE AREA, AND UP TO 72 PERCENT IF THE SOLAR HEATING FOR THE SWIMMING POOL IS INCLUDED. WITH LAND AND EQUIPMENT, A HOME LIKE ELECTRA III WOULD CURRENTLY COST ABOUT $200,000 ON THE REAL ESTATE MARKET. BUT ALMOST ALL OF THE HOME'S ENERGY SAVING IDEAS CAN BE BUILT INTO HOMES NOW GOING UP.

THE HOME HAS 10 FOOT CEILINGS, EXCEPT IN THE LIVING, DINING AND KITCHEN AREAS, WHERE THE CEILINGS ARE 12 1/2 FEET HIGH. THERE ARE 3 BEDROOMS, 2 1/2 BATHROOMS, LARGE LIVING AND DINING AREAS, A SNACK AREA, UTILITY SPACE, A PLAYROOM, AN EQUIPMENT ROOM, AND AN ENCLOSED TWO-CAR GARAGE. OUTDOORS, IT ALSO HAS A COVERED ENTRANCEWAY AND FOYER, A ROOFED GALLERY OUTSIDE THE LIVING AND DINING AREAS, AND OUTDOOR DINING AREAS ADJACENT TO THE KITCHEN. THE POOL AND PATIO ARE SCREENED.

AN HOUR-BY-HOUR COMPUTER STUDY SHOWED THE ELECTRA HOME'S MAXIMUM DEMAND FOR ELECTRIC POWER AT ANY TIME DURING A YEAR WOULD BE ABOUT 8.3 KILOWATTS, COMPARED WITH AN ESTIMATED 21.3 KILOWATTS FOR A SIMILAR CONVENTIONAL HOME.

TECHNIQUE	ENERGY SAVING IDEAS INCLUDE USE OF SOLAR ENERGY FOR HEATING WATER IN THE SWIMMING POOL AND FOR THE HOME'S HOT WATER SYSTEM; EXTRA WIDE OVERHANGS AROUND THE HOME; AND POSITIONING OF DOORS AND WINDOWS TO TAKE ADVANTAGE OF SOUTHEASTERLY BREEZES WHICH PREVAIL IN SOUTHERN FLORIDA IN SPRING, SUMMER AND FALL. USE OF HIGH ROOF HIPS INCREASES NATURAL ATTIC VENTILATION ABOUT 10 TIMES OVER CONVENTIONAL LOUVERED CONSTRUCTION AND REDUCES ATTIC TEMPERATURES CONSIDERABLY.

A HEAT PUMP, AN AIR CONDITIONING SYSTEM THAT OPERATES IN REVERSE TO EXTRACT HEAT FROM THE OUTDOOR AIR TO HEAT THE AIR INDOORS, ALSO CUTS DOWN THE NEED FOR AIR-CONDITIONING AND IS USED TO HELP HEAT WATER FOR DOMESTIC USE.

THE WALLS AND CEILINGS ARE INSULATED. WINDOWS AND GLASS DOORS USE BRONZE TINTED, DOUBLE INSULATED GLASS TO REDUCE THE HEATING AND COOLING NEEDS.

WHILE MOST AIR CONDITIONED HOMES RELY ON A THERMOSTAT THAT TURNS ON THE AIR CONDITIONER WHEN A SET TEMPERATURE IS REACHED INDOORS, THE ELECTRA HOME'S THERMOSTAT WORKS IN CONJUNCTION WITH A DEVICE THAT CHECKS THE OUTDOOR TEMPERATURE AND HUMIDITY. IF THE OUTSIDE AIR IS BELOW A CERTAIN TEMPERATURE AND HUMIDITY, FANS WILL DRAW IN OUTSIDE AIR TO COOL THE HOME INSTEAD OF TURNING ON THE AIR CONDITIONER.

MATERIAL	BRONZE TINTED DOUBLE INSULATING GLASS, SWIMMING POOL, HOT WATER SYSTEM, EXTRA WIDE OVERHANGS, HEAT PUMP, MODIFIED AIR CONDITIONING THERMOSTAT
AVAILABILITY	ADDITIONAL INFORMATION MAY BE OBTAINED FROM WESTINGHOUSE ELECTRIC CORP., INDUSTRY PRODUCTS CO., WESTINGHOUSE BUILDING, PITTSBURGH, PA 15222.
REFERENCE	PROFESSIONAL BUILDER HOW TO CUT ENERGY USE 72% PROFESSIONAL BUILDER, MARCH 1975, P 62

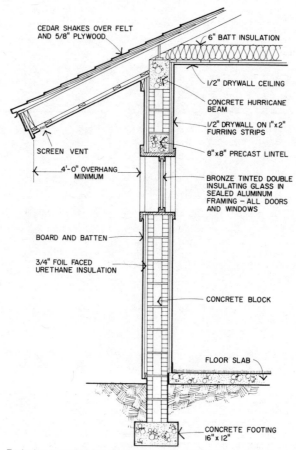

CEDAR SHAKES OVER FELT
AND 5/8" PLYWOOD

6" BATT INSULATION

1/2" DRYWALL CEILING

CONCRETE HURRICANE
BEAM

1/2" DRYWALL ON 1"x2"
FURRING STRIPS

SCREEN VENT

8"x8" PRECAST LINTEL

4'-0" OVERHANG
MINIMUM

BRONZE TINTED DOUBLE
INSULATING GLASS IN
SEALED ALUMINUM
FRAMING — ALL DOORS
AND WINDOWS

BOARD AND BATTEN

3/4" FOIL FACED
URETHANE INSULATION

CONCRETE BLOCK

FLOOR SLAB

CONCRETE FOOTING
16" x 12"

Typical cross section for Electra III by Coral Ridge Properties Inc.

362

TITLE	ENERGY CONSERVATION BUILDING
CODE	ENCON
KEYWORDS	OFFICE BUILDING LOUVER SOLAR COLLECTOR AWNING
AUTHOR	JOSEPH H. SOLOMON, PARTNER AND PROJECT ARCHITECT OF EMERY ROTH & SONS
DATE & STATUS	THE ENCON BUILDING WAS DESIGNED IN 1974, BUT HAS NOT BEEN BUILT.
LOCATION	A POTENTIAL SITE FOR THE ENCON BUILDING IS ON THIRD AVENUE IN NEW YORK CITY, NEW YORK
SCOPE	THE ENCON BUILDING WAS DESIGNED FOR A 200-BY-300 FOOT LOT, ONE MILLION SQUARE-FOOT, 43 STORY OFFICE BUILDING.

THE ENCON BUILDING CAN BE BUILT FOR $44 PER SQUARE FOOT - $4 PER SQUARE FOOT MORE THAN A TYPICAL CONVENTIONAL BUILDING WOULD COST. THE SAVINGS IN ENERGY COSTS ARE 90 CENTS PER SQUARE FOOT. THE AVERAGE MAJOR OFFICE BUILDING IN NEW YORK CITY, AT THE PRESENT RATES, COSTS ABOUT $1.90 PER SQUARE FOOT FOR ENERGY. $1.50 IS FOR ELECTRICITY AND 40 CENTS IS FOR STEAM. WITH THE ENCON BUILDING, THE COST IS FIGURED AT 90 CENTS FOR ELECTRICITY AND 10 CENTS FOR STEAM.

TECHNIQUE	THE BASIC FEATURE OF THE ENCON BUILDING IS THE FIVE-FOOT AWNINGS, SLANTED AT 45 DEGREES, OVERHANGING EACH LEVEL OF WINDOWS. THESE AWNINGS WILL BE MADE OF BLACK STEEL, ENCASED IN GLASS, WITH A CORE OF WATER CIRCULATION PIPES. WHEN THE SUN HITS THE AWNINGS, SOLAR ENERGY IS ABSORBED BY THE BLACK STEEL SURFACE AND HEATS THE WATER CIRCULATING INSIDE THE HALF INCH PIPE. THE WATER HEATS TO ABOUT 160 DEGREES IN THE SUMMER, 130 DEGREES IN THE WINTER.

THE HOT WATER WILL BE STORED IN LARGE BELOW GRADE STORAGE TANKS, AND DISTRIBUTED FOR HEATING AND COOLING. THE ENERGY PRODUCED WILL EFFECTIVELY REDUCE THE NEW YORK STEAM SUPPLIED TO THE BUILDING FOR THERMAL REQUIREMENTS BY APPROXIMATELY 30 PER CENT.

THE HOT WATER IS USED IN THE WINTER FOR HOT WATER HEATING; IN THE SUMMER, FOR THE ABSORPTION PORTION OF THE CENTRAL COOLING SYSTEM.

SINCE THE WEST FACADE OF THE BUILDING WILL ABSORB MAINLY A LATE, LOW HANGING SUN, THE SOLAR COLLECTOR AWNINGS WOULD NOT BE VERY EFFECTIVE. THEREFORE, THE VERTICALLY SLANTED LOUVER-LIKE PROJECTIONS OF THE BUILDING WALL WILL BE USED TO BLOCK OUT THE HEAT OF THE SUMMER SUN. THESE ARE DESIGNED, HOWEVER, TO REFLECT LIGHT INTO THE INTERIOR.

THE NORTH FACADE, WHICH IS FREE OF DIRECT SUN, IS FACED WITH SMALLER WINDOWS OF THERMOPANE GLASS TO REDUCE HEAT LOSS DURING THE WINTER.

THE ROOF OF THE BUILDING HAS A GIANT WEDGE-SHAPED SOLAR COLLECTOR WHICH GETS MOST OF THE EXPOSURE AND COLLECTS AS MUCH THERMAL ENERGY AS ALL THE WINDOW COLLECTORS COMBINED.

THE BASIC AIR CONDITIONING SYSTEM CONSISTS OF A CENTRAL REFRIGERATION PLANT IN THE PENTHOUSE MECHANICAL ROOM, WHICH DISTRIBUTES CHILLED WATER TO A SERIES OF INDIVIDUAL FAN ROOMS LOCATED AT THE PERIMETER OF EACH FLOOR. A COOLING TOWER FOR HEAT REJECTION WILL BE AT THE PENTHOUSE LEVEL. THERE WILL BE 3 FAN ROOMS SERVING EACH FLOOR STACKED VERTICALLY FOR THE ENTIRE HEIGHT OF THE BUILDING. EACH OF THE FAN ROOMS WILL CONTAIN A VARIABLE VOLUME SUPPLY AND RETURN FAN OF APPROXIMATELY 12,000 CU. FT. PER MINUTE. EACH FAN ROOM WILL DISTRIBUTE AIR THROUGH AN OVERHEAD DUCTED SUPPLY SYSTEM TO A PERIMETER AND INTERIOR ZONE IN THE VICINITY OF THE FAN ROOM.

INDIVIDUAL TEMPERATURE CONTROL IN EACH SPACE WILL BE OBTAINED
BY VARIABLE VALVE CONTROL OF THE AIR SUPPLIED THROUGH
THE CEILING OUTLETS. A PERIMETER BASEBOARD RADIATION SYSTEM
WILL PROVIDE THE HEATING TO OFFSET THE TRANSMISSION LOSSES
THROUGH THE COLD WEATHER.

MATERIAL   SOLAR COLLECTOR AWNINGS, BLACK STEEL, PIPES, WATER STORAGE
                TANK, LOUVERS, THERMOPANE, COOLING TOWER, FAN, BASEBOARD HEATER

AVAILABILITY   ADDITIONAL INFORMATION MAY BE OBTAINED FROM FLACK & KURTZ
                CONSULTING ENGINEERS, 29 WEST 38 ST., NEW YORK, N.Y. 10018.

REFERENCE   YUDIS, ANTHONY J.
                BROKER SHOWS OFF FUEL-SAVING OFFICE
                BOSTON SUNDAY GLOBE, OCTOBER 13, 1974, P A-49

                FLACK & KURTZ CONSULTING ENGINEERS
                ENCON BUILDING: ENERGY PLANT
                NEW YORK, N.Y., FLACK & KURTZ CONSULTING ENGINEERS, SEPTEMBER
                1974, 6 P

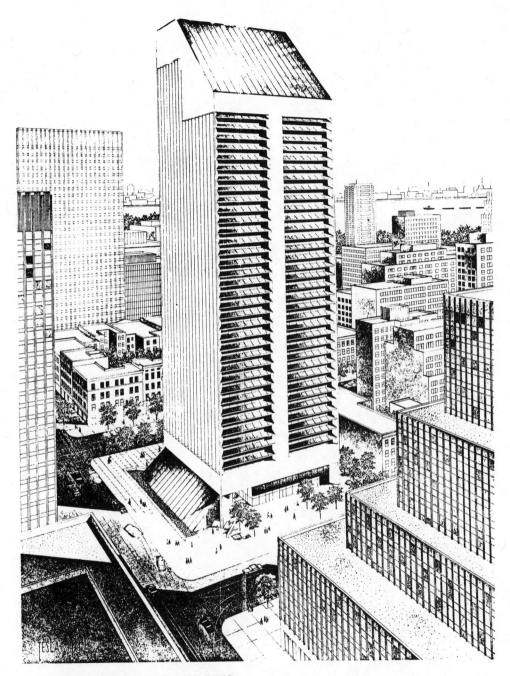

THE ENCON OFFICE BUILDING
A view of the west facade (at left) and the south facade
showing the solar collector and the deep louvers.

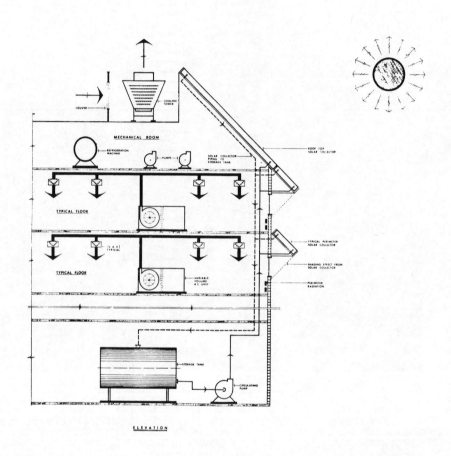

LOUVER

COOLING
TOWER

MECHANICAL ROOM

REFRIGERATION
MACHINE

PUMPS

SOLAR COLLECTOR
PIPING TO
STORAGE TANK

ROOF TOP
SOLAR COLLECTOR

TYPICAL FLOOR

TYPICAL PERIMETER
SOLAR COLLECTOR

(V.A.V)
TYPICAL

SHADING EFFECT FROM
SOLAR COLLECTOR

PERIMETER
RADIATION

TYPICAL FLOOR

VARIABLE
VOLUME
A.C UNIT

STORAGE TANK

CIRCULATING
PUMP

ELEVATION

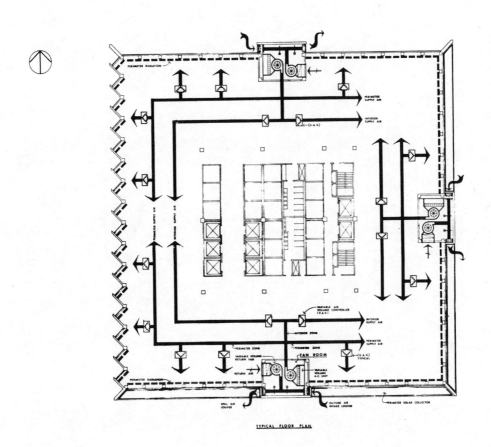

TYPICAL FLOOR PLAN

TITLE	ENERGY COST CUTTER HOUSE
CODE	
KEYWORDS	NATURAL VENTILATION HEAT LOSS BUFFER ENERGY LIGHTING
AUTHOR	PBD ARCHITECTS ASSOCIATED
DATE & STATUS	THE ENERGY COST CUTTER HOUSE WAS DESIGNED AS OF 1974.
LOCATION	THE DESIGN CAN BE ADAPTED TO VIRTUALLY ANY SITE.
SCOPE	THE ENERGY COST CUTTER HOUSE CAN CUT HEATING, COOLING AND ELECTRIC BILLS IN HALF. THE 1,900 SQUARE FOOT HOUSE CAN BE BUILT FOR APPROXIMATELY $40,000. WOOD FRAMED WINDOWS AND DOORS, PLYWOOD SYSTEMS FOR WALLS, FLOOR AND ROOF, AND COMPLETE INSULATION ARE SAID TO WORK TOGETHER TO CUT OPERATING COSTS AND TO HOLD CONSTRUCTION COSTS IN LINE.
	THE HOUSE IS SUPPORTED BY PRESSURE-TREATED 10 BY 10 POSTS WHICH CUT COSTS BOTH IN FOUNDATION AND SITE PREPARATION WORK. THE SYSTEM ALSO ADAPTS TO VIRTUALLY ANY SITE CONDITION, ACCORDING TO APA.
TECHNIQUE	IN WINTER, HEAT FROM THE SUN, THE FURNACE, AND THE FIREPLACE IS COLLECTED IN THE FAMILY ROOM LOFT AND IS REDISTRIBUTED THROUGH THE DUCTS WITH THE USE OF A BLOWER.
	FOR SUMMER COOLING, THE HOUSE WORKS LIKE A CHIMNEY. COOL AIR IS INTRODUCED THROUGH WINDOWS AND DOORS ON THE FIRST FLOOR AND DRAWN THROUGH THE HOUSE TO BE EXHAUSTED THROUGH OPEN WINDOWS IN THE SECOND STORY LOFT.
	VESTIBULES WITH INNER AND OUTER DOORS ARE USED TO PROTECT THE ENTRIES. THESE SMALL ANTEROOMS PROVIDE A BUFFER AREA OR AIR LOCK TO REDUCE THE FLOW OF OUTSIDE AIR INTO THE HOUSE AND PREVENT THE ESCAPE OF AIR WHICH HAS BEEN CONDITIONED TO COMFORTABLE TEMPERATURES.
	DUCTWORK IS PLACED WITHIN THE FLOOR PANELS SO THAT HEAT LOSS THROUGH THE DUCTS IS RADIATED UP INTO THE STRUCTURE. THE HOME IS SUITED TO ANY TYPE OF FURNACE OR HEAT PUMP; SELECTION DEPENDS ON THE ENERGY COSTS IN THE PARTICULAR AREA.
	ENERGY REQUIREMENTS OF THE LIGHTING SYSTEM HAVE BEEN REDUCED BY AT LEAST 1/3 THROUGH THE USE OF A SELECTIVE LIGHTING SCHEME WHICH INCORPORATES FLUORESCENT FIXTURES. THE LIGHTING SCHEME PROVIDES FOR EFFICIENT TASK-ORIENTED LIGHTING THROUGHOUT THE HOUSE. CONCENTRATED SUPPLEMENTAL LIGHTING IS ADDED IN PLACES SUCH AS STUDY AND MEAL PREPARATION AREAS.
MATERIAL	WOOD-FRAMED WINDOWS AND DOORS; PLYWOOD SYSTEMS FOR WALLS, FLOOR AND ROOF; PRESSURE-TREATED 10 BY 10 POSTS; FIREPLACE; DUCTS; BLOWER; FLUORESCENT FIXTURES
AVAILABILITY	ADDITIONAL INFORMATION MAY BE OBTAINED FROM THE AMERICAN PLYWOOD ASSOCIATION, 1119 ST., TACOMA, WASHINGTON 98401.
REFERENCE	ARCHITECTURAL RECORD ENERGY EFFICIENT HOME DEVELOPED BY THE AMERICAN PLYWOOD ASSOCIATION ARCHITECTURAL RECORD, OCTOBER 1974, P 35

TITLE	ENERGY HOUSE AT QUECHEE LAKE, VERMONT
CODE	
KEYWORDS	SOLAR COLLECTORS POST BEAM CONSTRUCTION THERMAL BARRIER
AUTHOR	DESIGNED BY BLUE/SUN LTD. OF FARMINGTON, CONN., BUILT BY TERROSI CONSTRUCTION INC. AT QUECHEE, VERMONT, SOLAR SYSTEM PROVIDED BY GRUMMAN AEROSPACE CORP. OF BETHPAGE, N.Y.
DATE & STATUS	THE QUECHEE, VERMONT HOUSE IS COMPLETE AS OF 1976.
LOCATION	QUECHEE, VERMONT
SCOPE	THE CEDAR-CLAD QUECHEE, VERMONT SOLAR HOUSE LOOKS LIKE AN ORDINARY VACATION HOUSE - IT DIFFERS IN THAT IT HAS GRUMMAN ROOF MOUNTED SOLAR COLLECTORS.
TECHNIQUE	THERE IS A LOT OF HEAT LOST IN CONVENTIONAL FRAMING THROUGH THE STUDS, WHICH, OF COURSE, CAN'T BE INSULATED. THE HOUSE USES POST-AND BEAM CONSTRUCTION AND OVERHUNG THE EXTERIOR WALL - 2X4 STUDS WITH CEDAR SIDING - 1-1/2 INCH OUTSIDE THE FRAMEWORK. IT WAS SPRAYED WITH 2-1/2 INCHES OF POLYURETHANE FROM THE INSIDE; THIS GIVES AN EFFECTIVE THERMAL BARRIER - NO STUDS GO "THROUGH" THE WALL FROM INSIDE TO OUTSIDE - AS WELL AS AN R-20 INSULATION RATING, THE EQUIVALENT OF SIX INCHES OF FIBERGLAS. THE ROOF HAS THREE INCHES OF POLYURETHANE AND 2-3/8 INCH HOMOSOTE DECKING.

THE HOME'S VERTICAL, THREE-LEVEL FLOOR PLAN PROVIDES NATURAL COOLING.

THE COLLECTORS DO NOT FACE DUE SOUTH; INSTEAD, THE HOUSE IS ORIENTATED 20 DEGREES WEST OF SOUTH TO TAKE ADVANTAGE OF PRE-VAILING SOUTHWESTERLY SUMMER WINDS. ALSO, THE COOL, HAZY MORNINGS AND WARMER, CLEARER AFTERNOONS IN THIS PART OF THE COUNTRY MAKE A SLIGHTLY WESTERLY-ORIENTED COLLECTOR MORE EFFECTIVE. |
| MATERIAL | CEDAR SIDING, POLYURETHANE, FIBERGLAS, HOMOSOTE DECKING, SOLAR COLLECTORS |
| AVAILABILITY | ADDITIONAL INFORMATION MAY BE OBTAINED FROM GRUMMAN AEROSPACE CORP., BETHPAGE, NEW YORK 11714. |
| REFERENCE | STEPLER, RICHARD SOLAR ARCHITECTURE - IT'S MORE THAN PUTTING COLLECTORS ON THE ROOF POPULAR SCIENCE, JULY 1976, PP 48-52, 96-97

DOODY, L.J. SOLAR HEATING AND COOLING OPTIONS ENGINEERING DIGEST, AUGUST 1976, PP 39-40 |

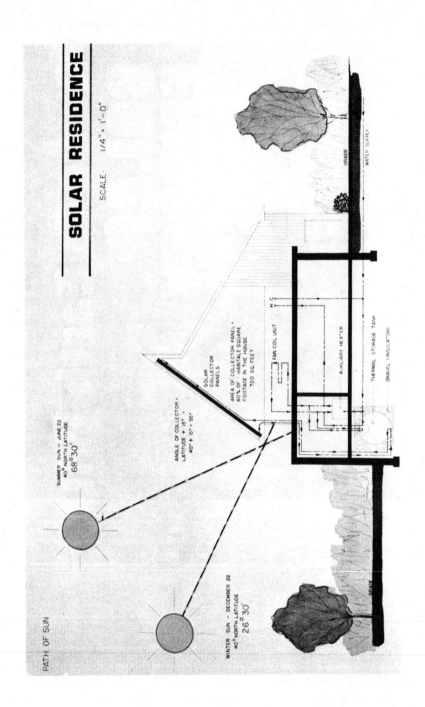

SOLAR RESIDENCE

SCALE: 1/4" = 1'-0"

PATH OF SUN

SUMMER SUN — JUNE 22
40° NORTH LATITUDE
68° 30'

WINTER SUN — DECEMBER 22
40° NORTH LATITUDE
26° 30'

ANGLE OF COLLECTOR =
LATITUDE + 15° =
40° + 15° = 55°

SOLAR
COLLECTOR
PANELS

AREA OF COLLECTOR PANEL =
60% OF HABITABLE SQUARE
FOOTAGE IN THE HOUSE
720 SQ. FEET

FAN COIL UNIT

H C

AUXILIARY HEATER

THERMAL STORAGE TANK

GRAVEL (INSULATOR)

GRADE

WATER SUPPLY

GRADE

372

TITLE	ENERGY SYSTEMS, INC. SOLAR HOUSES
CODE	
KEYWORDS	CUSTOM HOME COLLECTOR FLAT PLATE
AUTHOR	CASTER DEVELOPMENT CORP.
DATE & STATUS	TWO HOUSES OF THE ENERGY SYSTEMS, INC. SOLAR HOUSES WERE COM- PLETED IN 1975; THE REST ARE UNDER CONSTRUCTION.
LOCATION	EL CAJON (20 MI. EAST OF SAN DIEGO), CALIFORNIA
SCOPE	THE BUILDINGS INCLUDE A TOTAL OF 23 CUSTOM-BUILT DWELLINGS, ALL TO BE SOLAR HEATED.
TECHNIQUE	THE TYPICAL COLLECTOR AREA IS 300 TO 500 SQ. FT. THE COLLECTORS ARE FLAT-PLATE, WITH WATER-AND-ETHYLENE-GLYCOL. THEY CONSIST OF PANELS EMPLOYING BLACK ALUMINUM SHEET AND SINGLE GLAZING. THE PANELS ARE MOUNTED ON THE ROOF. THEY SLOPE 30 DEGREES FROM THE HORIZONTAL. HEAT EXCHANGERS ARE USED. THE STORAGE SYSTEM CONSISTS OF 1,500 TO 2,000 GALLONS OF WATER AND ETHYLENE GLYCOL.
MATERIAL	WATER AND ETHYLENE GLYCOL, BLACK ALUMINUM SHEET, SINGLE GLAZING
AVAILABILITY	ADDITIONAL INFORMATION MAY BE OBTAINED FROM THE CASTER DEVELOP- MENT CORP., 6626 MISSION GORGE RD., SAN DIEGO, CALIFORNIA.
REFERENCE	SHURCLIFF, W.A. SOLAR HEATED BUILDINGS, A BRIEF SURVEY CAMBRIDGE, MASSACHUSETTS, W.A. SHURCLIFF, MARCH 1976, 212 P

TITLE	ENGLE SOLAR HOME
CODE	
KEYWORDS	SOLAR SUN ANGLES HEATING COOLING ABSORPTION CHILLER
AUTHOR	ALAN LOWER, AIA
DATE & STATUS	THE ENGLE SOLAR HOME IS COMPLETE AS OF 1975.
LOCATION	WAGONER, OKLAHOMA
SCOPE	THE ENGLE SOLAR HOME INTEGRATES ITS SOLAR HEATING COMPONENTS WITH THE STRUCTURE RATHER THAN HAVING THEM APPEAR AS AN AFTER-THOUGHT.
TECHNIQUE	A MAJOR DESIGN ELEMENT OF THE STRUCTURE - THE STEEP, 50 DEGREE ROOF PLANE - POSITIONS THE COLLECTORS SO THAT THEY ARE PERPENDI-CULAR TO THE SUN'S RAYS AT LOW WINTER-SUN ANGLES. COLLECTORS ARE SELFDRAINING TO PREVENT FREEZE UP.
	ROOF OVERHANGS SHADE SOUTH-FACING WINDOWS DURING HOT OKLAHOMA SUMMERS, BUT LET IN SUN DURING THE WINTER MONTHS. ALL GLASS IS DOUBLE PANE, WALLS ABOVE AND BELOW GRADE HAVE SIX INCHES OF INSULATION; AND ROOF AND CEILING AREAS HAVE 12 INCHES.
	THE HOME'S HEATING AND COOLING SYSTEM USES ENERGY SYSTEMS, INC. LIQUID-TYPE COPPER COLLECTORS. THE HEAT STORAGE TANK IS COMPART-MENTALIZED; THE ENTIRE TANK - 2,600 GALLONS - STORES HEAT IN WINTER WHEN LOWER-TEMPERATURE WATER IS SUFFICIENT FOR SPACE HEATING. IN SUMMER, ONLY 1,000 GALLONS ARE USED SO THAT HIGHER TEMPERATURES CAN BE REACHED. THIS WATER - 195 DEGREES - WILL BE SUPPLIED TO A LITHIUM-BROMIDE ABSORPTION CHILLER FOR AIR CONDITIONING.
MATERIAL	DOUBLE PANE GLASS, INSULATION, LIQUID-TYPE COPPER COLLECTORS, HEAT STORAGE TANK, LITHIUM-BROMIDE ABSORPTION CHILLER
AVAILABILITY	ADDITIONAL INFORMATION MAY BE OBTAINED FROM ALAN LOWER, 3535 NW 58TH, OKLAHOMA CITY, OKLAHOMA. SEND SELF-ADDRESSED, STAMPED ENVELOPE.
REFERENCE	STEPLER, RICHARD SOLAR ARCHITECTURE - IT'S MORE THAN PUTTING COLLECTORS ON THE ROOF POPULAR SCIENCE, JULY 1976, PP 48-52, 96-97
	ARIZONA STATE UNIVERSITY, COLLEGE OF ARCHITECTURE ENGLE HOUSE WASHINGTON, D.C., SOLAR-ORIENTED ARCHITECTURE, PUBLISHED BY THE AIA RESEARCH CORP., JANUARY 1975, P 74

TITLE   EPA LABORATORY RESEARCH BUILDING

CODE

KEYWORDS   SOLAR PANEL LABORATORY BUILDING SYSTEMS CONTROL HEATING
                   COOLING WALL COMPACT ENCLOSURE WASTE RECOVERY ELECTRICAL

AUTHOR   ROE ASSOCIATES, ARCHITECTS-ENGINEERS

DATE & STATUS   THIS WAS A CONCEPTUAL STUDY DONE IN 1975.

LOCATION   NORTH CANADA

SCOPE   THIS CONCEPTUAL STUDY WAS FOR A 10-STORY BUILDING. THE
                   BUILDING WAS TO COVER 575,000 SQ. FT. AND HOUSE OFFICES,
                   LABORATORIES, ANIMAL RESEARCH FACILITIES, A CAFETERIA AND
                   AN AUDITORIUM.

TECHNIQUE   PROJECTED ENERGY SAVINGS FOR THIS CONCEPTUAL BUILDING WERE
                   ATTRIBUTED TO THE FOLLOWING DESIGN MEASURES: USE OF SOLAR
                   WALL PANELS FOR HEATING AND COOLING, COMPACT BUILDING ENCLOSURE,
                   REDUCED WINDOW AREAS, WASTE HEAT RECOVERY, AND MECHANICAL
                   AND ELECTRICAL SYSTEMS WITH BETTER CONTROLS.

MATERIAL   SOLAR PANELS, SYSTEMS CONTROLS

AVAILABILITY   ADDITIONAL INFORMATION MAY BE OBTAINED FROM ROE ASSOCIATES,
                   ARCHITECTS-ENGINEERS, 320 FULTON AVENUE, HEMPSTEAD, NEW YORK.

REFERENCE   ARCHITECTURAL RECORD
                   PROFESSION AND INDUSTRY FOCUS ON SOLAR ENERGY
                   ARCHITECTURAL RECORD, AUGUST 1975, P 35

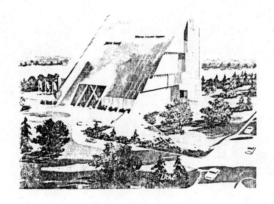

TITLE	ERIE-LAKAWANNA HOUSE
CODE	
KEYWORDS	SNOW PLOW TRAIN SOLAR AIR COOLED COLLECTOR GUEST HOUSE CLERESTORY GREENHOUSE BEADWALL
AUTHOR	DAN SCULLY
DATE & STATUS	THE ERIE-LAKAWANNA HOUSE IS DESIGNED AS OF 1975.
LOCATION	WOODBRIDGE, CONN., LAST SITED ALONG NEW HAMPSHIRE THROUGHWAY.
SCOPE	THE ERIE-LAKAWANNA HOUSE IS MODELED AFTER A BLUE SNOW PLOW TRAIN. THE TRAIN WAS THE IDEA FOR A COMBINATION GUEST HOUSE/STUDIO. THERE IS ONE LARGE ROOM WHICH SERVES AS THE LIVING ROOM AND STUDIO, AND ALSO EXTENDS INTO THE KITCHEN AREA. THERE IS ALSO A BEDROOM AND A BATHROOM. THE ENTIRE HOUSE IS SOLAR HEATED.
	THE LONG LIVING ROOM, STRETCHED OUT BEHIND THE COLLECTOR AND ALONG THE RIDGE, HAS WINDOWS ON THREE SIDES. THIS IS A NARROW ROOM, NO WIDER THAN A TRAIN. THE SOLAR COLLECTOR REQUIRES MAXIMUM SOUTHERN EXPOSURE, WITH A WIDE SOUTH FACE. THE BEDROOM AND BATH FALL INTO PLACE BEHING THE COLLECTOR, MAKING A VERY WORKABLE 'L'-SHAPED PLAN WITH THE INTERSECTION IN THE OPEN KITCHEN.
TECHNIQUE	THE PLOW OF THE TRAIN IS CONVERTED INTO A SOLAR COLLECTOR. THE CONTAINERIZED WATER STORAGE IS BEHIND THE COLLECTOR AND IN THE PLOW. THE CURVED TRAILING EDGE OF THE PLOW IS CONVERTED INTO A 2 FEET DEEP SOUTH FACING WINDOW OF CURVED PLEXIGLASS.
	THE TRAIN MAN'S CLERESTORY, HERE A TOKEN GREENHOUSE, IS INSULATED AT NIGHT WITH ZOMEWORKS' BEAD WALL – POLYSTYRENE BEADS BLOWN BETWEEN TWO PANES OF GLASS, AND THEN VACUUMED OUT IN THE MORNING, AND STORED IN METAL DRUMS ON THE ROOF.
MATERIAL	CORRUGATED ALUMINUM, SNOWPLOW TRAIN, WATER CONTAINERS, PLEXI-GLASS, SOLAR AIR COOLED COLLECTOR, GLASS BLOCK LINED ENTRYWAY, TRAIN MAN'S CLERESTORY, GREENHOUSE, BEADWALL
AVAILABILITY	ADDITIONAL INFORMATION ON THE ERIE-LAKAWANNA HOUSE MAY BE OBTAINED FROM DAN SCULLY, HARRISVILLE, NEW HAMPSHIRE 03450.
REFERENCE	JOURNAL OF ARCHITECTURAL EDUCATION ANIMUS JOURNAL OF ARCHITECTURAL EDUCATION, VOL. 29, NO. 1, SEPTEMBER 1975, PP 24-25

INSPIRATION

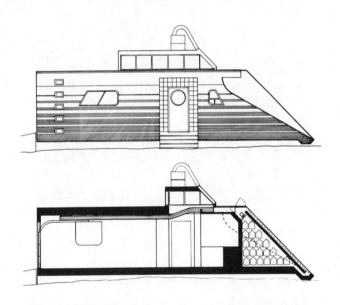

Combo Guest house/studio
Program: Large room—living room and studio, as extension of open
kitchen; one bedroom, bath—all solar heated
Site: Along crest of ledge hill, running N-S; access from main house
from the East; On the West side of hill, the road runs obliquely N-W.
views to East and West

DESIGN

TITLE	ERWIN HOUSE
CODE	
KEYWORDS	MOBILE HOME DETACHED COLLECTOR STORAGE
AUTHOR	TROY ERWIN
DATE & STATUS	THE ERWIN HOUSE WAS RETROFITTED WITH SOLAR COLLECTION EQUIPMENT IN APRIL 1975.
LOCATION	ROSEBURG, OREGON
SCOPE	THE BUILDING IS AN EXISTING 1970, DOUBLE-WIDTH, 40' X 24', BUDDY MOBILE HOME. IT WAS RETROFITTED WITH SOLAR HEAT FROM A DETACHED COLLECTOR.
TECHNIQUE	THE BUILDING IS WELL INSULATED. THE WINDOWS ARE DOUBLE GLAZED.

THE COLLECTOR CONTAINS 210 SQ. FT. OF WATER-AND ETHYLENE-GLYCOL COOLANT. THE COLLECTOR IS 21' X 10', AND SLOPES 60 DEGREES FROM THE HORIZONTAL. IT IS ON A SEPARATE STAND OR SHED, SITUATED 20' WEST OF THE HOUSE. THE HEART OF THE COLLECTOR IS A SHEET OF CORRUGATED GALVANIZED STEEL. THE CORRUGATIONS ARE HORIZONTAL. THE BLACK COATING IS NON-SELECTIVE. AFFIXED TO THE SHEET BY METAL BANDS OR CLIPS IS A SERPENTINE ARRAY OF 3/4" IN DIAMETER BLACK-PAINTED STEEL PIPE. THE PIPE SEGMENTS RUN PARALLEL TO THE CORRUGATIONS. THE SPACING OF THE SEGMENTS RANGES FROM 13" NEAR THE BOTTOM TO 9" NEAR THE TOP. THE GLAZING CONSISTS OF ONE LAYER OF GLASS. THE BACKING CONSISTS OF FOIL-FACED FIBERGLAS, 3-1/2" THICK, WITH THE FOIL TOWARD THE CORRUGATED SHEET. WHENEVER THE COLLECTOR TEMPERATURE EXCEEDS 105 DEGREES F, A WATER-PUMP CIRCULATES THE COOLANT THROUGH THE COLLECTOR AND THROUGH A 100' COIL OF 3/4" DIAMETER PLASTIC TUBING IN THE BOTTOM OF THE STORAGE TANK. AN ALUMINUM COVERED PLYWOOD REFLECTOR HAS BEEN ADDED TO INCREASE THE EFFICIENCY OF THE SOLAR COLLECTOR.

THE STORAGE SYSTEM CONTAINS 1,000 GALLONS OF WATER IN A VERTICAL CYLINDRICAL TANK, SITUATED INSIDE THE SHED. THE TANK IS INSULATED WITH 3-1/2" OF FIBERGLAS. THE TANK DIAMETER IS 7', AND THE HEIGHT IS 4'. WHEN THE ROOMS NEED HEAT, A SMALL WATER-PUMP CIRCULATES THE WATER FROM THE TANK TO A SMALL RADIATOR IN THE HOUSE. (THE RADIATOR IS FROM A 1950 DESOTO AUTOMOBILE.) A FAN BLOWS THE ROOM AIR THROUGH THE RADIATOR. THE FLOW OF WATER FROM THE SHED TO THE HOUSE AND VICE VERSA IS VIA INSULATED UNDERGROUND PIPES.

MATERIAL	WATER-AND-ETHYLENE-GLYCOL COOLANT; CORRUGATED GALVANIZED STEEL; NON-SELECTIVE BLACK COATING; 3/4" DIAMETER BLACK-PAINTED STEEL PIPE; GLASS; FOIL-FACED FIBERGLAS, 3-1/2" THICK; WATER PUMP; 100' COIL OF 3/4" DIAMETER PLASTIC TUBING; CYLINDRICAL TANK; 1,000 GALLONS OF WATER; 3-1/2" OF FIBERGLAS; SMALL WATER PUMP; 1950 DESOTO AUTOMOBILE RADIATOR; FAN; INSULATED UNDERGROUND PIPES; ALUMINUM COVERED PLYWOOD REFLECTOR
AVAILABILITY	ADDITIONAL INFORMATION MAY BE OBTAINED FROM TROY ERWIN, IDLEYLD RT. BOX 240, ROSEBURG, OREGON 97470.
REFERENCE	SHURCLIFF, W.A. SOLAR HEATED BUILDINGS, A BRIEF SURVEY CAMBRIDGE, MASSACHUSETTS, W.A. SHURCLIFF, MARCH 1976, 212 P

TITLE	ESSEN HOUSE
CODE	
KEYWORDS	HOUSE INSULATED COLLECTOR STORAGE
AUTHOR	DORNIER SYSTEM GMBH
DATE & STATUS	THE ESSEN HOUSE WAS COMPLETED IN 1975.
LOCATION	ESSEN, WEST GERMANY
SCOPE	THE BUILDING IS A THREE-STORY, WELL-INSULATED HOUSE, CONTAINING FOUR APARTMENTS.
TECHNIQUE	THE COLLECTOR CONSISTS OF 70 SQ.M OF NET AREA. IT IS MOUNTED ON THE STEEPLY SLOPING ROOF, AND EMPLOYS HEAT PIPES AND WATER. IT INCLUDES PANELS, 5.3 M X 0.2 M, THAT EXTEND UP AND DOWN THE ROOF. THE HEART OF THE PANEL IS AN ALUMINUM PLATE, SPECIALLY FORMED SO AS TO BE THICK NEAR THE INTEGRAL TUBE AND THIN AT LOCATIONS FAR FROM THE TUBE. THE TUBE OPERATES ON A HEAT-PIPE PRINCIPLE, CARRYING HEAT RAPIDLY AND EFFICIENTLY TO ONE END OF THE PANEL WHERE, VIA A HEAT EXCHANGER, THE HEAT IS TRANSFERRED TO A STREAM OF WATER. THE PANEL IS DOUBLE GLAZED WITH GLASS AND PLEXIGLAS, AND HAS AN INSULATING BACKING OF POLYURETHANE FOAM. THE STORAGE SYSTEM CONSISTS OF 7,200 KG OF WATER IN A 7.2 CUBIC METER, INSULATED TANK IN THE BASEMENT. THE SUPPLY OF HEAT FOR SPACE HEATING AND DOMESTIC HOT WATER HEATING IS ASSISTED BY A WATER-TO-WATER HEAT PUMP.
MATERIAL	WATER; ALUMINUM PLATE; HEAT EXCHANGER; GLASS; POLYURETHANE FOAM; PLEXIGLAS; 7,200 KG OF WATER; INSULATED TANK; WATER-TO-WATER HEAT PUMP
AVAILABILITY	ADDITIONAL INFORMATION MAY BE OBTAINED FROM DORNIER-SYSTEM GMBH, FRIEDRICHSHAFEN, WEST GERMANY.
REFERENCE	SHURCLIFF, W.A. SOLAR HEATED BUILDINGS, A BRIEF SURVEY CAMBRIDGE, MASSACHUSETTS, W.A. SHURCLIFF, MARCH 1976, 212 P

TITLE	EUREKA SOLAR HOUSE
CODE	
KEYWORDS	HOUSE WATER TYPE COLLECTOR STORAGE
AUTHOR	Y.B. SAFDARI AND T. LANDES, JR.
DATE & STATUS	THE EUREKA SOLAR HOUSE WAS COMPLETED IN APRIL 1975.
LOCATION	EUREKA, ILLINOIS
SCOPE	THE ONE-STORY BUILDING HAS AN ATTIC AND A WALK-OUT BASEMENT. THE LIVING AREA IS 2,000 SQ. FT. PLUS A WALK-OUT BASEMENT. THERE ARE A LIVING ROOM, A KITCHEN, AND TWO BEDROOMS ON THE GROUND FLOOR, AND TWO BEDROOMS AND A FAMILY ROOM IN THE ATTIC. DURING THE COLDEST WINTER MONTHS OF JANUARY AND FEBRUARY 1975-1976, THE SOLAR SYSTEM SUPPLIED 70% OF THE HEATING REQUIREMENT.
TECHNIQUE	THE FIBERGLAS INSULATION IS 3-1/2" THICK IN THE WALLS, AND 7" THICK IN THE CEILING OR ROOF.
	THE GROSS COLLECTOR AREA IS 800 SQ. FT. THE COLLECTOR IS OF THE WATER-TYPE, IT IS MOUNTED ON THE ROOF AND SLOPES 45 DEGREES FROM THE HORIZONTAL. IT IS INTEGRATED INTO THE ROOF. IT CONSISTS OF 36 OLIN BRASS CO. PANELS, EACH 3' X 8'. SOME PANELS EMPLOY BLACK ALUMINUM SHEETS; OTHERS EMPLOY BLACK COPPER SHEETS. EACH PANEL IS DOUBLE GLAZED AND HAS A FIBERGLAS BACKING. THE LIQUID IS WATER PLUS ETHYLENE GLYCOL AND INHIBITOR. THE LIQUID IS CIRCULATED BY A 1 HP CENTRIFUGAL PUMP.
	THE STORAGE SYSTEM CONSISTS OF 1,500 GALLONS OF WATER IN A FIBERGLAS-INSULATED STEEL TANK IN THE BASEMENT. THE TANK DIMENSIONS ARE 6' X 6' X 6'.
MATERIAL	FIBERGLAS INSULATION, 3-1/2" THICK IN THE WALLS, 7" THICK IN THE CEILING OR ROOF; 36 OLIN BRASS CO. PANELS, EACH 3' X 8'; BLACK ALUMINUM SHEETS; BLACK COPPER SHEETS; FIBERGLAS BACKING; WATER AND ETHYLENE GLYCOL; INHIBITOR; 1 HP CENTRIFUGAL PUMP; WATER; FIBERGLAS-INSULATED STEEL TANK
AVAILABILITY	ADDITIONAL INFORMATION MAY BE OBTAINED FROM SUN SYSTEMS, INC., P.O.BOX 155, EUREKA, ILLINOIS 61530.
REFERENCE	SOLAR ENGINEERING DR. SAFDARI - AN EDUCATOR WHO IS A PRACTITIONER SOLAR ENGINEERING, AUGUST 1976, PP 27-28

TITLE	EVANS HOUSE
CODE	
KEYWORDS	HOUSE INSULATION WINDOW AREA
AUTHOR	M.B. WELLS
DATE & STATUS	THE EVANS HOUSE WAS SCHEDULED TO BE COMPLETED IN LATE 1976.
LOCATION	MT. LAUREL, NEW JERSEY
SCOPE	THE BUILDING IS A TWO-STORY, 3-BEDROOM, 2,750 SQ. FT. (40' X 38') WOODFRAME HOUSE, WITH A SMALL ATTIC BUT NO BASEMENT. A 2-CAR GARAGE IS ATTACHED.
TECHNIQUE	THE HOUSE FACES SOUTHSOUTHEAST. THE WALL INSULATION IS 3-1/2" OF FOIL-FACED FIBERGLAS BATTS BETWEEN 5/8" OF PLYWOOD ON THE OUTSIDE AND 1/2" PLASTERBOARD ON THE INSIDE. THE OVERALL R-VALUES OF THE WALL AND ROOF ARE R-15 AND R-26, RESPECTIVELY. THERE IS A MODERATE AREA OF WINDOWS, WHICH ARE DOUBLE GLAZED. THEY ARE COVERED ON THE INSIDE AT NIGHT BY A 1" LAYER OF POLYURETHANE FOAM OR THE EQUIVALENT. THE BUILDING FOUNDATIONS ARE ALSO INSULATED.
	THE COLLECTOR IS A 600 SQ. FT., AIR TYPE COLLECTOR, MOUNTED ON THE ROOF, SLOPING 45 DEGREES FROM THE HORIZONTAL. THE ROOF SLOPES TOWARD THE SOUTHSOUTHEAST. THE COLLECTOR INCLUDES 38 PANELS, ARRANGED IN TWO ARRAYS, CONTAINING 19 PANELS EACH. EACH PANEL IS 8' X 2'. THE DETAILS OF THE PANEL DESIGN HAVE NOT YET BEEN DECIDED. THERE ARE WHITE MARBLE CHIPS ON THE HORIZONTAL ROOF OF THE GARAGE, JUST SOUTH OF THE COLLECTOR, WHICH DIRECT ADDITIONAL RADIATION TO THE COLLECTOR.
	THE STORAGE SYSTEM HAS 70 TONS (1,000 CUBIC FEET) OF 1-1/2" TO 3" DIAMETER STONES. THE STONES ARE IN TWO BINS, THE INSIDE DIMENSIONS OF WHICH ARE 15' X 11' X 3' HIGH. THE THICKNESS OF THE POLYURETHANE FOAM INSULATION ON THE INTERIOR OF THE BIN IS 4" ON THE SIDES, AND 2" ON THE BOTTOM. THE BINS ARE SITUATED BENEATH A 4" THICK FLOOR SLAB OF THE SOUTHSOUTHEAST PORTION OF THE HOUSE. THE HOT AIR FROM THE COLLECTOR FLOWS THROUGH THE BIN HORIZONTALLY, PARALLEL TO THE 15' EDGES. WHEN THE ROOMS NEED HEAT, A BLOWER CIRCULATES THE ROOM AIR THROUGH THE BIN IN THE OPPOSITE DIRECTION.
MATERIAL	FOIL-FACED FIBERGLAS BATT; 5/8" OF PLYWOOD; 1/2" PLASTERBOARD; DOUBLE GLAZED WINDOWS; 1" LAYER OF POLYURETHANE FOAM; WHITE MARBLE CHIPS; 70 TONS OF 1-1/2" TO 3" DIAMETER STONES; TWO BINS; BLOWER
AVAILABILITY	REFER TO THE FOLLOWING REFERENCE FOR ADDITIONAL INFORMATION.
REFERENCE	SHURCLIFF, W.A. SOLAR HEATED BUILDINGS, A BRIEF SURVEY CAMBRIDGE, MASSACHUSETTS, W.A. SHURCLIFF, MARCH 1976, 212 P

TITLE	FAMILY MUTUAL SAVINGS BANK
CODE	
KEYWORDS	SOLAR COLLECTOR PANEL HEATING COOLING HOT WATER
AUTHOR	KENNETH F. PARRY ASSOCIATES; CA CROWLEY ENGINEERING CO.
DATE & STATUS	THE FAMILY MUTUAL SAVINGS BANK WAS BUILT IN 1976.
LOCATION	CONCORD, NEW HAMPSHIRE
SCOPE	THIS BANK BUILDING HAS 2,500 SQ. FT. OF FLOOR SPACE.
TECHNIQUE	THIS BANK IS DESIGNED TO RECEIVE UP TO 30% OF ITS COOLING AND 70% OF ITS HEATING ENERGY FROM THE SUN THROUGH 30 SOLAR COLLECTORS, DESIGNED BY DAYSTAR CORP. OF BURLINGTON, MASSACHUSETTS. ALSO 100% OF ITS DOMESTIC HOT WATER WILL BE HEATED BY THE SUN YEAR ROUND.
MATERIAL	SOLAR COLLECTOR
AVAILABILITY	ADDITIONAL INFORMATION MAY BE OBTAINED FROM DAYSTAR CORP., 90 CAMBRIDGE STREET, BURLINGTON, MASSACHUSETTS 01803.
REFERENCE	DAYSTAR CORPORATION ALL-COPPER SOLAR COLLECTORS FOR THREE NEW ENGLAND BANKS BURLINGTON, MASSACHUSETTS, DAYSTAR CORP. PAMPHLET, 1976, 1 P
	DAYSTAR CORPORATION DAYSTAR 20 SOLAR COLLECTOR: TECHNICAL SPECIFICATIONS AND PERFORMANCE DATA BURLINGTON, MASSACHUSETTS, DAYSTAR CORP. PAMPHLET, 1976, 2 P

TITLE	FAUQUIER COUNTY PUBLIC HIGH SCHOOL
CODE	
KEYWORDS	HIGH SCHOOL SOLAR WATER TYPE COLLECTOR STORAGE
AUTHOR	INTERTECHNOLOGY CORP.
DATE & STATUS	THE SOLAR SPACE HEATING SYSTEM WAS COMPLETED IN 1974.
LOCATION	WARRENTON, VIRGINIA
SCOPE	THE FAUQUIER COUNTY PUBLIC HIGH SCHOOL DESIGN INCLUDES A SOLAR HEATING SYSTEM FOR A SET OF FIVE SEPARATE CLASSROOM BUILDINGS, COMPRISING A 4,100 SQ. FT. PORTION OF THE SCHOOL. THE OVERALL COST WAS $297,000.
TECHNIQUE	THE COLLECTOR IS A 126' X 26', 2,400 SQ. FT. NET, WATER-TYPE COLLECTOR, MOUNTED ON A 53-DEGREES-FROM-THE-HORIZONTAL SCAFFOLD ADJACENT TO THE SCHOOL. IT IS DOUBLE GLAZED WITH 1/8" DOUBLE-STRENGTH GLASS. RADIATION STRIKES A CHEMICALLY ETCHED SURFACE COVERED WITH SELECTIVE BLACK COATING ON AN OLIN BRASS CO. ROLL-BOND ALUMINUM SHEET. THE WATER, WITH CORROSION INHIBITOR, FLOWS IN THE INTEGRAL EXPANDED CHANNELS IN THE ALUMINUM SHEET. NO ANTI-FREEZE IS USED, BUT THE LIQUID CAN BE DRAINED FROM THE COLLECTOR INTO THE STORAGE TANK IN 1-1/2 MINUTES. THERE ARE 105 COLLECTOR PANELS, EACH 3-1/2' X 8'.
	THE STORAGE SYSTEM CONSISTS OF TWO SEPARATE, WATER-FILLED, UNDERGROUND, 5,500 GALLON TANKS. THE TOTAL CAPACITY IS 11,000 GALLONS. EACH TANK, AN ELECTRICAL TRANSFORMER BUNKER, IS OF REINFORCED CONCRETE WITH LENGTH, WIDTH, AND HEIGHT OF 11-1/2', 8-1/2', AND 7-1/2'.
MATERIAL	DOUBLE GLAZED WITH 1/8" DOUBLE-STRENGTH GLASS; SELECTIVE BLACK COATING; OLIN BRASS CO. ROLL-BOND ALUMINUM SHEET; WATER WITH CORROSION INHIBITOR: TWO WATER-FILLED, UNDERGROUND, 5,500 GALLON TANKS; REINFORCED CONCRETE
AVAILABILITY	ADDITIONAL INFORMATION MAY BE OBTAINED FROM THE INTERTECHNOLOGY CORP., BOX 340, WARRENTON, VIRGINIA 22186.
REFERENCE	SHURCLIFF, W.A. SOLAR HEATED BUILDINGS, A BRIEF SURVEY CAMBRIDGE, MASSACHUSETTS, W.A. SHURCLIFF, MARCH 1976, 212 P

TITLE   FEDERAL OFFICE BUILDING, MANCHESTER, NEW HAMPSHIRE

CODE

KEYWORDS   SOLAR COLLECTOR WINDOW FIN DOUBLE GLAZED OFFICE BUILDING

AUTHOR   DUBIN-MINDELL BLOOM ASSOCIATES

DATE & STATUS   THE FEDERAL OFFICE BUILDING WILL BE COMPLETED IN 1975.

LOCATION   THE BUILDING WILL BE LOCATED AT THE JUNCTION OF CHESTNUT
                   STREET AND MERRIMACK STREET, MANCHESTER, NEW HAMPSHIRE.

SCOPE   THE FEDERAL OFFICE BUILDING IS DESIGNED TO CONSUME AT LEAST
                   40 PERCENT LESS ENERGY THAN A COMPARABLE CONVENTIONAL
                   STRUCTURE. THIS BUILDING IS THE FIRST ENERGY CONSERVATION DEM-
                   ONSTRATION PROJECT OF THE GENERAL SERVICES ADMINISTRATION.
                   IT WILL BE A LIVING WORKING LABORATORY DESIGNED TO MINIMIZE
                   ENERGY CONSUMPTION AND RECOVER HEAT THAT WOULD ORDINARILY
                   BE LOST IN BUILDING OPERATIONS.

TECHNIQUE   THE SEVEN-STORY BUILDING WILL INCORPORATE MORE ENERGY-SAVING
                   FEATURES THAN ANY OTHER FEDERAL BUILDING. IT WILL INCLUDE A
                   10,000 SQUARE FOOT SOLAR ENERGY COLLECTOR THAT WILL FURNISH
                   NEARLY 30 PERCENT OF THE BUILDING'S ENERGY REQUIREMENT. THE
                   BUILDING'S WINDOWS WILL BE UNIQUELY FACED WITH FINS THAT WILL
                   ADMIT WARM SOLAR RAYS IN THE WINTER AND EXCLUDE THEM IN THE
                   SUMMER. THE DOUBLE-GLAZED WINDOWS WILL BE CAREFULLY LOCATED
                   TO AVOID ANY SENSE OF CONFINEMENT, WHILE GIVING THE FACADE A
                   HIGHLY-SCULPTURED AND DISTINCTIVE APPEARANCE.

                   THE EXTERIOR SHELL OF THE 175,000 SQUARE-FOOT BUILDING WILL
                   BE OF PRE-CAST RESIN-BASED PEBBLE AND INSULATED METAL PANELS
                   THAT WILL REDUCE THE HEAT TRANSFER BETWEEN THE INTERIOR AND
                   EXTERIOR.

MATERIAL   SOLAR COLLECTOR, INSULATED METAL PANELS, DOUBLE-GLAZED
                   WINDOWS, FINS

AVAILABILITY   FURTHER INFORMATION MAY BE OBTAINED BY WRITING TO DUBIN MINDELL
                   BLOOM ASSOCIATES, 42 WEST 39TH STREET, NEW YORK, N.Y.

REFERENCE   GENERAL SERVICES ADMINISTRATION
                   FEDERAL OFFICE BUILDING - MANCHESTER, N.H., A CASE STUDY:
                   DESIGNING AN ENERGY-EFFICIENT BUILDING
                   WASHINGTON, D.C., GENERAL SERVICES ADMINISTRATION DOCUMENT
                   NO. GSA DC 76-3360, 43 P

                   CONTI, FRAN
                   SOLAR ENERGY FORECAST - HAZY
                   INDUSTRY, JUNE 1974, PP 12-13, 40

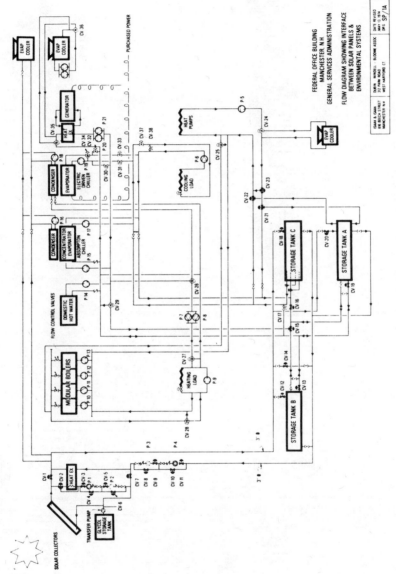

Schematic of mechanical heating and cooling plant serving top four floors of the GSA-Manchester building

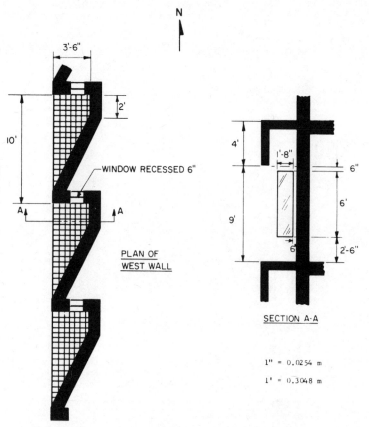

3'-6"

2'

10'

WINDOW RECESSED 6"

A   A

PLAN OF
WEST WALL

N

4'

1'-8"

6"

6'

6"

9'

6"

2'-6"

SECTION A-A

1" = 0.0254 m

1' = 0.3048 m

Schematic diagram of the west wall for one possible design

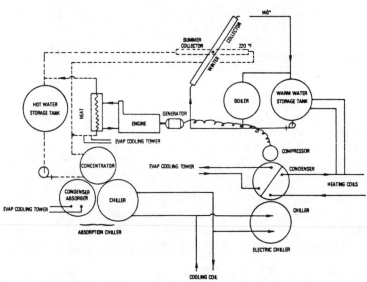

140°

SUMMER
COLLECTOR   220 °F

WINTER   COLLECTOR

HOT WATER
STORAGE TANK

HEAT

ENGINE

GENERATOR

BOILER

WARM WATER
STORAGE TANK

EVAP COOLING TOWER

CONCENTRATOR

EVAP COOLING TOWER

COMPRESSOR

CONDENSER

HEATING COILS

CONDENSER
ABSORBER

CHILLER

EVAP COOLING TOWER

CHILLER

ABSORPTION CHILLER

ELECTRIC CHILLER

COOLING COIL

Schematic representation of the way solar collectors will be utilized in the GSA-
Manchester Building

TITLE	FEDERAL BUILDING, SAGINAW, MICHIGAN
CODE	
KEYWORDS	FLAT PLATE SOLAR COLLECTOR OFFICE BUILDING HEAT PUMP UNDERGROUND
AUTHOR	SMITH, HINCHMAN & GRYLLS, ASSOCIATES, INC.
DATE & STATUS	THE FEDERAL BUILDING WAS PLANNED IN 1973.
LOCATION	SAGINAW, MICHIGAN
SCOPE	THE FEDERAL OFFICE BUILDING IS A ONE-STORY REINFORCED CONCRETE BUILDING, PARTLY RECESSED INTO THE GROUND TO ALLOW AN EXTENSION OF THE LANDSCAPED SITE ONTO THE ROOF. THREE LEVELS OF OFFICE SPACE ARE PROVIDED, AS WELL AS A LOADING DOCK LEVEL. THE MOST PROMINENT OF THE ENERGY CONSERVATION FEATURES IS AN 8,000 SQUARE FOOT FLAT PLATE SOLAR COLLECTOR WHICH SLANTS UPWARD FROM THE BUILDING AT THE OPTIMUM ANGLE TO THE SUN. ABOUT HALF OF THE ROOF IS LANDSCAPED. THE OTHER HALF IS PARKING AREA WHICH CAN BE USED AS A NEIGHBORHOOD PLAYGROUND AFTER BUSINESS HOURS. USE OF EARTH BERMS ALONG WITH ROOF LANDSCAPING WILL MAKE THE SITE A MAJOR PARK AND RECREATIONAL AREA.
TECHNIQUE	THE FEDERAL BUILDING USES AN 8,000 SQUARE FOOT FLAT PLATE SOLAR COLLECTOR, AND CLOSED LOOP HEAT PUMPS TO OBTAIN HEAT.
MATERIAL	SOLAR COLLECTOR, CLOSED LOOP HEAT PUMPS, SELF-CONTAINED MINERAL OIL TREATMENT SYSTEM FOR WASTES, EARTH BERMS
AVAILABILITY	FURTHER INFORMATION MAY BE OBTAINED BY WRITING TO SMITH, HINCHMAN & GRYLLS, ASSOCIATES, INC., 455 W. FORT ST., DETROIT, MICHIGAN 48226, OR THE GENERAL SERVICES ADMINISTRATION, REGION 5, 230 S. DEARBORN ST., CHICAGO, ILLINOIS 60604, ATTENTION: MR.RICHARD MCGINNIS, DIRECTOR, CONSTRUCTION MANAGEMENT DIVISION.
REFERENCE	PROGRESSIVE ARCHITECTURE OWENS-CORNING WINNERS ANNOUNCED PROGRESSIVE ARCHITECTURE, JANUARY 1975, P 32
	ARCHITECTURAL RECORD OWENS-CORNING FIBERGLAS CORPORATION ANNOUNCES ENERGY CONSERVATION WINNERS ARCHITECTURAL RECORD, JANUARY 1975, P 34
	CONTI, FRAN SOLAR ENERGY FORECAST - HAZY INDUSTRY, JUNE 1974, PP 12-13, 40

394

TITLE	FERMILAB HOUSE
CODE	
KEYWORDS	HOUSE INSULATED WINDOW AREA WATER TYPE COLLECTOR STORAGE
AUTHOR	J. O'MEARA
DATE & STATUS	THE FERMILAB HOUSE WAS COMPLETED IN JANUARY 1976.
LOCATION	BATAVIA, ILLINOIS
SCOPE	THE BUILDING IS AN EXISTING ONE-STORY, 3-ROOM, WOOD-FRAME HOUSE, 34' X 28'. THERE IS A SMALL ATTIC, AND A CRAWL SPACE UNDERNEATH.
TECHNIQUE	THE BUILDING IS WELL INSULATED. THE WINDOW AREA IS SMALL, AND THE WINDOWS ARE DOUBLE GLAZED.

THE COLLECTOR AREA IS 540 SQ. FT. GROSS, 480 SQ. FT. NET. THE COLLECTOR IS A WATER-TYPE COLLECTOR, MOUNTED ON A SEPARATE, 72' X 8' STAND, 20' SOUTHWEST OF THE HOUSE. THE TILT IS 53 DEGREES FROM THE HORIZONTAL. THE COLLECTOR CONSISTS OF ONE ROW OF 24 PANELS, EACH OF WHICH EMPLOYS AN 8' X 2-1/2' ALUMINUM ROLL-BOND SHEET WITH INTEGRAL PASSAGES FOR THE FLOW OF COOLANT. THE SURFACE IS NON-SELECTIVE BLACK, NEXTEL. THE COOLANT IS WATER. THERE IS NO ANTIFREEZE. THE LIQUID IS DRAINED BEFORE FREEZE-UP CAN OCCUR. THE CORROSION INHIBITOR IS ONE GALLON OF HOH CHEMICAL CO. #CS-42 PER 400 GALLONS OF WATER. A 1 HP CENTRIFUGAL PUMP PRODUCES, TYPICALLY, A FLOWRATE OF 24 GPM. GLAZING CONSISTS OF ONE TIGHTLY PACKED ARRAY OF 8' LONG, 1-1/2" OD TRANSPARENT GLASS TUBES, WHICH CONTAIN AIR AT ATMOSPHERIC PRESSURE. THERE IS NO CEMENT BETWEEN THE TUBES; A SLIGHT INLEAK OF RAIN WATER IS POSSIBLE. THE PANEL BACKING IS 4" OF FIBERGLAS. THE WATER FLOWS TO THE STORAGE SYSTEM VIA ABOVE-GROUND, 1-1/2" DIAMETER INSULATED PIPES.

THE STORAGE SYSTEM CONTAINS 1,300 GALLONS OF WATER IN A COLLAPSIBLE PLASTIC TANK, 12' X 8' X 2', SITUATED IN THE CRAWL SPACE BENEATH THE HOUSE. THE TANK IS INSULATED WITH 4" OF STYROFOAM. THE ROOMS ARE HEATED BY A FORCED HOT AIR SYSTEM.

MATERIAL	DOUBLE GLAZED WINDOWS; 8' X 2-1/2' ALUMINUM ROLL-BOND SHEET; WATER COOLANT; NEXTEL, NON-SELECTIVE BLACK; HOH CHEMICAL CO. #CS-42; 1 HP CENTRIFUGAL PUMP; 8' LONG, 1-1/2" OD TRANSPARENT GLASS TUBES; FIBERGLAS; 1-1/2" DIAMETER INSULATED PIPES; COLLAPSIBLE PLASTIC TANK; 4" OF STYROFOAM
AVAILABILITY	ADDITIONAL INFORMATION MAY BE OBTAINED FROM THE J. O'MEARA, PRESIDENT, CL5E, THE SOLAR ENERGY CLUB, FERMILAB NATIONAL ACCEL. LAB., BATAVIA, ILLINOIS 60510.
REFERENCE	O'MEARA, J.E. DESIGN AND CONSTRUCTION OF A RESIDENTIAL SOLAR HEATING SYSTEM AT FERMILAB BATAVIA, ILLINOIS, FERMI NATIONAL ACCELERATOR LABORATORY, 1976, 10 P

Collector array, 24 panels.

Testing of Storage Tank

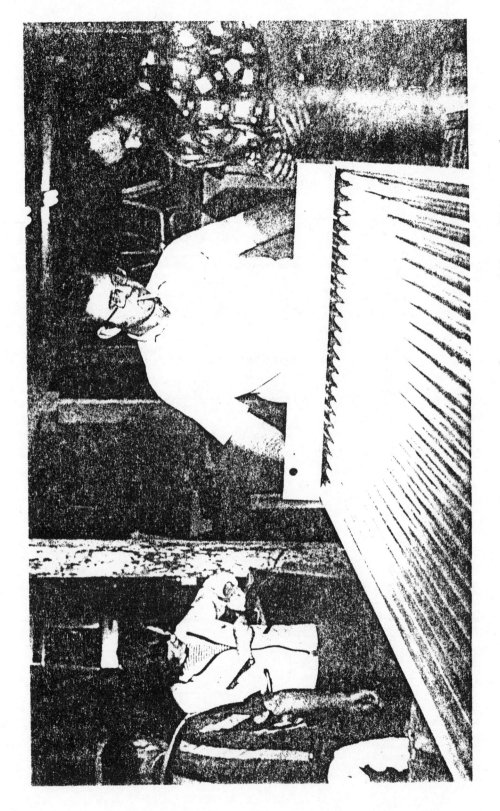

Fermilab Solar Energy Club members assembling collector panel

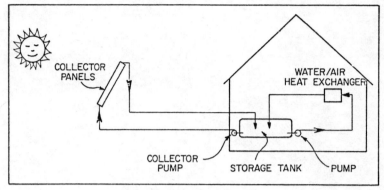

...Solar House heating schematic...

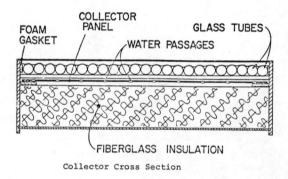

Collector Cross Section

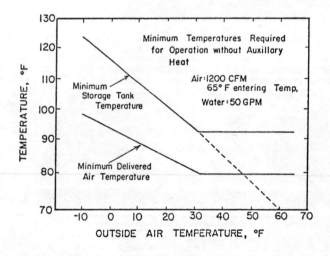

398

TITLE	FIRE STATION
CODE	
KEYWORDS	FIRE STATION SOLAR COLLECTION COMPOUNDED PARABOLIC COLLECTORS ABSORPTION GLAZING INSULATION OVERHANGS REFLECTED RERADIATED CLERESTORY LIGHTING
AUTHOR	BILL HIDELL
DATE & STATUS	THE FIRE STATION WILL BE UP FOR BIDS JULY, 1976.
LOCATION	DALLAS, TEXAS
SCOPE	THE DALLAS FIRE STATION #58 WAS DESIGNED TO COMPLY WITH THE ENERGY CONSERVATION GUIDELINES FOR THE CITY OF DALLAS, AND TO FULFILL THE UNIQUE CHARACTERISTICS OF A FIRE STATION AS WELL AS PROVIDE A LIVEABLE 24-HOUR OCCUPANCY AREA FOR 15 FIREFIGHTERS.
TECHNIQUE	A SOLAR COLLECTION SYSTEM IS THE FOCAL POINT OF ENERGY CONSERVING FEATURES OF THE FIRE STATION. THE SYSTEM WILL CONSIST OF 1,203 SQUARE FEET OF COMPOUNDED PARABOLIC SOLAR COLLECTORS MOUNTED ON A SUPPORT STRUCTURE ABOVE THE STATION'S APPARATUS ROOM. THE SUPPORT STRUCTURE WILL FACE 31 DEGREES WEST OF SOUTH, AT AN ANGLE OF 17 DEGREES HORIZONTAL. THERE WILL BE 65 COLLECTOR PANELS, EACH 36" X 78", IN FIXED MOUNTINGS.

THE SOLAR COLLECTORS WILL FURNISH HOT WATER TO SPACE HEATING COILS IN THE SUPPLY AIR DUCTS SERVING ALL AREAS OF THE PROJECT EXCEPT THE APPARATUS ROOM. THIS ROOM WILL BE HEATED WITH RADIANT HEATING COILS EMBEDDED IN THE FLOOR SLAB. THESE COILS WILL ALSO RECEIVE HOT WATER FROM THE SOLAR COLLECTORS WITH SUPPLEMENTAL HEAT SUPPLIED WHEN NECESSARY BY A GAS FIRED HOT WATER HEATER TO BOTH THE APPARATUS ROOM AND THE REMAINING AREAS OF THE BUILDING.

THE SOLAR COLLECTORS WILL ALSO FURNISH HOT WATER TO A 3-TON WATER-FIRED ABSORPTION AIR CONDITIONER. THIS UNIT WILL PARALLEL TWO OTHER 3-TON ELECTRIC DRIVEN AIR CONDITIONING UNITS TO FURNISH SPACE COOLING TO THE DAYTIME LIVING AREAS.

THE HEAT STORAGE FOR THE SYSTEM WILL BE PROVIDED BY AN INSULATED 2,000 GALLON WATER STORAGE TANK BURIED IN THE GROUND NEAR THE BUILDING. THIS STORAGE WILL BE SUFFICIENT TO OPERATE THE BUILDING DOMESTIC HOT WATER SYSTEM AND THE SPACE HEATING SYSTEM TWO JANUARY DAYS WITHOUT SUN.

THE GOAL IS TO PROVIDE THE OCCUPANTS WITH 75% OF THE DOMESTIC HOT WATER, 40% OF THE SPACE HEATING (AVERAGE FOR WINTER MONTHS OF OCTOBER - FEBRUARY), AND 50% OF THE COOLING IN THE LIVING QUARTERS.

THE LIVING AREAS AND DORMITORY AREAS WILL BE ORIENTED TO TAKE ADVANTAGE OF PREVAILING SOUTHERLY BREEZES. OPERABLE WINDOWS, SMALLER ON THE INCOMING AIR SIDE THAN ON THE OUTLET SIDE, WILL GIVE THE GREATEST POSSIBLE AIR FLOW.

GLAZING IN THE AIR CONDITIONED AREAS WILL BE DOUBLE PANE INSU-LATION GLASS; ALL GLASS AREAS WILL BE ORIENTED AWAY FROM DIRECT EASTERN OR WESTERN SUN. OTHER GLASS AREAS WILL BE ORIENTED TO THE SOUTH AND SHADED BY OVERHANGS TO BLOCK SUMMER SUN WHILE ADMITTING WINTER SUN.

PLANTING AREAS NEXT TO THE BUILDING WILL PREVENT UNDUE HEAT GAIN FROM REFLECTED TO RE-RADIATED HEAT. OTHER LANDSCAPING IN-CLUDES DECIDUOUS TREES ON THE SOUTH SIDE OF THE STRUCTURE AS A NATURAL SUN SHADE.

CLERESTORY LIGHTING IN THE APPARATUS ROOM, THE BUILDING'S LARGEST AREA, WILL PROVIDE SUFFICIENT NATURAL LIGHT FOR DAYTIME OPERATIONS.

CONTROLS OR WATER FLOW RESTRICTORS IN THE SHOWER AND LAVATORIES WILL REDUCE WATER WASTE, WHICH IN TURN WILL HELP REDUCE FUEL COSTS. MINIMUM AMOUNTS OF OUTSIDE AIR FOR AIR CONDITIONING AND HEATING WILL BE USED TO AVOID EXCESS LOADS ON THE MECHANICAL SYSTEM.

MATERIAL   COMPOUNDED PARABOLIC SOLAR COLLECTORS, HOT WATER, SPACE HEATING COILS, SUPPLY AIR DUCTS, RADIANT HEATING COILS, GAS FIRED HOT WATER HEATER, ABSORPTION AIR CONDITIONER, ELECTRIC DRIVEN AIR CONDITIONING UNITS, OPERABLE WINDOWS, DOUBLE PANE INSULATION GLASS, OVERHANGS, PLANTING AREAS, DECIDUOUS TREES, STORAGE TANK

AVAILABILITY   ADDITIONAL INFORMATION MAY BE OBTAINED FROM BILL HIDELL, AIA, DALLAS, TEXAS.

REFERENCE   AIA ENERGY NOTEBOOK
                  ENERGY AIA ENERGY NOTEBOOK AN INFORMATION SERVICE ON ENERGY AND THE BUILT ENVIRONMENT
                  AMERICAN INSTITUTE OF ARCHITECTS, WASHINGTON, D.C., 1975

TITLE	FIRST CONNECTICUT SOLAR HOUSE
CODE	
KEYWORDS	SOLAR COLLECTOR HOUSE STORAGE AUXILIARY HEATING AIR CONDITIONING NATURAL VENTILATION
AUTHOR	DONALD WATSON, EVERETT M. BARBER, JR., AND SUNWORKS INC., GUILFORD, CONN.
DATE & STATUS	THE HOUSE WAS COMPLETED FOR HEATING IN 1973-74.
LOCATION	LONG ISLAND SOUND, CONNECTICUT
SCOPE	THE CONNECTICUT HOUSE IS THE FIRST SOLAR HEATED HOUSE IN THE STATE. THE SOLAR ENERGY SYSTEM IN THE THREE BEDROOM, $60,000 HOUSE WILL SUPPLY ABOUT TWO-THIRDS OF THE HEAT FOR THE HOUSE. LOCAL ZONING REQUIREMENTS ON HEIGHT LIMITED THE AREAS OF THE PANELS TO 20% OF THE HOME'S 19,000 SQ. FT. LIVING AREA. THE SYSTEM COST AN ESTIMATED $3,000 MORE THAN A CONVENTIONAL HEATING SYSTEM, AND REPRESENTS ABOUT 10% OF THE OVERALL COST OF THE HOUSE. IT WILL PAY FOR ITSELF IN 15 YEARS. THERE RE NO MOVING PARTS TO THE SYSTEM. THE AVERAGE TEMPERATURE OF THE WATER IN THE STORAGE TANK WILL BE 120 F - 140 F IN THE WINTER. THE TANK IS CAPABLE OF STORING HEAT FOR 4-5 DAYS OF CLOUDY WEATHER.
TECHNIQUE	3 LONG SLANTED ROWS OF GLASS COVERED PANELS ARE MOUNTED ON THE ROOF, FACING SOUTH. THE NORTH SIDES OF THE RAISED ROOF SHEDS FORMED BY THE COLLECTORS ARE USED FOR CLERESTORY WINDOWS IN ORDER TO ENCOURAGE NATURAL VENTILATION AND SUMMER COOLING. THE PANELS CONTAIN SUNWORKS COLLECTORS WHICH ABSORB SUNLIGHT AND HEAT AND TRANSFER IT TO SELECTIVE SURFACE COPPER PIPES CARRYING WATER FOR USE IN THE HOME'S SPACE-HEATING AND WATER HEATING SYSTEMS. TO TRAP AS MUCH HEAT AS POSSIBLE, THE COLLECTOR IS LINED WITH 3" FIBERGLAS INSULATION. CIRCULATION OF WATER THROUGH PIPING TRAVELS TO A 2,000 GALLON STORAGE TANK IN THE BASEMENT. A HOT WATER COIL IN A HOT AIR DUCT THAT BLOWS HEAT INTO THE HOUSE. AUXILIARY HEATING FOR THE HOUSE IS PROVIDED BY AN OVERSIZED OIL-FIRED WATER HEATER, CONNECTED TO THE HOUSE AIR DUCT SYSTEM.
MATERIAL	SHEET GLASS, FIBERGLAS INSULATION, NON TOXIC PROPOLENE GLYCOL (ANTIFREEZE), WATER, COPPER PIPES, 2,000 GALLON WATER TANK, HOT WATER COIL, HOT AIR DUCT BLOWER, BLACKENED COPPER SOLAR COLLECTOR PANELS
AVAILABILITY	ADDITIONAL INFORMATION MAY BE OBTAINED FROM DONALD WATSON, AIA, BOX 401, GUILFORD, CONN. 06437.
REFERENCE	SUNWORKS INC. ENERGY CONSERVATION IN ARCHITECTURE GUILFORD, CONNECTICUT, SUNWORKS INC., REPRINT #04, 1974 PULLEN, JOHN J. ENERGY FROM THE SUN COUNTRY JOURNAL, JUNE 1975, PP 32-37 CLARK, WILSON SOLAR HOUSE PRIMER, A NEW YORK TIMES MAGAZINE, APRIL 7, 1974, PP 92-94

403

TITLE	FIRST INTERNATIONAL BUILDING
CODE	
KEYWORDS	INSULATING REFLECTIVE GLASS SOLAR RECLAIM ALL AIR MECHANICAL SYSTEM OFFICE BUILDING
AUTHOR	HARWOOD K. SMITH & PARTNERS, ARCHITECTS
DATE & STATUS	THE FIRST INTERNATIONAL BUILDING HAS BEEN OCCUPIED SINCE THE SUMMER OF 1974.
LOCATION	DALLAS, TEXAS
SCOPE	THE FIRST INTERNATIONAL BUILDING IS A BRILLIANT EXAMPLE OF THE EFFICIENCY OF GLASS BUILDINGS. IN SPITE OF THE FACT THAT ITS SKIN IS NEARLY ALL GLASS, IT IS ONE OF THE MOST ENERGY EFFICIENT BUILDINGS IN DALLAS.

IN PLANNING THIS BUILDING THE DESIGN TEAM SAW THAT ABOUT 50% OF THE ENERGY WOULD GO TO LIGHT IT, ANOTHER 14% TO RUN THE FANS, ELEVATORS AND VARIOUS OFFICE MACHINES, ABOUT 7% TO HEAT IT, AND BECAUSE IT IS IN DALLAS, 29% TO COOL IT.

SOLARBAN 480 INSULATING REFLECTIVE GLASS WAS SELECTED TO PROVIDE AN INNOVATIVE, ENERGY-CONSERVING AIR-CONDITIONING SYSTEM. THE MANUFACTURER CLAIMS THAT SOLARBAN INSULATING REFLECTIVE GLASS CAN BE 50-75% MORE EFFECTIVE THAN REGULAR INSULATING GLASS. SOLARBAN INSULATING REFLECTIVE GLASS, CALLED HIGH-PERFORMANCE GLASS, CAN REDUCE HEAT GAIN BY UP TO 80% COMPARED TO CLEAR SINGLE GLASS.

THE INNOVATIVE, ALL-AIR MECHANICAL SYSTEM SAVES BOTH ENERGY AND MONEY. IT RECLAIMS HEAT FROM LIGHTING AND LARGE INTERIOR SPACES; AND REDISTRIBUTES IT FOR PERIMETER HEATING WHEN NEEDED.

IN GENERAL, INSTALLATION COSTS FOR REFLECTIVE INSULATING GLASS ARE ALMOST THREE TIMES GREATER THAN FOR CLEAR OR TINTED SINGLE GLASS, BUT THIS IS A ONE TIME COST AND MUST BE CONSIDERED IN TERMS OF A BUILDING'S LIFE CYCLE OPERATING COSTS AND THE FUEL SAVINGS THAT CAN BE REALIZED THROUGH THE USE OF HIGH-PERFORMANCE GLASS.

A BUILDING'S OUTDOOR WALL AREA, THAT IS 70% SOLARBAN INSULATING GLASS, IS MORE ENERGY EFFICIENT THAN THE SAME BUILDING USING CRAMPED LITTLE CLEAR GLASS WINDOWS TOTALING ONLY 20% VISION AREA.

TOTAL INSTALLATION COSTS FOR REFLECTIVE INSULATING GLASS UNITS MAY BE PERHAPS THREE TIMES GREATER THAN FOR SINGLE GLAZED CLEAR OR TINTED GLASS. HOWEVER, THIS ONE TIME COST WHEN CONSIDERED IN TERMS OF BUILDING LIFE CYCLE OPERATING COSTS IS EASILY RECOVERED BY THE FUEL SAVINGS REALIZED WITH HIGH-PERFORMANCE GLASS.

TECHNIQUE	INSULATING REFLECTIVE GLASS IS USED TO REDUCE HEAT GAIN OR LOSS THROUGH THE SKIN OF THE BUILDING. AN ALL-AIR MECHANICAL SYSTEM IS USED TO RECLAIM HEAT FROM LIGHTING.
MATERIAL	SOLARBAN INSULATING REFLECTIVE GLASS, ALL-AIR MECHANICAL SYSTEM
AVAILABILITY	FURTHER INFORMATION MAY BE OBTAINED FROM DONALD G. HEGNES, PPG INDUSTRIES INC., ONE GATEWAY CENTER, PITTSBURGH, PENNSYLVANIA 15222.
REFERENCE	ENGINEERING NEWS-RECORD INSULATING REFLECTIVE GLASS CALLED AN ENERGY SAVER ENGINEERING NEWS-RECORD, FEBRUARY 7, 1974, P 14

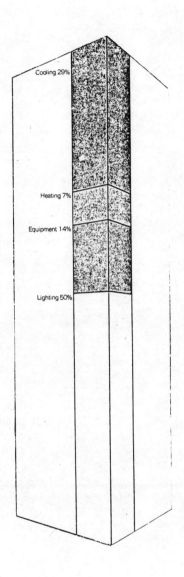

Cooling 29%

Heating 7%

Equipment 14%

Lighting 50%

TITLE	FLIP TOP MOBILE HOME
CODE	
KEYWORDS	FLIP TOP MOBILE HOME SOLAR COLLECTOR CELL BATTERY ELECTRICITY STORAGE HEAT EXCHANGER TANK TRENCH PANEL WATER
AUTHOR	THOMAS J. MILLERBAUGH; W. MARION BAMMAN
DATE & STATUS	THIS MOBILE HOME WAS DESIGNED IN 1976.
LOCATION	MOBILE, CAN BE LOCATED ANYWHERE.
SCOPE	WHEN TRAVELING, THE FLIP TOP MOBILE HOME IS 12' WIDE AND 58' LONG. IT IS READY FOR UNFOLDING AND ASSEMBLY AT THE ARRIVAL SITE.
TECHNIQUE	SPECIAL GLASS IS USED FOR EXTRA ILLUMINATION AND INSULATION. THE ENTIRE ROOF OF THIS MOPILE HOME IS A SOLAR COLLECTOR. DURING THE DAY, WATER FROM THE STORAGE TANK IS CIRCULATED THROUGH THE PANELS AND STORED OVER NIGHT, OR CIRCULATED THROUGH THE HOME VIA HEAT EXCHANGERS TO DUCT CHANNELS FOR SPACE HEATING. ROOFTOP SOLAR CELLS USE THE SUN'S ENERGY TO CREATE ELECTRICITY, WHICH IS STORED IN BATTERIES. TO PROVIDE A STABLE FOUNDATION FOR THE CHASIS OF THE HOME, A BASEMENT TRENCH HIDES UNSIGHTLY PORTIONS, AND AT THE SAME TIME GIVES A CAPACIOUS AREA FOR WATER STORAGE. IN ADDITION, THIS MOBILE HOME INCORPORATES A UNIQUE SOUND SYSTEM, WHICH IS SWITCHED ON WHEN ONE PASSES NEAR PROXIMITY SWITCHES.
MATERIAL	SOLAR PANELS, SOLAR CELLS, BATTERIES, HEAT EXCHANGERS, DUCT CHANNELS, STORAGE TRENCH, SPECIAL GLASS
AVAILABILITY	ADDITIONAL INFORMATION MAY BE OBTAINED FROM FRED RICE PRODUCTIONS, INC., 48780 EISENHOWER DRIVE, LA QUINTA, CALIFORNIA.
REFERENCE	RICE, FREDERICK H. NEW CONCEPT IN RECREATIONAL LIVING: THE SOLAR/SONIC FLIP TOP MODULAR MOBILE HOME, A PALM SPRINGS, CALIFORNIA, FREDERICK H. RICE PAMPHLET, 1976, 10 P

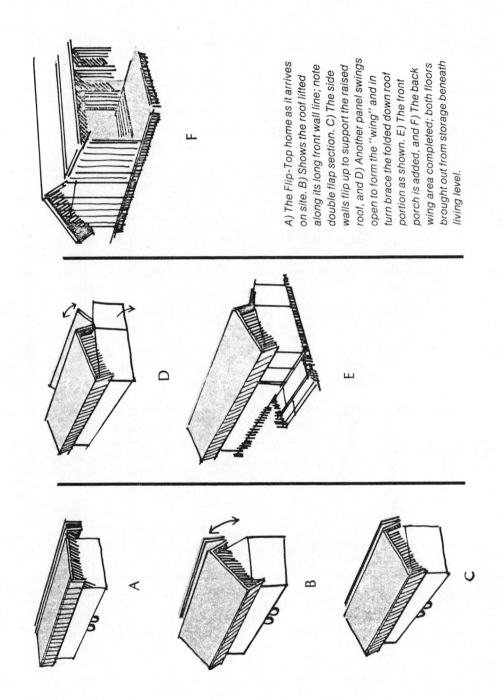

A) The Flip-Top home as it arrives on site. B) Shows the roof lifted along its long front wall line; note double flap section. C) The side walls flip up to support the raised roof, and D) Another panel swings open to form the "wing" and in turn brace the folded down roof portion as shown. E) The front porch is added, and F) The back wing area completed; both floors brought out from storage beneath living level.

F

D

E

A

B

C

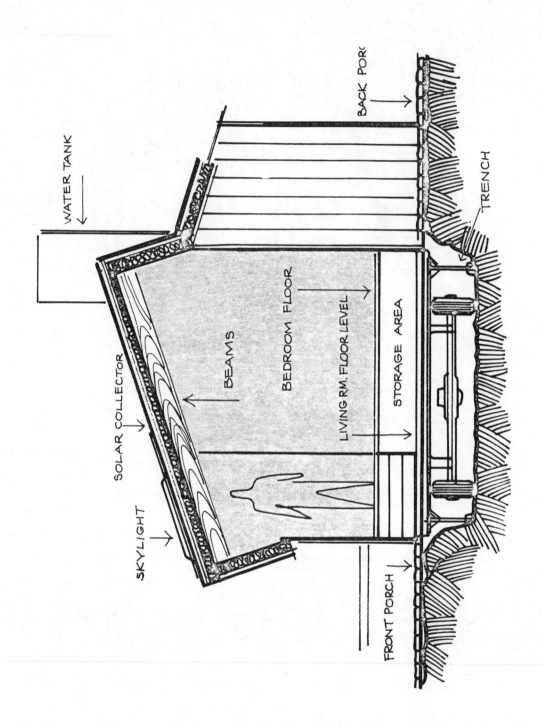

WATER TANK

DRAIN

BACK PORCH

TRENCH

SOLAR COLLECTOR

BEAMS

BEDROOM FLOOR

LIVING RM. FLOOR LEVEL

STORAGE AREA

SKYLIGHT

FRONT PORCH

407c

TITLE	FLORIDA ENERGY SAVING OFFICE
CODE	
KEYWORDS	COMPUTER PROGRAMS
AUTHOR	LEWIS & BURKE ASSOCIATES, INC.
DATE & STATUS	CONSTRUCTION OF THE FLORIDA ENERGY SAVING OFFICE BEGAN AS OF APRIL, 1976.
LOCATION	ORLANDO, FLORIDA
SCOPE	THE FLORIDA ENERGY SAVING OFFICE IS THE REGIONAL SERVICE CENTER, CONSOLIDATING PUBLIC SERVICE AGENCIES FOR SEVEN FLORIDA COUNTIES. THE ENERGY CONSUMPTION OF THE BUILDING IS ESTIMATED TO BE 13 KWH PER SQ. FT. PER YEAR, OR 21% UNDER THE LIMIT OF 16 KWH PER SQ. FT. PER YEAR SET BY THE STATE UNDER THE LAW. THE FLORIDA BUILDING'S PROJECTED CONSUMPTION OF 13 KWH PER SQ. FT. PER YEAR IS EQUIVALENT TO 44,000 BTU PER SQ. FT. PER YEAR. GSA SETS A LIMIT OF 55,000 BTU PER SQ. FT. PER YEAR ON ITS BUILDINGS.
TECHNIQUE	THE DESIGNERS USED TWO COMPUTER PROGRAMS TO OPTIMIZE DESIGN AND OPERATION FOR ENERGY EFFICIENCY. ONE WAS THE "E CUBE" PROGRAM DEVELOPED BY THE AMERICAN GAS ASSOCIATION, AND THE OTHER, REQUIRED BY THE STATE, IS CALLED FLEET FOR FLORIDA LIFECYCLE ENERGY EVALUATION TECHNIQUE.
MATERIAL	COMPUTER PROGRAMS
AVAILABILITY	ADDITIONAL INFORMATION MAY BE OBTAINED FROM LEWIS & BURKE ASSOCIATES, INC., 5750 MAJOR BOULEVARD, ORLANDO, FLORIDA 32805.
REFERENCE	ENGINEERING NEWS RECORD ALL BIDS LOW FOR ENERGY-SAVING OFFICE ENGINEERING NEWS RECORD, MARCH 25, 1976, P 11

408½

TITLE	FLORIDA SOLAR HOUSE
CODE	
KEYWORDS	SOLAR COLLECTOR AIR CONDITIONING INSULATED TANK HEAT PUMP FAN ABSORPTION EVAPORATING INSULATION HOUSE
AUTHOR	H.A. INGLEY
DATE & STATUS	THE FLORIDA SOLAR HOUSE HAS BEEN HEATED BY SOLAR HEAT SINCE 1958. IT HAS BEEN COOLED BY THE SUN SINCE 1975.
LOCATION	GAINESVILLE, FLORIDA
SCOPE	THE SOLAR ENERGY SYSTEM AT THE UNIVERSITY OF FLORIDA COSTS ABOUT $10,000. THE COST OF THE AIR CONDITIONING SYSTEM IS ESTIMATED AT ABOUT $4,000, SOMEWHAT MORE THAN THE COST OF A CONVENTIONAL SYSTEM. BUT THE SOLAR AIR CONDITIONING SYSTEM USES THE SAME SOLAR COLLECTORS AND STORAGE TANK USED FOR HEATING AND HOT WATER. ALSO IT DOESN'T INCLUDE THE COST OF AIR DUCTS ALREADY IN THE HOUSE FOR THE CONVENTIONAL AIR CON- DITIONING SYSTEM. IT WOULD PROBABLY COST ABOUT $10,000 TO INSTALL A COMPARABLE COOLING SYSTEM WITHOUT ANY PRE-INSTALLED PARTS.
TECHNIQUE	TWELVE SOLAR COLLECTORS ON THE ROOF, EACH 48 SQUARE FEET, GATHER ALL THE ENERGY NEEDED TO PROVIDE HOT WATER, HEAT AND AIR CONDITIONING FOR THE THREE-BEDROOM TEST HOUSE. THE HOT WATER FROM THE CONVENTIONAL COLLECTORS IS STORED IN A 3,000 GALLON TANK COVERED WITH 4 INCHES OF FOAM, GLASS AND INSULATION. THE TANK HOLDS ENOUGH WATER TO POWER ALL SYSTEMS FOR A FULL THREE DAYS WITHOUT BEING RECHARGED.

THE SOLAR AIR CONDITIONING PACKAGE CONSISTS OF TWO INSULATED TANKS, TWO SMALL ELECTRIC PUMPS AND A SMALL FAN FOR CIRCU- LATING COOL AIR THROUGHOUT THE HOUSE. THIS EQUIPMENT HAS BEEN ADDED TO A CONVENTIONALLY SOLAR HEATED HOME.

THE FIRST TANK, CALLED A GENERATOR, CONTAINS A SOLUTION OF 60% AMMONIA AND 40% WATER. THE SOLUTION AT THE BEGINNING OF THE COOLING CYCLE IS UNDER PRESSURE AT 160 POUNDS PER SQUARE INCH. AT THE BOTTOM OF THIS TANK IS A LOT OF TUBING CONNECTED TO THE CENTRAL HOT WATER TANK, WHICH POWERS ALL SYSTEMS IN THE HOUSE. THE PIPES ARE ALSO JOINED TO AN OUTSIDE COOL-WATER SOURCE.

THE SECOND TANK, FOR HEAT ABSORPTION, CONTAINS SEPARATE SETS OF PIPES AT BOTH TOP AND BOTTOM. WATER IN THE TOP PIPES IS AT 80 DEGREES AND DRAWN FROM THE OUTSIDE COOLING POND. THE PIPES AT THE BOTTOM HOLD WATER THAT ALSO COMES FROM THE COOLING POND, THEN FLOWS FROM THE HEAT ABSORPTION TANK INTO THE HOUSE, AND LATER BACK TO THE COOLING POND.

AT THE BEGINNING OF THE SYSTEM'S CYCLE, THE AMMONIA-WATER IN THE GENERATOR IS HEATED TO 140 DEGREES WITH HOT WATER FROM THE CENTRAL ENERGY TANK. THIS CAUSES THE SOLUTION TO BOIL, EVAPORATING THE AMMONIA WHICH IS PIPED INTO THE HEAT ABSORPTION TANK.

THE COOLING PROCESS BEGINS HERE. WATER CIRCULATING THROUGH THE TOP PIPES AT 80 DEGREES COOLS AND CONDENSES THE AMMONIA VAPOR (WHICH IS UNDER 180 POUNDS PER SQUARE INCH OF PRESSURE), AND IT FALLS TO THE BOTTOM OF THE TANK. THE PROCESS CONTINUES UNTIL 10% OF THE GENERATOR'S AMMONIA HAS BEEN EVAPORATED AND CONDENSED IN THE HEAT ABSORPTION TANK.

AS THE AMMONIA VAPOR CONDENSES, THE PRESSURE IN THE HEAT ABSORPTION TANK DROPS TO 75 POUNDS PER SQUARE INCH. THIS PRESSURE DROP COOLS THE AMMONIA TO 45 DEGREES, JUST UNDER ITS BOILING POINT.

WHEN THE THERMOSTAT IN THE HOUSE TURNS ON, COOL WATER IS CIRCU- LATED THROUGH THE GENERATOR TO LOWER THE TEMPERATURE OF THE AMMONIA WATER SOLUTION, WHICH IS NOW A 50-50 MIXTURE. |

AS THE SOLUTION COOLS, THE PRESSURE DROPS TO 75 POUNDS PER SQUARE INCH, EQUALIZING THE PRESSURE IN BOTH TANKS. AT THE SAME TIME, WATER CIRCULATES IN THE LOWER PIPES OF THE HEAT ABSORPTION TANK, HEATING THE PURE LIQUID AMMONIA TO 50 DEGREES, ITS BOILING POINT AT 75 LBS PSI.

WHEN AMMONIA BOILS IT ABSORBS HEAT. AS THE AMMONIA BOILS, IT COOLS THE WATER IN THE LOWER PIPES TO 50 DEGREES. THE WATER FLOWS INTO THE HOUSE, WHERE IT CIRCULATES THROUGH A WIDE, CIRCULAR COIL IN FRONT OF A FAN. THE FAN BLOWS AIR OVER THE COIL, COOLING IT TO 60 DEGREES, AND CIRCULATES IT THROUGH AIR DUCTS IN THE HOUSE.

THE SYSTEM IS NOT WITHOUT FLAWS, AT LEAST FOR THE PRESENT. AFTER RUNNING FOR THREE HOURS, THE SYSTEM IS COMPLETELY DRAINED AND HAS TO BE SHUT DOWN AND RECHARGED BY EVAPORATING THE AMMONIA FROM THE GENERATOR AND CONDENSING IT AGAIN IN THE HEAT ABSORPTION UNIT. THE RECHARGE CYCLE NOW TAKES TWENTY MINUTES, BUT DURING THAT TIME THE CONVENTIONAL AIR CONDITIONING SYSTEM IN THE HOUSE TAKES OVER, IF NEEDED.

MATERIAL   SOLAR COLLECTOR, 3,000 SQUARE FOOT TANK INSULATED WITH FOAM AND GLASS, TWO SMALL TANKS, AMMONIA-WATER SOLUTION, SMALL FAN, CIRCULAR COIL, PIPING

AVAILABILITY   ADDITIONAL INFORMATION ON THE FLORIDA SOLAR HOUSE MAY BE OBTAINED FROM THE SOLAR ENERGY AND ENERGY CONVERSION LABORATORY, DEPARTMENT OF MECHANICAL ENGINEERING, UNIVERSITY OF FLORIDA, GAINESVILLE, FLORIDA 32611.

REFERENCE   BUTCHER, LEROY
                AIR CONDITIONING WITH SUN POWER
                MECHANIX ILLUSTRATED, AUGUST 1975, PP 27-29

                ARIZONA STATE UNIVERSITY, COLLEGE OF ARCHITECTURE
                UNIVERSITY OF FLORIDA SOLAR HOUSE
                WASHINGTON, D.C., SOLAR-ORIENTED ARCHITECTURE, PUBLISHED BY
                THE AIA RESEARCH CORPORATION, JANUARY 1975, P 37

                VILLECCO, MARGUERITE
                SUN POWER
                ARCHITECTURE PLUS, SEPTEMBER/OCTOBER 1974, PP 85-99

THE University of Florida Solar House at Gainesville, with 12 roof collectors, each 48 sq. ft., supplying the cooling system.

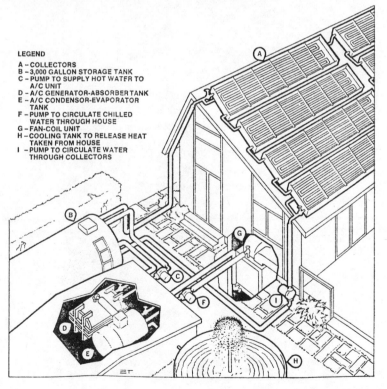

LEGEND

A – COLLECTORS
B – 3,000 GALLON STORAGE TANK
C – PUMP TO SUPPLY HOT WATER TO
     A/C UNIT
D – A/C GENERATOR-ABSORBER TANK
E – A/C CONDENSOR-EVAPORATOR
     TANK
F – PUMP TO CIRCULATE CHILLED
     WATER THROUGH HOUSE
G – FAN-COIL UNIT
H – COOLING TANK TO RELEASE HEAT
     TAKEN FROM HOUSE
I – PUMP TO CIRCULATE WATER
     THROUGH COLLECTORS

STORAGE tank, above, has 3,000-gal. capacity, is covered with 4 inches of insulation all around. Collector for solar water heater is in right, foreground.

AIR conditioning system, above, is an intermittent ammonia-absorption type producing 30,000 BTU/hr. One of the two circulating pumps is shown right, foreground.

THERE are 12 collectors, right, totaling 576 sq. ft. Their square aluminum tubing differs from round tubing of the collector for water heater which is in the left foreground.

FAN-COIL assembly is coupled to conventional heating and cooling system. By opening and closing the valves, system changes as needed.

411

TITLE	FREESE HOUSE
CODE	
KEYWORDS	HOUSE INSULATION COLLECTOR STORAGE
AUTHOR	E-M ARCHITECTS
DATE & STATUS	THE FREESE HOUSE WAS COMPLETED IN 1975.
LOCATION	CONCORD, NEW HAMPSHIRE
SCOPE	THE BUILDING IS AN EXISTING FARMHOUSE WHICH HAS BEEN REMODELED, AND A NEW WING HAS BEEN ADDED.
TECHNIQUE	THE THICKNESS OF THE FIBERGLAS INSULATION IN THE NEW WING IS 6" FOR THE WALL, AND 12" FOR THE CEILING. THE COLLECTOR AREA IS 500 SQ. FT. THE COLLECTOR CONSISTS OF KALWALL SUN-LITE AND ALSO A BEADWALL. IT IS SITUATED IN THE ATTIC OF THE NEW WING.
	THE MAIN PORTION OF THE STORAGE SYSTEM CONSISTS OF WATER-FILLED BAGS, OF VINYL PLASTIC, ON THE FLOOR OF THE ATTIC. THE SUPPLEMENTARY PORTION CONSISTS OF GRANITE FOUNDATIONS, WHICH ARE EXTERNALLY INSULATED WITH 4" OF FOAM. THIS PORTION RECEIVES HEAT VIA BLOWER-DRIVEN AIR, CIRCULATED FROM THE ATTIC SPACE TO THE CRAWL SPACE ADJACENT TO THE FOUNDATION.
MATERIAL	6" AND 12" FIBERGLAS; KALWALL SUN-LITE; BEADWALL
AVAILABILITY	ADDITIONAL INFORMATION MAY BE OBTAINED FROM JACKSON D. FREESE, 97 FISKE ROAD, CONCORD, NEW HAMPSHIRE.
REFERENCE	SHURCLIFF, W.A. SOLAR HEATED BUILDINGS, A BRIEF SURVEY CAMBRIDGE, MASSACHUSETTS, W.A. SHURCLIFF, MARCH 1976, 212 P

TITLE	FREMONT SCHOOL
CODE	
KEYWORDS	BERM RECESSED UNDERGROUND
AUTHOR	RALPH ALLEN OF ALLEN AND MILLER ARCHITECTS
DATE & STATUS	THE FREMONT SCHOOL WAS COMPLETED IN 1973.
LOCATION	SANTA ANA, CALIFORNIA
SCOPE	THIS SCHOOL IS A $1.5 MILLION STRUCTURE, UTILIZING RECESSED CONSTRUCTION AS ITS ENERGY SAVING DEVICE. THIS TECHNIQUE ALLOWS UTILIZATION OF THE ROOF FOR A PLAY AREA, BUT DEMANDS HEAVIER REINFORCING STEEL BARS IN THE CONCRETE. THIS UPPED CONSTRUCTION COSTS TO $28 PER SQ. FT. OR $2 PER SQ. FT. MORE THAN USUAL. PAYBACKS ARE EXPECTED IN THE FORM OF LOWER OPERATING COSTS, AND THE FACT THAT ADDITIONAL ACREAGE DID NOT HAVE TO BE PUR- CHASED FOR RECREATIONAL USE.
TECHNIQUE	THE MAJOR PART OF THIS STRUCTURE IS RECESSED INTO THE GROUND, ALLOWING FOR NATURAL INSULATION. EARTH BERMS ALSO SURROUND THE BUILDING, ADDING FURTHER INSULATION.
MATERIAL	EARTH BERMS, RECESSED CONSTRUCTION
AVAILABILITY	ADDITIONAL INFORMATION MAY BE OBTAINED FROM ALLEN AND MILLER ARCHITECTS, 1606 BUSH STREET, SANTA ANA, CALIFORNIA 92701.
REFERENCE	PROGRESSIVE ARCHITECTURE THREE UNDERGROUND SCHOOLS PROGRESSIVE ARCHITECTURE, OCTOBER 1975, P 30

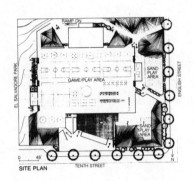

414

TITLE	GANANOQUE HOUSE
CODE	
KEYWORDS	HOUSE GREENHOUSE INSULATING SHUTTERS FIREPLACE EARTH BERMS SODS ROOF WATER TYPE COLLECTOR STORAGE FIREPLACE
AUTHOR	G. ALLEN
DATE & STATUS	THE GANANOQUE HOUSE WAS COMPLETED IN THE FALL OF 1975.
LOCATION	GANANOQUE, ONTARIO, CANADA

SCOPE　　　THE BUILDING IS A 1-1/2-STORY, HEXAGONAL HOUSE, WITH A FLOOR AREA OF 2,200 SQ. FT., SITUATED ON A SOUTH-FACING LIMESTONE CLIFF. THE HALF-STORY IS THE GROUND FLOOR, WHICH IS 60' EASTWEST AND 16' NORTHSOUTH. IT CONTAINS A LIVING ROOM AND A MASSIVE FIREPLACE AT THE WEST END. A SMALL GREENHOUSE AREA AND, AT THE EAST END, THERE IS A GUEST ROOM, BATHROOM AND SAUNA. THE FULL STORY IS THE UPPER STORY, 60' EASTWEST AND 30' NORTHSOUTH. THE ENTRANCE FOYER IS AT THE CENTER OF THE NORTH SIDE. THE KITCHEN AND DINING AREA ARE AT THE WEST END. AT THE CENTER, THERE IS AN ORNAMENTAL 8' DIAMETER POOL WITH FLOWING WATER, AND A STAIRCASE. AT THE EAST END, THERE ARE THREE BEDROOMS. THE EAST AND WEST WALLS, INCLUDING VERTICAL CEDAR LOGS, MORTAR AND TAR, ARE WELL INSULATED.

TECHNIQUE　　THE SOUTH FACE IS ALL-GLAZED (SINGLE GLAZED) WITH A 15 DEGREE LEAN-BACK FROM THE VERTICAL. THE INTERNAL INSULATING SHUTTERS ARE MADE OF WOOD, AND COVER THE WINDOWS AT NIGHT. THE NORTH SIDE CONSISTS OF THE FACE OF THE CLIFF, WITH A FIREPLACE PARTIALLY RECESSED THEREIN, AND A MASSIVE CHIMNEY RESTING THEREON. EARTH BERMS COVER PORTIONS OF THE EAST, NORTH AND WEST WALLS. SODS COVER THE LOW-PITCH PORTIONS OF THE EAST, NORTH AND WEST ROOF AREAS. THE ROOF INSULATION IS 6".

THE COLLECTOR IS A 240 SQ. FT., 32' X 7-1/2', WATER-TYPE COLLECTOR, MOUNTED ALONG THE CENTRAL EASTWEST RIDGE. IT SLOPES 75 DEGREES FROM THE HORIZONTAL. IT INCLUDES EIGHT PANELS, EACH 7-1/2' X 4', AND EACH EMPLOYING A STEEL SHEET WITH A BLACK CHROME COATING. DOUBLE GLAZING IS USED. THE INNER SHEET IS OF 1/8" GLASS, AND THE OUTER SHEET IS OF KALWALL PREMIUM SUN-LITE (FIBERGLAS AND POLYESTER). THERE IS A 1/2" AIR-SPACE BETWEEN THE TWO LAYERS. THE FLUID IS WATER AND ETHYLENE GLYCOL. THE TRANSPORT PIPES ARE OF 1-1/4" COPPER. ADJACENT TO THE BASE OF THE COLLECTOR IS A NEAR-HORIZONTAL ROOF AREA THAT IS REFLECTIVE (BAKED WHITE ENAMEL COATING ON AN ALUMINUM SHEET) AND DIRECTS ADDITIONAL RADIATION TO THE COLLECTOR PROPER. THE LARGE SOUTH WINDOWS OF THE UPPER AND LOWER STORIES ALSO COLLECT MUCH RADIATION.

THE STORAGE SYSTEM CONSISTS OF 2 TONS, 4,000 LBS, OF WAX THAT HAS A M.P. OF 120 DEGREES F. THE WAX IS INSIDE 1-3/4" X 2' X 3' MYLAR BAGS SPACED BY POLYETHYLENE CHANNELS IN WHICH WATER FLOWS. THE WATER CARRIES HEAT FROM THE COLLECTOR OR TO THE BASEBOARD RADIATORS IN THE ROOMS. THERE IS A HEAT EXCHANGER BETWEEN THIS WATER AND THE COLLECTOR FLUID. THE WAX IS INSIDE TWO TANKS, ONE AT EACH END OF THE ROOF RIDGE. SOME HEAT IS SCAVENGED FROM THE FIREPLACE.

A LARGE CENTRAL FIREPLACE PROVIDES AUXILIARY HEAT. THE FLUE IS OF STAINLESS STEEL AND IS ENCLOSED IN A WATER JACKET CONNECTED TO THE STORAGE SYSTEM. A 40 KW ELECTRIC HOT WATER HEATER IS ALSO USED.

MATERIAL　　STEEL SHEET WITH BLACK CHROME COATING; 1/8" GLASS; KALWALL PREMIUM SUN-LITE (FIBERGLAS AND POLYESTER); WATER; ETHYLENE GLYCOL; 1-1/4" COPPER PIPES; BAKED WHITE ENAMEL COATING ON ALUMINUM SHEET; 4,000 LBS WAX; 1-3/4" X 2' X 3' MYLAR BAGS; FIREPLACE; STAINLESS STEEL; 40 KW ELECTRIC HOT WATER HEATER

AVAILABILITY　ADDITIONAL INFORMATION MAY BE OBTAINED FROM LARRY SOUTH, RR 3, GANANOQUE, ONTARIO, CANADA.

REFERENCE　　SHURCLIFF, W.A.
SOLAR HEATED BUILDINGS, A BRIEF SURVEY
CAMBRIDGE, MASSACHUSETTS, W.A. SHURCLIFF, MARCH 1976, 212 P

TITLE   GENERAL ELECTRIC CO. SOLAR HEATED BUILDING

CODE

KEYWORDS   CAFETERIA KITCHEN COLLECTOR

AUTHOR   GENERAL ELECTRIC CO.

DATE & STATUS   THE GENERAL ELECTRIC CO. SOLAR HEATED BUILDING WAS COMPLETED IN
                 MAY 1975.

LOCATION   VALLEY FORGE, PENNSYLVANIA

SCOPE   THE BUILDING IS AN EXISTING BUILDING, BUILT IN 1960. THE
                 SOLAR HEATING SYSTEM IS INTENDED TO HEAT THE 20,000 SQ. FT.
                 CAFETERIA-AND-KITCHEN AREA. THE COST WAS APPROXIMATELY $400,000.

TECHNIQUE   THE COLLECTOR IS A 4,900 SQ. FT., WATER-TYPE COLLECTOR,
                 MOUNTED ON THE HORIZONTAL ROOF. THE COLLECTOR CONSISTS OF
                 203 PANELS, EACH 8' X 3', AND 5" THICK, TILTED 45 DEGREES
                 FROM THE HORIZONTAL. THE HEART OF THE PANEL IS A SERIES 1100
                 ROLL-BOND ALUMINUM SHEET WITH 240' OF INTEGRAL PASSAGES. FOR
                 COMPARISON, THREE KINDS OF SELECTIVE BLACK COATINGS ARE USED
                 AND SEVERAL GLAZING SCHEMES, EMPLOYING THE G.E. PLASTIC LEXAN.
                 THE INSULATED BACKING INCLUDES 0.001" OF ALUMINUM FOIL, A LAYER
                 OF FOAM-TYPE INSULATION AND, IN ADDITION, A THIN LAYER OF
                 FIBERGLAS. THE FLUID IS 500 GALLONS OF WATER (70%) AND ETHYLENE
                 GLYCOL (30%) SOLUTION IN A CLOSED SYSTEM. CIRCULATION IS BY A
                 5 HP CENTRIFUGAL PUMP.

                 THE STORAGE SYSTEM CONSISTS OF 8,000 GALLONS OF WATER IN SIX
                 TANKS. HEAT EXCHANGERS ARE USED. THE ROOMS ARE HEATED BY FAN-
                 AND-COIL SYSTEMS. WHEN AND IF THE TANK TEMPERATURE REACHES
                 200 DEGREES F, ADDITIONAL HEAT IS "VENTED" TO THE COLD WATER.

MATERIAL   SERIES 1100 ROLL-BOND ALUMINUM SHEET; THREE KINDS OF SELECTIVE
                 BLACK COATING; G.E. PLASTIC LEXAN; 0.001" ALUMINUM FOIL; FOAM-
                 TYPE INSULATION; THIN LAYER OF FIBERGLAS; WATER (70%) AND
                 ETHYLENE GLYCOL (30%) SOLUTION; 5 HP CENTRIFUGAL PUMP; 8,000
                 GALLONS OF WATER; SIX TANKS; HEAT EXCHANGERS

AVAILABILITY   ADDITIONAL INFORMATION MAY BE OBTAINED FROM THE GENERAL ELECTRIC
                 CO., SPACE DIVISION, KING OF PRUSSIA PARK (VALLEY FORGE),
                 P.O.BOX 8661, PHILADELPHIA, PENNSYLVANIA 19101.

REFERENCE   SHURCLIFF, W.A.
                 SOLAR HEATED BUILDINGS, A BRIEF SURVEY
                 CAMBRIDGE, MASSACHUSETTS, W.A. SHURCLIFF, MARCH 1976, 212 P

417

TITLE	GENERAL ELECTRIC MOBILE HOME
CODE	
KEYWORDS	MOBILE HOME SOLAR COLLECTOR STORAGE WATER TYPE FORCED HOT AIR
AUTHOR	GENERAL ELECTRIC COMPANY, SPACE DIVISION
DATE & STATUS	BY SPRING OF 1975 CONSTRUCTION WAS COMPLETED. IN NOVEMBER, DECEMBER, AND JANUARY OF THE WINTER OF 1975-1976, THE DEVICE WAS OPERATED IN CHICAGO, LOS ANGELES, AND LOUISVILLE.
LOCATION	THE LOCATION IS VARIABLE. THE MOBILE HOME IS TO BE TAKEN TO MANY DIFFERENT LOCATIONS FOR TESTING. TYPICAL LOCATIONS ARE PHILADELPHIA AND WASHINGTON, D.C.
SCOPE	THE BUILDING IS A 60' X 12' MOBILE HOME MADE BY SKYLINE CORP. IT IS OF THE STANDARD TYPE EXCEPT THAT IT HAS BETTER INSULATION, AND HAS BEEN MODIFIED TO INCLUDE A SOLAR COLLECTOR AND STORAGE SYSTEM.
TECHNIQUE	THE COLLECTOR IS A 450 SQ. FT., WATER-TYPE COLLECTOR. THERE ARE THREE ARRAYS OF PANELS. THE ARRAYS ARE MOUNTED ON THE ROOF. THE TILT IS USUALLY 40 DEGREES FROM THE HORIZONTAL, BUT THAT IS VARIABLE. THE ARRAYS CAN BE FOLDED DOWN WHEN THE MOBILE HOME IS TRAVELING ON THE HIGHWAY. THERE ARE 19 G.E. TYPE SOLAR COLLECTOR PANELS. EACH IS 3' X 8', AND EMPLOYS A ROLL-BOND ALUMINUM SHEET WITH INTEGRAL PASSAGES FOR THE WATER-AND-ETHYLENE-GLYCOL FLUID. THE BLACK COATING IS SELECTIVE. THE GLAZING IS DOUBLE, OF LEXAN.
	THE STORAGE SYSTEM HAS 400 GALLONS OF WATER IN TEN 40 GALLON HORIZONTAL CYLINDRICAL STEEL TANKS, MOUNTED BENEATH THE FLOOR. THE HEAT EXCHANGER IS BETWEEN THE COLLECTOR AND THE TANK SYSTEM. DISTRIBUTION OF HEAT IS BY FORCED HOT AIR THROUGH FAN COIL SYSTEMS.
	THE DOMESTIC HOT WATER IS PREHEATED BY THE SOLAR HEATING SYSTEM. 95% OF THE HEAT NEEDED FOR DOMESTIC HOT WATER IS TO BE SUPPLIED BY THE SOLAR HEATING SYSTEM.
	COOLING IN SUMMER IS ACCOMPLISHED BY AN ARKLA DX LIBR ABSORPTION COOLER. IT HAS BEEN MODIFIED FOR POWERING BY HOT WATER, AT ABOUT 190 DEGREES F, AND HAS BEEN DERATED FROM 3 TONS TO 2 TONS CAPACITY. REJECTED HEAT IS SENT TO A HALSTEAD & MITCHELL 91,000 BTU COOLING TOWER, MODEL GCKA-7-1/2.
MATERIAL	MOBILE HOME BY SKYLINE CORP.; ROLL-BOND ALUMINUM SHEET WITH INTEGRAL PASSAGES; WATER-AND-ETHYLENE-GLYCOL FLUID; SELECTIVE BLACK COATING; LEXAN DOUBLE GLAZING; 400 GALLONS OF WATER; TEN 40 GALLON HORIZONTAL CYLINDRICAL STEEL TANKS; HEAT EXCHANGER; ARKLA DX LIBRIUM BROMIDE ABSORPTION COOLER; HALSTEAD & MITCHELL 91,000 BTU COOLING TOWER
AVAILABILITY	ADDITIONAL INFORMATION MAY BE OBTAINED FROM THE GENERAL ELECTRIC CO., SPACE DIVISION, KING OF PRUSSIA PARK (VALLEY FORGE), P.O. BOX 8661, PHILADELPHIA, PENNSYLVANIA 19101.
REFERENCE	SHURCLIFF, W.A. SOLAR HEATED BUILDINGS, A BRIEF SURVEY CAMBRIDGE, MASSACHUSETTS, W.A. SHURCLIFF, MARCH 1976, 212 P

419

TITLE	GEORGE A. TOWNS ELEMENTARY SCHOOL
CODE	
KEYWORDS	ELEMENTARY SCHOOL WATER TYPE SOLAR COLLECTOR REFLECTOR SHADE RETROFIT
AUTHOR	DUBIN - BLOOME ASSOCIATES
DATE & STATUS	THE GEORGE A. TOWNS ELEMENTARY SCHOOL WAS BUILT IN 1962, AND RETROFITTED WITH SOLAR COLLECTOR EQUIPMENT IN OCTOBER 1975.
LOCATION	ATLANTA, GEORGIA
SCOPE	THE BUILDING IS A ONE-STORY, 32,000 SQ. FT. ELEMENTARY SCHOOL WITH A HORIZONTAL ROOF. IT IS USED YEAR ROUND BY 660 STUDENTS. THE WINTER HEATING LOAD EQUALS THE SUMMER COOLING LOAD.
TECHNIQUE	THE COLLECTOR AREA IS 10,000 SQ. FT. THE COLLECTOR IS OF THE FLAT-PLATE, WATER-TYPE, AND IS MOUNTED ON THE ROOF. IT HAS 576 PPG INDUSTRIES, INC., PANELS, IN 12 ROWS. THE SLOPE OF THE PANEL IS 45 DEGREES. THE PANEL AREA IS 34" X 76". THE ABSORBING SHEET IS OF ALUMINUM, OLIN BRASS CO., ROLL-BOND, WITH A SHEET THICKNESS OF 0.060". IN THE REGION BETWEEN, THERE ARE INTEGRAL PASSAGES FOR WATER; THE CROSS SECTION OF THE PASSAGES IS 3/8" X 0.065". THE ABSORBING SURFACE IS COVERED WITH ALCOA SELECTIVE BLACK COATING. THE GLAZING CONSISTS OF TWO LAYERS OF 1/8" PPG HERCULITE TEMPERED GLASS IN STANDARD INSULATED FRAMES. THE FLUID IS WATER AND A SPECIAL INHIBITOR; THE INHIBITOR WILL REDUCE THE ALUMINUM CORROSION RATE TO LESS THAN 0.0001"/YEAR. THERE IS NO ANTIFREEZE. THE LIQUID WILL BE DRAINED AT THE END OF THE DAY AND REPLACED BY NITROGEN.
	ADJACENT TO EACH ROW OF COLLECTOR PANELS IS AN OPPOSITELY SLOPING (36 DEGREES FROM THE HORIZONTAL) REFLECTOR. THE REFLECTOR DIRECTS ADDITIONAL RADIATION TO THE COLLECTOR, AND HELPS SHADE THE ROOF IN SUMMER. THE REFLECTOR CONSISTS OF A SHEET OF ALUMINIZED MYLAR BONDED BETWEEN THE MYLAR SHEETS AND SUPPORTED BY A 1/8" SHEET OF MASONITE. THE TOTAL REFLECTOR AREA IS 12,000 SQ. FT. THE REFLECTANCE IS 0.74. THE IRRADIATION GAIN FACTOR PROVIDED BY THE REFLECTOR IS NEAR-UNITY IN MIDWINTER, AND NEAR 1.3 IN MIDSUMMER. THE COLLECTION TEMPERATURE IS 140 DEGREES F IN MIDWINTER, AND 200 DEGREES F IN MIDSUMMER.
	THE STORAGE SYSTEM CONTAINS 45,000 GALLONS OF WATER, STORED IN THREE 15,000 GALLON INSULATED UNDERGROUND STEEL TANKS. ALL THE TANKS STORE HOT WATER IN WINTER.
	AUXILIARY HEAT IS PROVIDED BY AN EXISTING GAS FURNACE. COOLING IN SUMMER IS DONE BY A 100-TON ARKLA LIBR ABSORPTION CHILLER, POWERED BY WATER AT 195 TO 200 DEGREES F, AND HEATED BY SOLAR RADIATION OR, WHEN THAT IS INSUFFICIENT, BY THE AUXILIARY HEAT SOURCE.
MATERIAL	576 PPG INDUSTRIES, INC., PANELS; ALUMINUM, OLIN BRASS CO. ROLL-BOND, WITH A SHEET THICKNESS OF 0.060"; 1/8" PPG HERCULITE TEMPERED GLASS IN STANDARD INSULATED FRAMES; WATER; ALUMINUM CORROSION INHIBITOR; NITROGEN; SHEET OF ALUMINIZED MYLAR BONDED BETWEEN MYLAR SHEETS AND SUPPORTED BY A 1/8" SHEET OF MASONITE; THREE 15,000 GALLON INSULATED UNDERGROUND STEEL TANKS; 100-TON ARKLA LIBR ABSORPTION CHILLER
AVAILABILITY	ADDITIONAL INFORMATION MAY BE OBTAINED FROM DUBIN - BLOOME ASSOCIATES, 42 WEST 39TH STREET, NEW YORK, NY.
REFERENCE	SYMMONDS, BARRY D. ATLANTA SCHOOL PROJECT: INTERFACING THE SOLAR AND EXISTING MECHANICAL SYSTEMS SPECIFYING ENGINEER, NOVEMBER 1975, 7 P
	RITTELMANN, RICHARD P. ATLANTA SCHOOL PROJECT: WHY SOLAR WAS CHOSEN AND HOW THE SYSTEM WILL WORK SPECIFYING ENGINEER, NOVEMBER 1975, 5 P

421

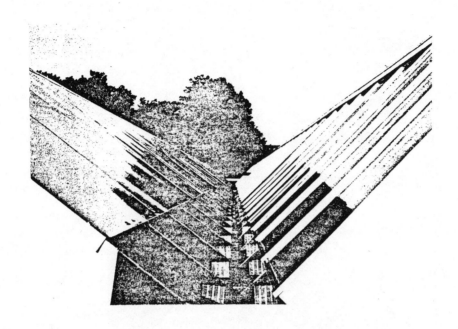

ZONING DIAGRAM

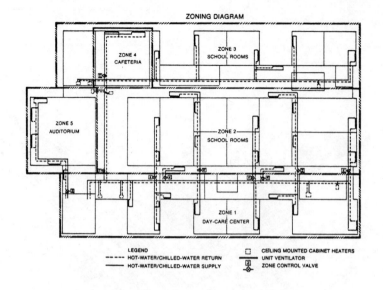

LEGEND
- – – – HOT-WATER/CHILLED-WATER RETURN
- ——— HOT-WATER/CHILLED-WATER SUPPLY
- ▢ CEILING MOUNTED CABINET HEATERS
- UNIT VENTILATOR
- ⊠ ZONE CONTROL VALVE

FRAMING DETAIL

GASKET ASSEMBLY

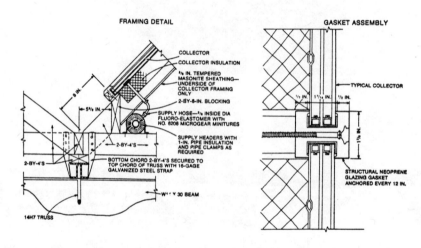

COLLECTOR AND REFLECTOR FRAMING

TYPICAL FRAMING SECTION

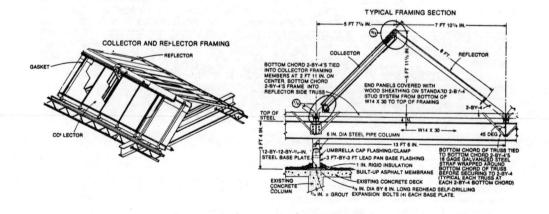

423

TITLE	GOOSEBROOK SOLAR HOME
CODE	
KEYWORDS	HOUSE PASSIVE SOLAR COLLECTOR ROOF MOUNTED TRIPLE GLAZING WARM AIR DELIVERY WATER SAVING FIXTURE INSULATION
AUTHOR	TOTAL ENVIRONMENTAL ACTION, INC.
DATE & STATUS	THE GOOSEBROOK SOLAR HOME WILL BE OFFERED FOR SALE IN THE SPRING OF 1977.
LOCATION	GOOSEBROOK, NEW HAMPSHIRE
SCOPE	THIS IS A SINGLE FAMILY RESIDENCE WITH THREE BEDROOMS, ONE AND A HALF BATHS, LIVING, DINING AND COOKING AREAS, A FULL BASEMENT, A 2-CAR GARAGE, AND A GREENHOUSE. FUEL BILLS FOR THE GOOSEBROOK HOME WILL BE LESS THAN 1/5 OF MOST HOME HEATING BILLS.
TECHNIQUE	THE COMPACT DESIGN, WITH ITS JUDICIOUS PLACEMENT OF WINDOWS AND DOORS, EXTRA-HEAVY INSULATION, TRIPLE-GLAZING AND CAREFUL SEALING OF JOINTS, USES NORMAL WOOD-FRAMED CONSTRUCTION METHODS. PASSIVE SOLAR HEATING COLLECTION FROM THE LARGE AMOUNTS OF SOUTH GLASS IS ADDED TO THE ENERGY FROM THE INEXPENSIVE, ROOF-MOUNTED SOLAR COLLECTORS. HEAT FROM THE UNIQUE WATER STORAGE SYSTEM IS DELIVERED TO THE HOUSE BY A GENTLE, EFFECTIVE WARM-AIR DELIVERY SYSTEM.
MATERIAL	TRIPLE-GLAZING; PASSIVE SOLAR HEATING COLLECTORS; WATER STORAGE SYSTEM; WARM AIR DELIVERY SYSTEM; WATER SAVING PLUMBING FIXTURES
AVAILABILITY	FURTHER INFORMATION MAY BE OBTAINED FROM TOTAL ENVIRONMENTAL ACTION, INC., CHURCH HILL, HARRISVILLE, NEW HAMPSHIRE 03450.
REFERENCE	TOTAL ENVIRONMENTAL ACTION, INC. GOOSEBROOK: AN ENERGY CONSERVING SOLAR HOME DESIGNED AND BUILT BY TOTAL ENVIRONMENTAL ACTION, INC., HARRISVILLE, N.H. HARRISVILLE, NEW HAMPSHIRE, TOTAL ENVIRONMENTAL ACTION, INC. PAMPHLET, 1976, 1 P

TITLE	GOVERNMENT EMPLOYEES INSURANCE COMPANY BUILDING
CODE	GEICO
KEYWORDS	HEAT RECOVERY ELECTRIC HVAC SYSTEM AUTOMATIC COMPRESSOR PREHEAT VENTILATION PRECOOL CONDENSER EVAPORATIVE COOLER WHEEL
AUTHOR	KLING PARTNERSHIP
DATE & STATUS	THE GEICO BUILDING WAS BUILT AND FUNCTIONING IN JANUARY 1973. THE GEICO BUILDING IN MACON, GEORGIA WAS OPENED IN AUGUST 1974.
LOCATION	THE PROTOTYPE IS IN WOODBURY, NEW YORK. THERE ARE ALSO PLANS FOR ANOTHER BUILDING ALMOST IDENTICAL TO THE GEICO BUILDING IN MACON, GEORGIA.
SCOPE	THE BUILDING (IN WOODBURY) IS PRESENTLY L-SHAPED, PARTIALLY ENCLOSING THE ENTRANCE COURT. A 150,000 SQUARE FOOT WING TO BE ADDED IN A SECOND PHASE OF CONSTRUCTION WILL ENCLOSE THE COURT ON A THIRD SIDE, AND THE BUILDING WILL TAKE THE FORM OF A U. THE CENTER PORTION OF THE PLAZA HAS BEEN GRADED TO FORM A GENTLE SLOPE BETWEEN THE SECOND LEVEL OF THE BUILDING AND THE PARKING AREA. THE MOUNDED EARTH IS WELL PLANTED. THIS ENABLED THE ARCHI-TECTS TO PUT THE MAIN ENTRANCE ON THE SECOND FLOOR AT MID-POINT, THUS ENCOURAGING PEOPLE TO USE THE STAIRS. ANY LEVEL OF THE BUILDING CAN BE REACHED EASILY BY CLIMBING AT MOST TWO FLIGHTS OF STAIRS. THIS BUILDING COULD SURVIVE AS A WALK-UP.

SCOPE (continued)

A MAJOR INNOVATION MADE TO ACHIEVE OPENNESS IS THE STRATEGIC BREAK-UP OF THOSE SERVICES THAT ARE USUALLY CONCENTRATED IN A SOLID CORE BUILT AT DEAD CENTER. THREE CURVING BRICK TOWERS EXTEND OUT FROM THE BUILDING AT INTERVALS ALONG THE PERIMETER. THESE OUTBOARD CORES CONTAIN ELEVATOR SHAFTS, STAIRWELLS, REST-ROOMS, AND TELEPHONE AND ELECTRICAL CLOSETS. THE PENTHOUSES ABOVE THE TOWERS ARE FOR MISCELLANEOUS MECHANICAL EQUIPMENT SUCH AS VENTILATION FANS, HEAT RECOVERY WHEELS AND EVAPORATE COOLERS. TWO ADDITIONAL CORES WILL BE ERECTED WHEN EXPANSION PLANS MATERIALIZE.

TECHNIQUE	THE ENGINEERS SELECTED AN ELECTRIC HEAT RECOVERY HVAC SYSTEM PRIMARILY TO OPTIMIZE ENERGY CONSUMPTION. THE SYSTEM EMPLOYS ELECTRIC WATER-TO-AIR HEAT PUMPS WORKING INTO A CLOSED LOOP OF CIRCULATING WATER, PLUS AN ARRANGEMENT OF TWO HEAT WHEELS IN THE EXHAUST AND INLET PASSAGES. AS IN ANY HEAT RECOVERY APPLICATION THE DESIGN OBJECTIVE HERE WAS TO SALVAGE HEAT THAT WOULD NORMALLY BE WASTED AND REUSE IT ELSEWHERE.

THE MAJOR PORTIONS OF THE BUILDING, BOTH INTERIOR AND PERIMETER, ARE SERVED BY A TOTAL OF 185 CEILING-MOUNTED HEAT PUMP UNITS RATED AT THREE OR FOUR TONS. EACH OPERATES INDEPENDENTLY OF THE OTHERS AND CAN BE ON HEATING OR COOLING AT ANYTIME REGARDLESS OF WHAT IS HAPPENING IN THE REST OF THE SYSTEM. IN THE COOLER MONTHS, THE SYSTEM REMOVES EXCESS HEAT FROM THE INTERIOR ZONES AND, BY THE MEDIUM OF THE WATER CIRCULATING IN THE PIPES CON-NECTING THE HEAT PUMPS, MAKES IT IMMEDIATELY AVAILABLE FOR USE AT THE PERIMETER.

EACH HEAT PUMP UNIT IS CONTROLLED FROM A SPACE THERMOSTAT WITH AUTOMATIC CHANGEOVER FROM HEATING TO COOLING AND VICE VERSA. INTERIOR ZONE THERMOSTATS ARE PRESENT FOR A 68F TO 76F DEADBAND, WHICH PERMITS REDUCED COMPRESSOR OPERATION WITHOUT SIGNIFICANTLY AFFECTING THE COMFORT OF THE OCCUPANTS. HAD THERMOSTATS IN THESE AREAS BEEN SET A FIXED 68 F, COMPRESSORS WOULD BE OPERATED ALMOST CONTINUOUSLY - ON HEATING OR COOLING - WITH HIGHER ENERGY CON-SUMPTION. THE WIDE DEADBAND ALSO PRECLUDES THE POSSIBILITY OF ADJACENT UNITS ATTEMPTING TO COOL WHILE ITS NEIGHBOR IS TRYING TO HEAT THE SAME SPACE.

DURING WINTER, THE HEAT WHEELS RECOVER UP TO 80 PERCENT OF THE HEAT CONTENT OF THE AIR EXHAUSTED FROM THE BUILDING AND USE IT TO PREHEAT THE VENTILATION AIR BEING BROUGHT INTO THE BUILDING. IN SUMMER, WHEN THE EXHAUST AIR IS COOLER THAN THE OUTSIDE AIR, THE WHEELS SERVE AS PRECOOLERS.

MATERIAL   CEILING MOUNTED HEAT-PUMPS, STORAGE TANK WITH IMMERSION HEATERS,
                    PUMPS, CONDENSER, EVAPORATIVE COOLER WHEEL

AVAILABILITY   ADDITIONAL INFORMATION ABOUT THE CLOSED-LOOP HVAC SYSTEM USED
                    IN THE GEICO BUILDING AT WOODBURY, NEW YORK, MAY BE OBTAINED BY
                    WRITING TO THE GOVERNMENT EMPLOYEES INSURANCE COMPANY BUILDING,
                    BUILDING MAINTAINANCE DEPARTMENT, WOODBURY, NEW YORK.

REFERENCE   AIA JOURNAL
                    OWNER TESTS HEAT RECOVERY CONCEPTS FOR FIVE SIMILAR OFFICE
                    BUILDINGS IN A $13.1 MILLION PROTOTYPE
                    AIA JOURNAL, FEBRUARY 1975, PP 11-14

Three existing outboard "cores" will be joined by two more when new wing is added.

Attractively planted courtyard in Woodbury leads to main entrance on second floor.

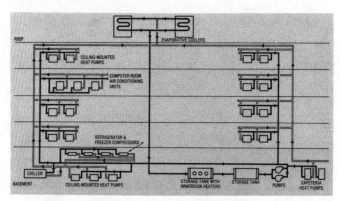

HEAT RECOVERY WATER LOOP

This flow diagram points up the wide extent to which the closed loop of circulating water is used for recovering heat from many sources in the GEICO building. The bulk of the connections are made, of course, to the 185 ceiling-mounted heat pump units distributed throughout the general office areas. Some other spaces require special refrigeration and air conditioning considerations. The equipment serving those areas was selected for compatibility with the overall energy conservation concept.

All units are water cooled and, as indicated, on the flow diagram, have their condensers connected into the heat pump water loop. The special areas and the equipment serving them are: computer room with three 16-ton packaged air conditioning units; kitchen/cafeteria with air handlers and a 125-ton centrifugal chiller; and the food storage area with five compressors for walk-in freezers and refrigerators.

Excess heat generated during the day is stored in the two 10,000-gallon tanks which are part of the loop. The stored heat is available for use at night, supplemented by three 90-kw immersion heaters installed in one of the tanks. The temperature of the condenser water loop is maintained between 70F (winter low) and 92F (summer high). Below 70F the immersion heaters are energized and above 92F the evaporative coolers are gradually phased into operation.

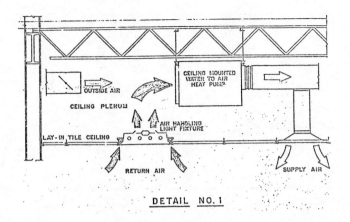

DETAIL NO. 1

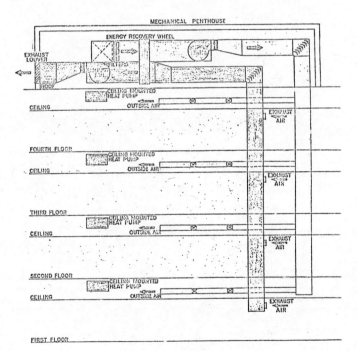

VENTILATION AND EXHAUST AIR SYSTEM

DETAIL NO. 2

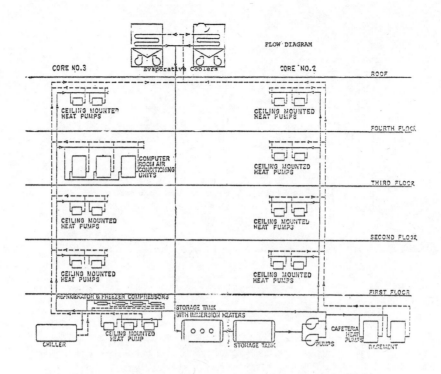

FLOW DIAGRAM

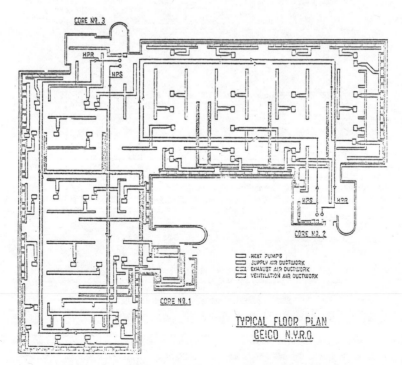

TYPICAL FLOOR PLAN
GEICO N.Y.R.O.

431

TITLE   GRAS HOUSE

CODE

KEYWORDS   HOUSE COLLECTOR AIR TYPE STORAGE MASSIVE WALLS FLOORS THERMAL
               CAPACITY

AUTHOR   R.W. GRAS

DATE & STATUS THE GRAS HOUSE WAS COMPLETED IN 1960.

LOCATION   LINCOLN, MASSACHUSETTS

SCOPE   THE BUILDING IS A SPLIT-LEVEL HOUSE WITH 2 STORIES ON THE
               SOUTH SIDE, AND 3,000 SQ. FT. OF FLOOR AREA. IT IS SITED ON
               THE SOUTHEAST BROW OF A HILL, ON THE LEDGE OF A ROCK.

TECHNIQUE   MOST OF THE WALLS ARE MADE OF PUMMICE BLOCKS. THE FLOORS ARE
               MASSIVE, MADE OF PRECAST CONCRETE FLOOR SLABS WITH INTEGRAL
               DUCTS. THE WINDOWS ARE DOUBLE GLAZED. THE HOUSE IS 70' X 28'.

               THE COLLECTOR AREA IS 1,300 SQ. FT. IT IS AN AIR TYPE COLLECTOR,
               OCCUPYING MOST OF THE SOUTH VERTICAL FACE OF THE BUILDING.
               RADIATION PASSES THROUGH THE DOUBLE GLAZING, WITH PANES 1" APART,
               THEN THROUGH A 1-1/2" AIRSPACE, AND THEN STRIKES A CORRUGATED
               ALUMINUM SHEET THAT HAS A NON-SELECTIVE BLACK COATING. BEHIND
               THIS SHEET THERE IS A 7" AIRSPACE, AND A MASSIVE WALL. THE
               COLLECTOR IS TWO STORIES HIGH. A FAN DRIVES THE AIR UPWARD IN
               THE 7" AIRSPACE TO A 60' LONG PLENUM AT THE TOP OF THE COLLECTOR,
               THEN DOWN TO THE ROOMS OR TO THE PLENUM BENEATH THE SECOND
               STORY FLOOR, THEN DOWN TO THE PLENUM BENEATH THE FIRST STORY
               FLOOR, AND THEN BACK TO THE BASE OF THE COLLECTOR.

               THE STORAGE SYSTEM CONSISTS OF THE BUILDING ITSELF, ESPECIALLY
               THE MASSIVE WALLS AND FLOORS, AND ALSO THE EXPOSED LEDGE ROCK
               ASSOCIATED WITH THE LOWEST PLENUM. THE THERMAL CAPACITY IS
               VERY LARGE.

MATERIAL   PUMMICE BLOCKS; PRECAST CONCRETE FLOOR SLABS WITH INTEGRAL
               DUCTS; DOUBLE GLAZED WINDOWS; CORRUGATED ALUMINUM SHEET; NON-
               SELECTIVE BLACK COATING; FAN; EXPOSED LEDGE ROCK

AVAILABILITY   ADDITIONAL INFORMATION MAY BE OBTAINED FROM R.W. GRAS, LAUREL
               DRIVE, LINCOLN, MASSACHUSETTS.

REFERENCE   SHURCLIFF, W.A.
               SOLAR HEATED BUILDINGS, A BRIEF SURVEY
               CAMBRIDGE, MASSACHUSETTS, W.A. SHURCLIFF, MARCH 1976, 212 P

TITLE   GRASSY BROOK VILLAGE

CODE

KEYWORDS   SOLAR COLLECTOR PANEL STORAGE TANK HEAT HOT WATER WIND FAN
               COIL OIL BURNER EARTH

AUTHOR   RICHARD D. BLAZEJ; DUBIN, MINDELL ET AL; ROBERT F. SHANNON

DATE & STATUS IN JANUARY OF 1976 A GRANT WAS RECEIVED FROM HUD, AND FURTHER
               FUNDING WAS AWAITED AS OF MID-1976.

LOCATION   BROOKLINE, VERMONT

SCOPE   GRASSY BROOK VILLAGE IS A DEVELOPMENT OF 10 HOUSES. A 4,500
               SQ. FT. ARRAY OF SOLAR COLLECTOR PANELS WILL PROVIDE HEAT AND
               HOT WATER TO RESIDENTS. PHASE TWO OF THE PROJECT WILL SEEK
               WAYS IN WHICH TO PROVIDE ON SITE ELECTRICAL GENERATION; WIND
               POWER IS THE MOST LIKELY CANDIDATE FOR PROVIDING NEEDED ELEC-
               TRICITY. INITIALLY, SOLAR POWER WILL PROVIDE 60% TO 75% OF
               THE ANNUAL HEATING LOAD PER YEAR. THE EVENTUAL GOAL OF THE
               PROJECT IS TO BECOME SELF SUFFICIENT IN TERMS OF ENERGY NEEDS.

TECHNIQUE   HEAT DERIVED FROM THE SUN AND COLLECTED IN THE SOLAR PANELS
               WILL BE STORED IN TANKS UNDER EACH UNIT. A CENTRAL ELECTRIC
               HEAT PUMP WILL ACT AS A BACKUP. FAN COILS WILL DISTRIBUTE HEAT
               AMONG THE INDIVIDUAL HOUSES. EACH UNIT WILL CONTAIN AN OIL
               BURNER AND A WOODBURNING STOVE FOR SUPPLEMENTAL HEAT. THE
               ROOF OF EACH HOUSE IS FLAT AND SUPPORTS A GARDEN, WHICH INSU-
               LATES THE UNIT AND ADDS TO THE BEAUTY OF THE SURROUNDING AREA.

MATERIAL   SOLAR PANELS, WOOD STOVES, STORAGE TANKS, HEAT PUMP, FAN COILS,
               OIL BURNERS, EARTH

AVAILABILITY   ADDITIONAL INFORMATION MAY BE OBTAINED FROM GRASSY BROOK
               VILLAGE, INC., R.F.D. #1, BOX 39, NEWFANE, VERMONT 05345.

REFERENCE   BLAZEJ, RICHARD D. / MORIARTY, PHILLIP M.
               LIVING WITHIN OUR MEANS
               NEWFANE, VERMONT, GRASSY BROOK VILLAGE PAMPHLET, 1976, 8 P

TITLE	GROVER CLEVELAND JUNIOR HIGH SCHOOL
CODE	
KEYWORDS	SOLAR COLLECTOR SCHOOL HOT AIR SYSTEM STORAGE HEAT PUMP
AUTHOR	SPACE DIVISION OF GENERAL ELECTRIC CO.
DATE & STATUS	THE BUILDING IS PRESENTLY BEING TESTED. THE SYSTEM WAS BUILT IN EARLY 1974.
LOCATION	DORCHESTER, MASSACHUSETTS

SCOPE
THE PROJECT USES SOLAR ENERGY TO PROVIDE FOR ABOUT 65 PERCENT OF THE HEATING NEEDS FOR A 20,000 SQ. FT. CLASSROOM SECTION IN THE THREE STORY SCHOOL BUILDING. IT IS PART OF AN EXPERIMENTAL ENERGY RESEARCH AND DEVELOPMENT ADMINISTRATION PROGRAM (SOLAR ENERGY). 144 G.E. PANELS, COVERING 4,600 SQ. FT., HAVE BEEN INSTALLED ON THE SCHOOL ROOF, SOUTH FACING, 45 DEGREE ANGLE FROM HORIZONTAL. THE PROJECT COSTS $354,000.00.

TECHNIQUE
AS SUNLIGHT PASSES THROUGH A PANEL'S CLEAR PLASTIC COVER WINDOWS, THE SOLAR ENERGY IS ABSORBED BY THE BLACK ALUMINUM ABSORBER SURFACE AND CONVERTED INTO HEAT. THIS HEAT IS TRANS-FERRED TO A LIQUID, WHICH IS BEING PUMPED THROUGH THE FLUID PASSAGE NETWORK INTEGRAL TO THE ALUMINUM ABSORBERS TO A PAIR OF SPECIAL SOLAR ENERGY HEAT EXCHANGERS. THE HEAT EXCHANGERS PROVIDE THE HEAT NEEDED TO MAINTAIN A COMFORTABLE TEMPERATURE IN THE SERVED SECTION OF THE SCHOOL BUILDING. IN PERIODS WHERE ROOM TEMPERATURE NEEDS DON'T USE ALL THE HEAT OF THE SYSTEM, IT IS PUMPED TO A 2,000 GALLON THERMAL STORAGE TANK FOR USE AT NIGHT OR ON DAYS WHEN THERE IS LITTLE OR NO SUN. COOL WATER FROM EXCHANGERS AND/OR TANKS THEN IS RECIRCULATED BACK TO COLLECTORS TO BE REHEATED BY SOLAR ENERGY.

THERE ARE 3 BASIC MODES OF OPERATION: THE FIRST MODE IS USED WHENEVER THE AMOUNT OF HEAT COLLECTED EQUALS OR IS LESS THAN THE AMOUNT NEEDED IN THE CLASSROOM. THE SCHOOL ROOMS ARE HEATED USING THE ENERGY PROVIDED BY THE SOLAR ENERGY COLLECTOR PANELS. ONCE THE LIQUID IN THE PANELS IS HEATED, IT FLOWS TO EITHER SET OF SOLAR HEAT EXCHANGERS. THE COOLED WATER IS THEN PUMPED BACK TO THE SOLAR COLLECTORS TO BE REHEATED. ANY DEFICIENCY IN HEATING REQUIREMENTS IS PROVIDED BY THE ORIGINAL HEATING.

THE SECOND MODE IS USED WHERE MORE HEAT IS BEING COLLECTED BY THE COLLECTORS THAN IS NEEDED FOR HEATING THE CLASSROOMS. IN THIS MODE, AS MUCH HEATED LIQUID AS IS NEEDED IS PASSED TO THE HEAT EXCHANGERS, AND THE REST IS PASSED TO THE THERMAL STORAGE TANK. THE COOLED WATER FROM THE HEAT EXCHANGERS AND THE THERMAL STORAGE TANK IS THEN RETURNED TO THE COLLECTOR. IF NO HEAT IS NEEDED IN THE CLASSROOMS, IN THIS MODE, HEATED LIQUID FROM THE COLLECTORS IS PUMPED INTO THE STORAGE TANK. THE COOL WATER IN THE TANK IS RETURNED TO THE COLLECTOR.

IN THE THIRD MODE, THE SCHOOL ROOMS ARE HEATED USING ONLY THE HEAT STORED IN THE THERMAL STORAGE TANK. IF THE WATER FROM THE COLLECTORS IS TOO COOL TO HEAT THE BUILDING, THEN THE HEATED WATER FROM THE THERMAL STORAGE TANK IS PUMPED THROUGH THE HEAT EXCHANGERS.

MATERIAL
TWO LEXAN PLASTIC WINDOWS, INSULATION, SOLAR COLLECTOR PANELS, ALUMINUM ABSORBERS, COPPER TUBING, STEEL STORAGE TANK, FORCED AIR DUCTWORK, PUMPS, EXPANSION TANK, VALVES AUTOMATIC CONTROL

AVAILABILITY
FURTHER INFORMATION MAY BE OBTAINED FROM GENERAL ELECTRIC CO., P.O. BOX 8555, PHILADELPHIA, PENNSYLVANIA 19101.

REFERENCE
BUILDING DESIGN & CONSTRUCTION
SOLAR BUILDINGS RISING FAST
BUILDING DESIGN & CONSTRUCTION, AUGUST 1974, P 98

ENGINEERING NEWS RECORD
SOLAR COLLECTORS REACH PRODUCTION, BUT MARKET IS CLOUDY
ENGINEERING NEWS RECORD, JULY 4, 1974, P 12

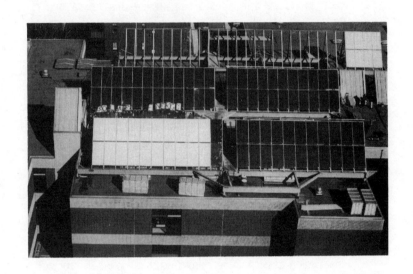

TITLE	HAMDEN HOUSING FOR THE ELDERLY
CODE	
KEYWORDS	HOUSING INSULATION WINDOWS EARTH BERMS THERMAL CAPACITIES
AUTHOR	MCHUGH & ASSOCIATES - ARCHITECT
DATE & STATUS	CONSTRUCTION OF THE HAMDEN HOUSING FOR THE ELDERLY WILL BEGIN IN MARCH 1977, AND WILL BE COMPLETED IN ONE YEAR.
LOCATION	HAMDEN, CONNECTICUT
SCOPE	THE BUILDING COMPLEX HAS AN OPEN ARRANGEMENT WITH ONE COMMUNITY BUILDING AND FIVE RESIDENTIAL BUILDINGS, TO BE HEATED CONVENTIONALLY, AND FIVE RESIDENTIAL BUILDINGS TO BE SOLAR HEATED. EACH OF THE RESIDENTIAL BUILDINGS INCLUDES 4 APARTMENTS.
	A TYPICAL BUILDING IS A ONE-STORY, 4-APARTMENT, MODIFIED-U-SHAPE BUILDING, WITH SOME ATTIC SPACE AND NO BASEMENT. THE FLOOR AREA IS 1,800 SQ. FT., I.E. 400 TO 500 SQ. FT. PER APARTMENT. THE COURTYARD IS NOT ROOFED-OVER.
TECHNIQUE	EXCELLENT INSULATION IS USED; THE THICKNESS OF FIBERGLAS IN THE WALLS IS 6", IN THE CEILING, 9". THERE IS 3" OF RIGID INSULATION BENEATH THE FLOOR SLAB. THE WINDOWS ARE OF WOOD-FRAME TYPE AND SEAL TIGHTLY. THE DOORS WILL BE INSULATED, STEEL-CLAD, AND WILL SEAL TIGHTLY. THERE ARE FEW WINDOWS ON THE EAST, NORTH, AND WEST. THERE ARE EARTH BERMS ON THE NORTH. THE MASONRY WALLS HAVE 3" OF INSULATION ON THE EXTERIOR. THE BUILDING FACES 6 DEGREES WEST OF SOUTH. EACH BUILDING HAS ITS OWN COLLECTION, STORAGE, AND DISTRIBUTION SYSTEM.
	THE COLLECTOR AREA IS ABOUT 500 SQ. FT. THE COLLECTORS ARE OF THE WATER-TYPE, IN THREE ARRAYS, MOUNTED AT 42 DEGREES FROM THE HORIZONTAL ON THE THREE ROOF AREAS. THE LIQUID IS WATER AND ETHYLENE GLYCOL. THE HEAT IS TRANSFERRED TO THE DOMESTIC HOT WATER SYSTEM VIA A HEAT EXCHANGER, AND TO THE STORAGE SYSTEM VIA A HEAT EXCHANGER. THE COLLECTOR COVER IS FIBERGLAS.
	THE STORAGE SYSTEM CONSISTS OF 2,000 GALLONS OF WATER IN A CONCRETE STORAGE TANK IN THE GROUND UNDER THE COURT. THE LARGE THERMAL CAPACITIES OF THE FLOOR AND WALLS CONTRIBUTE TO THE STORAGE. THE DOMESTIC HOT WATER IS PREHEATED BY THE SOLAR HEATING SYSTEM IN WINTER, AND FULLY HEATED BY THAT SYSTEM IN SUMMER. A 60 GALLON, ELECTRICALLY-HEATED TANK IS PROVIDED FOR EACH APARTMENT.
MATERIAL	FIBERGLAS; 3" OF RIGID INSULATION; WATER AND ETHYLENE GLYCOL; HEAT EXCHANGER
AVAILABILITY	ADDITIONAL INFORMATION MAY BE OBTAINED FROM MCHUGH & ASSOCIATES, 790 FARMINGTON AVENUE, FARMINGTON, CONNECTICUT 06032.
REFERENCE	SHURCLIFF, W.A. SOLAR HEATED BUILDINGS, A BRIEF SURVEY CAMBRIDGE, MASSACHUSETTS, W.A. SHURCLIFF, MARCH 1976, 212 P

TITLE	HANSBERGER APARTMENT COMPLEX
CODE	
KEYWORDS	APARTMENT COMPLEX EAVES WATER TYPE COLLECTOR COMPRESSION EXPANSION COMPRESSORS
AUTHOR	E.L. HANSBERGER, JR.
DATE & STATUS	THE SOLAR HEATING SYSTEM WAS 95% COMPLETE IN MARCH 1975. THE SOLAR COOLING SYSTEM WAS TO BE COMPLETED BY THE SUMMER OF 1975.
LOCATION	YUMA, ARIZONA
SCOPE	THIS REGION HAS AN EXCEPTIONAL NUMBER OF SUNNY HOURS (I.E. 4,000 PER YEAR).

SCOPE

THE BUILDING COMPLEX CONSISTS OF AN EXISTING 10-APARTMENT COMPLEX, AND TWO ADJACENT BUILDINGS, EACH ONE-STORY AND EACH 34 FT. X 126 FT. TOTAL FLOOR AREA IS 8,400 SQ. FT.

BECAUSE THE FAN AND COIL SYSTEM IS OF LARGE CAPACITY, THE BUILDINGS CAN BE HEATED ADEQUATELY, EVEN ON THE COLDEST DAYS, AND EVEN WHEN THE TEMPERATURE OF THE WATER IN THE STORAGE TANK IS ONLY 110 DEGREES F. SOLAR ENERGY PROVIDES 25% OF THE COOLING NEEDED IN SUMMER. THE PERCENTAGE WILL BE MUCH LARGER WHEN THE COLLECTOR AREA IS INCREASED.

TECHNIQUE

THE BUILDING COMPLEX IS OF BLOCK CONSTRUCTION, WITH VERMICULITE INSULATION POURED IN THE BLOCK WALLS. THERE ARE 6" OF FIBERGLAS IN THE CEILING. 80% OF THE SOUTHERN WALLS ARE SINGLE GLAZED WINDOWS. THE DRAPERIES PROVIDE SOME INSULATION AT NIGHT. THERE ARE 4 FT. EAVES WHICH BLOCK THE SUMMER SUN. THE SAME SOLAR HEATING AND COOLING SYSTEM SERVES BOTH BUILDINGS.

THE 500 SQ. FT., WATER-TYPE COLLECTOR IS MOUNTED ON THE ROOF OF THE SOUTH BUILDING. THE ROOF SLOPES 7 DEGREES FROM THE HORIZONTAL. THE COLLECTOR CONSISTS OF 22 PANELS (MADE SPECIALLY, LOCALLY), EMPLOYING 3 FT. X 8 FT. OLIN BRASS CO. ROLL-BOND ALUMINUM SHEETS WITH INTEGRAL PASSAGES. THE BLACK COATING IS 3M BLACK VELVET, NON-SELECTIVE. THE SINGLE GLAZING IS OF 1/8" PPG TEMPERED GLASS. THE PANEL HAS A 4" FIBERGLAS BACKING. THE LIQUID IS WATER, WITH NO ANTIFREEZE OR INHIBITOR; BUT PH IS MONITORED AND CONTROLLED. THE LIQUID IS DRAINED AT THE END OF THE DAY, AND IS REPLACED BY NITROGEN. THE LIQUID IS CIRCULATED BY A 1/4 HP CENTRIFUGAL PUMP. THE COLLECTOR AREA MAY LATER BE DOUBLED, IN ORDER TO PROVIDE MORE COOLING IN SUMMER.

THE STORAGE SYSTEM CONSISTS OF 800 GALLONS OF WATER IN A VERTICAL, CYLINDRICAL, NON-PRESSURIZED, TANK OF FIBERGLAS AND EPOXY RESIN, 6 FT. IN DIAMETER, 8 FT. HIGH, SITUATED ABOVE THE GROUND BESIDE THE EAST END OF THE SOUTHERN BUILDING. THE TANK IS INSULATED WITH 3" OF POLYURETHANE FOAM (SPRAYED) AND FIBERGLAS. THE TANK IS WITHIN A CONCRETE-BLOCK ENCLOSURE. THE PIPES ARE ENCLOSED IN 2" DIAMETER, PREFORMED, SNAP-ON FIBERGLAS INSULATION; A VAPOR-BARRIER WRAPPING ENCLOSES THE INSULATION. THE ROOMS ARE HEATED BY INDIVIDUAL FAN-COIL SYSTEMS FED WITH HOT WATER FROM THE COPPER HEAT-EXCHANGER COIL WITHIN THE STORAGE TANK. THERE IS ONE FAN-COIL SYSTEM PER APARTMENT.

ROOMS ARE COOLED WITH THE SAME FAN-COIL SYSTEMS MENTIONED ABOVE: THEY ARE FED WITH CHILLED WATER, CHILLED BY A COMPRESSION-EXPANSION SYSTEM. THIS SYSTEM IS SERVED BY TWO COMPRESSORS, CONNECTED IN PARALLEL. COMPRESSOR A OPERATES STEADILY AND PROVIDES 25% OF NEEDED CHILLING; IT IS DRIVEN BY A RANKINE ENGINE, WHICH IS POWERED BY R-114 THAT HAS BEEN HEATED TO 180-200 DEGREES F BY PASSING THROUGH SPECIAL BOILER-COIL AT THE TOP OF THE STORAGE TANK. COMPRESSOR B OPERATES DURING PERIODS OF HIGH LOAD; IT IS LARGER; IT IS POWERED BY AN ELECTRIC MOTOR, AND HAS A LARGER OUTPUT THAN COMPRESSOR A.

MATERIAL   VERMICULITE INSULATION, BLOCK WALLS, SINGLE GLAZED WINDOWS,
                  DRAPERIES, WATER TYPE COLLECTOR, HEAT EXCHANGER, FAN-COIL SYSTEM

AVAILABILITY   ADDITIONAL INFORMATION MAY BE OBTAINED FROM THE BUILDING MAINTE-
                  NANCE OR SUPERINTENDENT, 669 AVE. B, YUMA, ARIZONA 85364.

REFERENCE   SHURCLIFF, W.A.
                  SOLAR HEATED BUILDINGS, A BRIEF SURVEY
                  CAMBRIDGE, MASSACHUSETTS, W.A. SHURCLIFF, MARCH 1976, 212 P

TITLE   HARTFORD INSURANCE GROUP OFFICE COMPLEX

CODE

KEYWORDS   SOLAR CONTROL LANDSCAPING SHADE REFLECTIVE GLASS SITE

AUTHOR   GUIREY, SRNKA, ARNOLD & SPRINKLE

DATE & STATUS   THE FIRST STRUCTURE IS SCHEDULED FOR COMPLETION IN JULY 1975.

LOCATION   PHOENIX, ARIZONA

SCOPE   THE FIRST PHASE INCLUDES A PAIR OF FOUR-STORY BUILDINGS OF
                 80,000 SQUARE FEET, EACH FLANKING A TWO-STORY, 40,000 SQUARE
                 FOOT BUILDING. BANKS, RESTAURANTS, AND OTHER FACILITIES WILL
                 FOLLOW. ENERGY SAVINGS OF EIGHT PERCENT ARE EXPECTED FOR THE
                 PROJECT AS A RESULT OF A COMPREHENSIVE SET OF METHODS.

                 DESIGN FOR ENERGY CONSERVATION BEGINS WITH THE BUILDING SITE.
                 IN THE HARTFORD COMPLEX, THIS CONCEPT WILL BE EXPRESSED WITH
                 BUILDINGS ORIENTED TO SHADE EACH OTHER, WITH LANDSCAPING AS A
                 SHADING DEVICE, WITH REFLECTIVE GLASS, AND WITH THE INTERNAL
                 BUILDING SYSTEMS THEMSELVES.

                 THE COMPLEX ALSO WILL INCORPORATE ADVANCED FIRE-SAFETY DEVICES,
                 INCLUDING SPRINKLERS AND MONITORS WHICH DETECT SMOKE, TRIGGER
                 EVACUATION SIGNALS, AND ACTIVATE THE AIR CONDITIONING SYSTEM
                 FOR SMOKE REMOVAL.

TECHNIQUE   LANDSCAPING IS USED TO SHADE BUILDINGS FROM THE SUN. THE TALLER
                 BUILDINGS ARE USED TO SHADE THE SHORTER ONES.

MATERIAL   LANDSCAPING

AVAILABILITY   ADDITIONAL INFORMATION MAY BE OBTAINED FROM GUIREY, SRNKA,
                 ARNOLD & SPRINKLE, 3122 NORTH 3RD AVE., PHOENIX, ARIZONA 85013.

REFERENCE   BUILDING DESIGN & CONSTRUCTION
                 FLASH GORDON & THE REALISTS
                 BUILDING DESIGN & CONSTRUCTION, JANUARY 1975, PP 11-12

TITLE   HAWAIIAN ENERGY HOUSE

CODE

KEYWORDS   NATURAL VENTILATION LIGHT COLOR WIND GENERATOR RECYCLED SEWAGE
                 SYSTEM LANAI MOVABLE SCREEN OVERHANG SOLAR COLLECTOR PANELS
                 SKYLIGHT THERMOSYPHON SUBTROPICAL HOUSE

AUTHOR   JAMES E. PEARSON, AIA

LOCATION   UNIVERSITY OF HAWAII, MANOA CAMPUS, HAWAII

DATE & STATUS   THE HAWAIIAN ENERGY HOUSE WAS COMPLETED IN JUNE, 1976.

SCOPE   THE HAWAIIAN ENERGY HOUSE WAS BUILT TO TAKE ADVANTAGE OF
                 MAXIMUM NATURAL VENTILATION AND LIGHT. SOLAR COLLECTOR PANELS
                 ARE INSTALLED ON THE ROOF. A WIND GENERATOR TO PROVIDE ELECTRI-
                 CITY, ENERGY-SAVING APPLIANCES, A SEWAGE SYSTEM, AS WELL AS A
                 SYSTEM TO COLLECT RAIN WATER, ARE INCLUDED IN THE DESIGN.

                 THE HAWAIIAN ENERGY HOUSE HAS THREE BEDROOMS, TWO BATHS, A
                 LIVING ROOM, DINING ROOM, KITCHEN, LAUNDRY AREA, AND LARGE LANAI.
                 THERE WILL BE A CARPORT FOR AN ELECTRIC CAR AND BIKE.

                 THE COST OF THE HAWAIIAN ENERGY HOUSE IS ESTIMATED AT $50,000.
                 (THE RESEARCH GRANT TOTAL WAS $75,000.)

TECHNIQUE   THE LANAI WILL HAVE MOVABLE SCREENS, AND WILL ENCIRCLE THE HOUSE.
                 IT WILL BE COVERED BY A 6 FT ROOF OVERHANG ON THE EAST, SOUTH,
                 AND WEST WALLS, AND A 3 FT OVERHANG ON THE NORTH SIDE. THE
                 OVERHANG IS NECESSARY TO KEEP THE SUN'S RAYS FROM HEATING THE
                 INTERIOR OF THE HOUSE.

                 THE SLANTED ROOF WILL BE OF CORRUGATED ALUMINUM, AND PAINTED A
                 LIGHT COLOR FOR MAXIMUM HEAT REFLECTION. THE SOUTHERN FACE OF THE
                 ROOF WILL HAVE A 30 DEGREE SLANT, AND THE NORTHERN FACE WILL
                 SLANT 60 DEGREES. THIS PEAKED TYPE ROOF WILL PERMIT THE HEAT TO
                 RISE AND BE FLUSHED OUT THROUGH VENTS BY THE TRADE WINDS BLOWING
                 THROUGH VENTS BY THE TRADE WINDS THROUGH MANOA VALLEY.

                 THERE WILL BE A 6 FT X 6 FT SKYLIGHT ON THE NORTHERN FACE OF THE
                 ROOF FOR LIGHT. DURING THE TWO MONTHS OF THE YEAR THAT THE SUN
                 WILL SHINE DIRECTLY ON THE SKYLIGHT FROM THE NORTH, A PLASTIC
                 SOLAR FILM WILL BE UNROLLED TO COVER THE SKYLIGHT. THE FILM WILL
                 STILL ALLOW LIGHT TO ENTER THE HOUSE, BUT WILL KEEP OUT MOST OF
                 THE SOLAR HEAT, ALLOWING THE INTERIOR TO REMAIN COMFORTABLE.

                 THE SLANTED ROOF WILL MAKE IT POSSIBLE TO CATCH RAIN WATER IN
                 GUTTERS. THE WATER WILL PASS THROUGH A SAND FILTER TO PURIFY IT.
                 THE HOT WATER SUPPLY WILL COME FROM WATER THAT WILL BE HEATED BY
                 SOLAR COLLECTORS ON THE ROOF. IT IS A THERMOSYPHON SYSTEM WITH
                 THE HOT WATER STORAGE TANK ELEVATED ABOVE THE COLLECTORS.

                 A WIND GENERATOR WILL GENERATE PART OF THE NECESSARY ELECTRIC
                 POWER FOR THE HOUSE. CONVENTIONAL PUBLIC UTILITY SYSTEMS WILL BE
                 CONNECTED TO THE HOUSE AS A BACK-UP.

MATERIAL   SOLAR COLLECTOR PANELS, WIND GENERATOR, MOVABLE SCREENS, ROOF
                 OVERHANG, LIGHT COLORED CORRUGATED ALUMINUM, VENTS, SKYLIGHT,
                 PLASTIC SOLAR CONTROL FILM, SLANTED ROOF, ROOF GUTTERS, SAND
                 FILTER

AVAILABILITY   ADDITIONAL INFORMATION MAY BE OBTAINED FROM JAMES E. PEARSON,
                 THE HAWAIIAN ENERGY HOUSE, DEPARTMENT OF ARCHITECTURE, UNIVERSITY
                 OF HAWAII, GEORGE ANNEX B4, 2560 CAMPUS ROAD, MANOA, HONOLULU,
                 HAWAII 96822.

REFERENCE   BENSON, BRUCE
                 ENERGY HOUSE: A PLACE IN THE SUN
                 HONOLULU ADVERTISER, SAT, JAN 10, 1976, P 20

                 KA LEO O HAWAII THE VOICE OF HAWAII
                 WIND, SUN TO POWER ENERGY HOUSE
                 KA LEO O HAWAII THE VOICE OF HAWAII, VOL 54, NO 54,
                 FRI, JANUARY 23, 1976, P 1

443

TITLE	HEALTH AND POLICE SCIENCE CLASSROOM BUILDING
CODE	
KEYWORDS	SITE REDUCED EXTERIOR SURFACE TASK LIGHTING RECOVERY SYSTEM HEAT PUMP EXHAUST AIR CLASSROOM BUILDING SCHOOL
AUTHOR	THE STUDY WAS PREPARED BY D.W. BROWN & ASSOCIATES FOR THE HANDREN ASSOCIATES, ARCHITECTS.
DATE & STATUS	THE HEALTH AND POLICE SCIENCE BUILDING WAS DESIGNED IN JANUARY 1975, BUT HAS NOT BEEN BUILT.
LOCATION	STATE UNIVERSITY AGRICULTURE AND TECHNICAL COLLEGE, FARMINGDALE, NEW YORK
SCOPE	THE DESIGN AIMS TO PROVIDE THE ENVIRONMENT TO WHICH PEOPLE HAVE BECOME ACCUSTOMED IN NON-ENERGY CRISIS DAYS, YET SAVE NEARLY 50 PERCENT IN ENERGY CONSUMPTION AND DO SO WITHIN THE ORIGINAL BUDGET.
	THE BUILDING IS PROGRAMMED FOR 2,000 STUDENTS AND 170 FACULTY, AND CONTAINS 37 CLASSROOMS, 12 LABORATORIES AND SUPPORT FACILITIES, TWO 150-STUDENT LECTURE HALLS, AND FACULTY OFFICES.
TECHNIQUE	IN A COMPACT SCHEME THAT REDUCES THE EXTERIOR SURFACE OF THE BUILDING WHERE THERE IS THE GREATEST HEAT GAIN AND LOSS, THE WALLS AND THE ROOF WILL SIGNIFICANTLY REDUCE LOAD REQUIREMENTS. IN ADDITION, WITH PROPER MECHANICAL SYSTEMS, THE HEAT FROM LIGHTS WHICH IS REMOVED FROM THE INTERIOR WILL BE USED TO SUPPLEMENT HEAT REQUIREMENTS FOR THE PERIMETER SPACE, AND BY PLACING THE MAJOR MECHANICAL EQUIPMENT ON THE ROOF, HVAC LOADS WILL BE REDUCED FOR SPACES ON THE TOP FLOOR DIRECTLY BELOW THE EQUIP- MENT ROOM. TO FURTHER REDUCE EXTERIOR SURFACES, A FULL FLOOR BELOW GRADE WILL BE ENTIRELY OCCUPIED BY OFFICES, EXCEPT FOR SOME MECHANICAL SPACES.
	GLASS WILL BE USED SPARINGLY IN THE BUILDING DESIGN TO REDUCE HEAT LOSS IN THE WINTER AND HEAT GAIN IN THE SUMMER FROM THE SUN. THE GLASS USED IN THE DESIGN IS FOR THE CLASSROOMS AND LABORATORIES, RATHER THAN FOR THE OFFICES WHICH GET RELATIVELY LESS USE THAN THE CLASSROOMS. THUS, AS THE FINAL ARCHITECTURAL CONCEPT AND FINAL SCHEMATIC WAS DEVELOPED, THE GLASS USED FOR CLASSROOM AND LAB AREAS WAS REDUCED FROM THE GENERAL STANDARD OF 40-50 PERCENT OF WALL AREA TO ONLY 14 PERCENT GLASS FOR THE ENTIRE BUILDING. THE RESULT ALLOWS NATURAL AIR VENTILATION BY MEANS OF EXHAUST FANS DRAWING AIR THROUGH THE RIBBON-LIKE PERIMETER WINDOWS, YET REDUCING HEAT LOSS AND GAIN.
	HEATING, VENTILATING, AND AIR CONDITIONING WAS THE OTHER MAJOR AREA FOR EXAMINATION TO REDUCE ENERGY USE. A FOUR-PIPE PERIMETER FAN COIL SYSTEM WITH AN ECONOMIZER CYCLE WITH A SMALL SEPARATE INTERNAL-SOURCE HEAT RECOVERY SYSTEM (HEAT PUMP) AND VARIABLE AIR VOLUME ARRANGEMENT WAS SELECTED. IT IS THE FIRST HEAT RECOVERY SYSTEM OF THIS TYPE WHICH WITHDRAWS HEAT FROM EXHAUST AIR TO HEAT WATER, WHICH THEN GOES TO THE BUILDING FOR HEATING USE.
MATERIAL	FOUR-PIPE PERIMETER FAN COIL SYSTEM, HEAT PUMP, EXHAUST AIR, RIBBON-LIKE PERIMETER WINDOWS, 14 PERCENT GLASS FOR THE ENTIRE EXTERIOR OF THE BUILDING
AVAILABILITY	REFER TO THE FOLLOWING REFERENCE FOR FURTHER INFORMATION.
REFERENCE	BUILDING DESIGN & CONSTRUCTION FLASH GORDON & THE REALISTS BUILDING DESIGN & CONSTRUCTION, JANUARY 1975, PP 11-12

TITLE	HILL HOUSE
CODE	
KEYWORDS	SOLAR ASSISTED HEAT PUMP WATER TYPE DETACHED COLLECTOR
AUTHOR	R.C. HILL AND N. SMITH
DATE & STATUS	THE HILL HOUSE WAS COMPLETED IN 1975.
LOCATION	ORONO, MAINE
SCOPE	THE BUILDING IS AN EXISTING 1966 BUILDING WITH TWO STORIES, BUT NO ATTIC OR BASEMENT. THE LOWER STORY IS PARTLY BELOW GRADE. THE FLOOR AREA IS 2,100 SQ. FT.
TECHNIQUE	THE WINDOW AREA IS SMALL. THE SOUTH EAVES HAVE A LARGE OVERHANG.
	THE COLLECTOR AREA IS 1,200 SQ. FT. IT IS A WATER TYPE COLLECTOR, OF THE DETACHED TYPE. IT IS ABOUT 50 FEET SOUTHEAST OF THE HOUSE, ON THE LAWN. IT CONSISTS OF TWO 10' X 60' ARRAYS, EACH SLOPING 61 DEGREES FROM THE HORIZONTAL. EACH ARRAY INCLUDES SIX 10' X 10' PANELS.
	THE STORAGE SYSTEM CONTAINS 22,000 GALLONS OF PLAIN WATER, IN TWO TANKS: A LARGE TANK OF 20,000 GALLONS, AND A SMALL TANK OF 2,000 GALLONS. THE SMALL TANK IS A VERTICAL CYLINDER, ABOUT 7' IN DIAMETER, AND 7' HIGH. IT IS SITUATED INSIDE THE LARGE TANK. THE TANKS ARE LARGELY UNDERGROUND OUTSIDE THE HOUSE, CLOSE TO THE SOUTHERN SIDE OF THE HOUSE. EACH TANK INCLUDES A STEEL ROD REINFORCING AND GUNITE CONCRETE, AND IS INSULATED – EXCEPT FOR THE LARGE TANK WHICH FACES ADJACENT TO THE HOUSE – WITH TWO INCHES OF SPRAYED-ON POLYURETHANE OR 2 INCHES OF POLYSTYRENE PLANK. THE PIPES CONNECTING THE COLLECTOR AND THE STORAGE SYSTEM ARE HOUSED IN AN INSULATED, ABOVE-GROUND CONDUIT. THERMAL STRATIFICATION WITHIN THE TANKS IS ENCOURAGED AND EXPLOITED. IN MIDWINTER, THE COLLECTOR OUTPUT GOES TO THE LARGE TANK, WHICH IS THEN NOT VERY HOT, BUT IS EXPECTED NEVER TO GET SO COLD AS TO FREEZE EVEN IN THE MOST SEVERE WINTER. A CARRIER 5F-20 8-TON HEAT PUMP IS USED TO TRANSFER ENERGY FROM THE LARGE TANK TO THE SMALL TANK, THE TEMPERATURE OF WHICH IS 120 DEGREES F OR SOMEWHAT LESS. THE HEAT IS DISTRIBUTED TO THE ROOMS BY A FAN-COIL SYSTEM, THE COIL BEING FED FROM THE SMALL TANK. THE HEAT PUMP OPERATES ABOUT 8 HOURS PER DAY, TYPICALLY A 10 HOUR MAXIMUM IS PREDICTED. THE TYPICAL PREDICTED COP OF THE HEAT PUMP IS 5. IN THE FALL AND SPRING, THE COLLECTOR OUTPUT TEMPERATURE IS HIGHER, AND THE OUTPUT FLOWS DIRECTLY TO THE SMALL TANK. AT THIS TIME THE HEAT PUMP REMAINS OFF.
MATERIAL	WATER; 20,000 GALLON TANK; 2,000 GALLON TANK; STEEL ROD REINFORCING; GUNITE CONCRETE; 2" OF SPRAYED-ON POLYURETHANE; 2" OF POLYSTYRENE PLANK; INSULATED CONDUIT; CARRIER 5F-20 8-TON HEAT PUMP
AVAILABILITY	ADDITIONAL INFORMATION MAY BE OBTAINED FROM R. C. HILL, 495 COLLEGE AVE., ORONO, MAINE.
REFERENCE	SHURCLIFF, W.A. SOLAR HEATED BUILDINGS, A BRIEF SURVEY CAMBRIDGE, MASSACHUSETTS, W.A. SHURCLIFF, MARCH 1976, 212 P

TITLE   HINESBURG HOUSE

CODE

KEYWORDS   ORIENTATION SOLAR COLLECTORS RETROFITTING

AUTHOR   GARDEN WAY LABORATORIES

DATE & STATUS   THE HINESBURG HOUSE HAS BEEN RETROFITTED WITH SOLAR COLLECTORS
                  AS OF 1975.

LOCATION   HINESBURG, VERMONT

SCOPE   ONE OF THE DIFFICULTIES IN EQUIPPING EXISTING BUILDINGS WITH
                  SOLAR HEATING IS ORIENTATION. IN ADDITION, THE SLOPE OF THE
                  ROOFS IS GENERALLY SHALLOWER THAN THE ANGLE SUGGESTS FOR
                  EFFICIENT COLLECTOR OPERATION.

TECHNIQUE   THESE DIFFICULTIES WERE OVERRIDEN IN THE HINESBURG HOUSE BY
                  INSTALLING A DUAL COLLECTOR ORIENTED EAST AND WEST. THE COLLEC-
                  TORS ARE MOUNTED TO CONFORM TO THE EXISTING SLOPE. DESPITE THESE
                  VIOLATIONS OF ACCEPTED PRACTICE, THE SYSTEM WORKS EFFECTIVELY,
                  SUPPLYING ABOUT 80% OF DOMESTIC HOT WATER REQUIREMENTS.

MATERIAL   SOLAR COLLECTORS

AVAILABILITY   ADDITIONAL INFORMATION MAY BE OBTAINED FROM DR. DOUGLAS TAFF,
                  GARDEN WAY LABORATORIES, P.O.BOX 66, CHARLOTTE, VERMONT 05445.

REFERENCE   ARCHITECTURAL RECORD
                  EAST-WEST COLLECTOR ORIENTATION PROVES EFFECTIVE
                  ARCHITECTURAL RECORD, MID-OCTOBER 1975, P 25

TITLE	HOFFMAN HOUSE
CODE	
KEYWORDS	HOUSE FLAT PLATE WATER TYPE COLLECTOR
AUTHOR	ERICH W. HOFFMAN
DATE & STATUS	THE HOFFMAN HOUSE WAS DESIGNED FOR AND FITTED WITH A SOLAR HEATING SYSTEM IN 1971. IN 1971.
LOCATION	SURREY, B.C., CANADA
SCOPE	THE BUILDING CONSISTS OF ONE STORY, 1,400 SQ. FT., PLUS A PARTIALLY HEATED BASEMENT, 1,300 SQ. FT.
TECHNIQUE	THE COLLECTOR CONSISTS OF 460 SQ. FT. GROSS, 413 SQ. FT. NET. IT IS A FLAT PLATE, WATER-TYPE COLLECTOR, MOUNTED ON THE ROOF, AND SLOPING 58 DEGREES FROM THE HORIZONTAL. RADIATION STRIKES THE 0.005" COPPER SHEET WITH A NON-SELECTIVE BLACK COATING. THE 1/4" BLACK CU TUBES ARE SPACED 6" FROM EACH OTHER, AND SOLDERED TO THE SHEET. THERE IS DOUBLE GLAZING. THE INNER LAYER IS 3/4" FROM THE COPPER SHEET, AND IS OF SINGLE-STRENGTH GLASS. THE PANE SIZE IS 3' X 18'. THE OUTER LAYER IS 3/4" FROM THE INNER LAYER, AND IS OF DOUBLE STRENGTH GLASS. THE COLLECTOR BACKING IS 6" OF FIBERGLAS. WATER IS CIRCULATED THROUGH THE COLLECTOR BY A CENTRIFUGAL PUMP, AND IS DRAINED AUTOMATICALLY AT THE END OF THE DAY. THE MAXIMUM COLLECTOR TEMPERATURE IS 270 DEGREES F WHEN THE PUMP SHUTS DOWN. THE STORAGE SYSTEM CONSISTS OF 800 US GALLONS OF WATER IN TWO VERTICAL CYLINDRICAL UNINSULATED TANKS, OF 300 GALLONS AND 500 GALLONS, IN AN INSULATED BASEMENT ROOM. THE MAXIMUM TEMPERA-TURE ACHIEVED IN THE TANKS IN SUMMER IS 170 DEGREES F.
MATERIAL	0.005" COPPER SHEET; NON-SELECTIVE BLACK COATING; 1/4" BLACK CU TUBES; SINGLE-STRENGTH GLASS; DOUBLE-STRENGTH GLASS; 6" OF FIBERGLAS; CENTRIFUGAL PUMP; 800 US GALLONS OF WATER; TWO VERTICAL CYLINDRICAL UNINSULATED TANKS OF 300 GALLONS AND 500 GALLONS
AVAILABILITY	ADDITIONAL INFORMATION MAY BE OBTAINED FROM ERICH HOFFMAN, 5511 128 ST., SURREY, B.C., CANADA W3W 485.
REFERENCE	SHURCLIFF, W.A. SOLAR HEATED BUILDINGS, A BRIEF SURVEY CAMBRIDGE, MASSACHUSETTS, W.A. SHURCLIFF, MARCH 1976, 212 P

TITLE	HOLLIS MIDDLE SCHOOL
CODE	
KEYWORDS	OUTDOOR FLOODLIGHT INDIRECT ILLUMINATION LAMP SCHOOL LIGHTING
AUTHOR	MICHAEL B. INGRAM, ARCHITECT, KAUFMANN ASSOCIATES, CONSULTING ENGINEERS
DATE & STATUS	THE HOLLIS MIDDLE SCHOOL WAS BUILT IN 1974.
LOCATION	HOLLIS, NEW HAMPSHIRE
SCOPE	THE HOLLIS MIDDLE SCHOOL IS SITUATED ON A SLOPING SITE, WHICH IS THE REASON FOR ITS STEPPED PROFILE. IT IS HEATED AND VEN-TILATED BY MEANS OF AIR-HANDLING UNITS USING ELECTRIC HEAT. THE BUILDING HAS A MINIMUM AREA OF GLASS IN FRONT AND BACK, AND THE ROOF USES URETHANE-INSULATED CEMENT-FIBER PLANK, HAVING A 0.08 U-FACTOR. THE STEEL-FRAMED BUILDING COST $730,000 OR ABOUT $22 PER SQUARE FOOT. THE SCHOOL IS DESIGNED FOR 415 PUPILS. THE HVAC CONTRACT WAS $80,000, AND THE ELECTRICAL CONTRACT, $75,000. OUTDOOR-TYPE MERCURY FLOODLIGHTS ARE USED FOR INDIRECT LIGHTING. THIS APPROACH IS USED BECAUSE IT TAKES THE FIXTURES AWAY FROM THE CEILING PLANE, AVOIDING THE IMAGE OF BARE-LAMP SUPERMARKET TYPE LIGHTING. THE INDIRECT APPROACH HELPED TO REALIZE THE CONCEPT OF A LARGE SPACE UNDER ONE LID.
TECHNIQUE	THE INDIRECT LIGHTING SYSTEM, USING PHOSPHOR-COATED, 1,000 WATT METAL-HALIDE LAMPS, PROVIDES AN ILLUMINATION LEVEL OF 50 FOOT CANDLES AT DESK HEIGHT FOR AN ENERGY EXPENDITURE OF ONLY 1.5 WATTS PER SQUARE FOOT. IN THE CLASSROOM AREAS, ONE LUMINAIRE IS USED FOR EVERY 26 BY 28 FOOT CLASSROOM UNIT. LAMPS OF 400, 750 AND 1,000 WATTS SIZES, OPERATING AT 277 VOLTS, WERE CONSIDERED. THE 400 WATT SIZE WOULD HAVE REQUIRED TWICE AS MANY LUMINAIRIES; THE 1,000 WATT SIZE WOULD HAVE MEANT HEAVIER MOUNTINGS AT THE WALL, BUT YIELDED OBVIOUS SAVINGS OVER-ALL. THEORETICALLY IN FAVOR OF THE LARGER NUMBER OF LUMINAIRES WAS THE POSSIBILITY OF A MORE EVEN BAND OF BRIGHTNESS ON THE CEILING. WITH THE 1,000 WATT LAMPS, THE LUMINAIRES NEEDED CAN BE LOCATED AS FAR DOWN THE BALCONY FACES AS POSSIBLE TO EVEN OUT THE BRIGHTNESS. SCOOP-TYPE SHIELDS AROUND LUMINAIRES SHUT OUT THE GLARE OF THE 1,000 WATT LAMPS FROM SIDE ANGLES. BUT THE BRIGHT LAMPS CAN BE SEEN WHEN LUMINAIRES ARE VIEWED FROM ACROSS THE ROOM.
MATERIAL	PHOSPOR-COATED, 1,000 WATT METAL-HALIDE LAMPS, RAKED CEILING
AVAILABILITY	ADDITIONAL INFORMATION ABOUT THE ENERGY SAVING LIGHTING SYSTEM USED IN THE HOLLIS MIDDLE SCHOOL MAY BE OBTAINED FROM MORRIS KAUFMANN OF KAUFMANN ASSOCIATES, 19 MT. AUBURN ST., CAMBRIDGE MASS. 02138
REFERENCE	ARCHITECTURAL RECORD FLOODLIGHTS IN AN OPEN-PLAN SCHOOL YIELD LOW-ENERGY ILLUMINATION ARCHITECTURAL RECORD, JANUARY 1974, PP 157-160

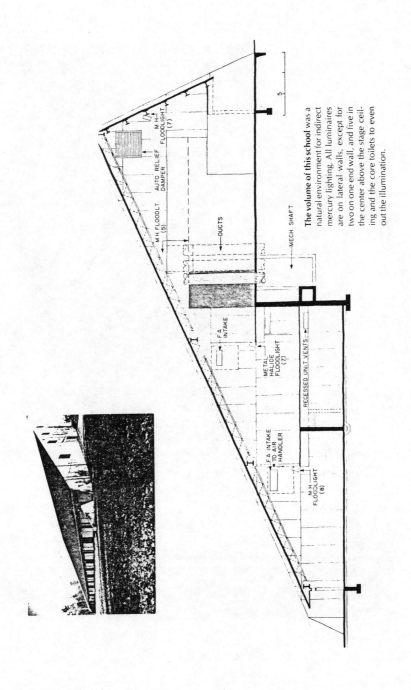

The volume of this school was a natural environment for indirect mercury lighting. All luminaires are on lateral walls, except for two on one end wall, and five in the center above the stage ceiling and the core toilets to even out the illumination.

452

The lighting situations at the
upper level, lower level, and the
west side wall are shown above
and across page. The concrete
block structure at left is the resource center office, and over it
is part of the faculty lounge area.
Notice that luminaire shields in
the classroom areas are only
partial on the sides. The lighting
level is 50 footcandles at only
1.5 watts per sq. ft.

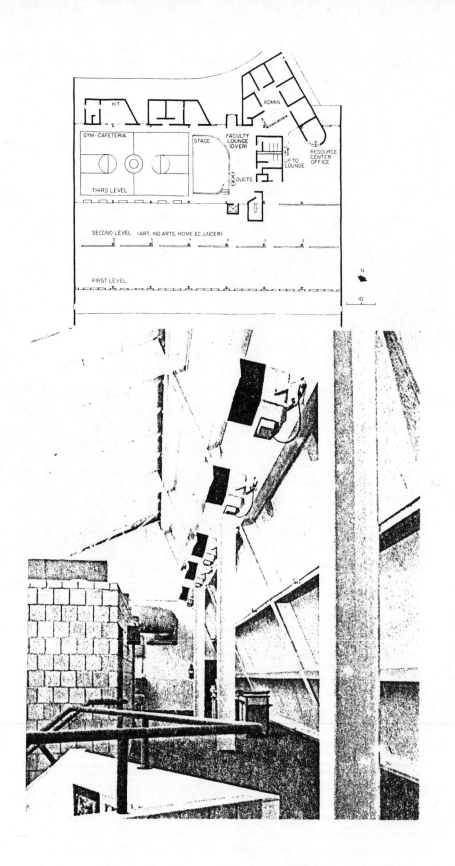

454

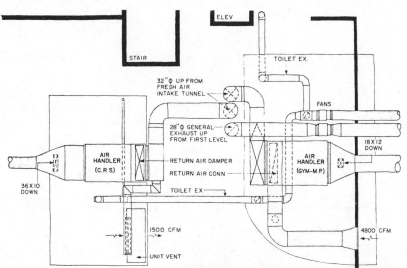

Air-handling units for heating and ventilating the upper level of the school and the gymnasium are shown in the plan. Fresh-air risers for these air handlers and an exhaust riser for the interior classrooms emerge at the third level, and are made a colorful feature. The faculty lounge area is conditioned by a 1500-cfm unit ventilator. The pneumatic controls can be seen in the photo bottom, left. All ductwork is left exposed, except where there is none in the interior classroom spaces at the first level. These have positive exhaust by means of in-duct exhausters. Main classroom areas, on the other hand, have automatically-controlled relief dampers set high in the end walls to allow vitiated air out.

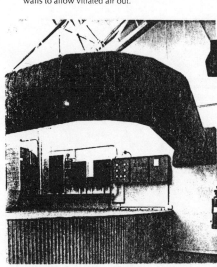

TITLE	HONEYWELL MOBILE SOLAR LABORATORY
CODE	
KEYWORDS	SOLAR LABORATORY THERMAL LOAD FLAT PLATE WATER TYPE COLLECTOR STORAGE
AUTHOR	HONEYWELL, INC.
DATE & STATUS	THE HONEYWELL MOBILE SOLAR LABORATORY WAS COMPLETED IN 1974.
LOCATION	MOBILE, CAN BE LOCATED ANYWHERE
SCOPE	THE SOLAR-HEATED-AND-COOLED LABORATORY CONSISTS OF AN 8' X 9' X 45' TRAILER TO WHICH TWO LARGE AREAS OF COLLECTOR ARE ATTACHED. A SECOND TRAILER, AN OFFICE-TYPE TRAILER WITH A 12' X 50' FLOOR, SERVES AS THERMAL LOAD AND A VISITOR RECEPTION AREA.
TECHNIQUE	THE COLLECTOR HAS 570 SQ. FT. OF NET AREA. IT IS A FLAT-PLATE, WATER-TYPE COLLECTOR, WHICH COMPRISES TWO SIMILAR LARGE PLANAR RECTANGULAR PORTIONS. THE PORTIONS CAN BE TILTED TO ANY DESIRED SLOPE, AND CAN BE FOLDED CLOSE TO THE TRAILER WHEN THE TRAILER IS TO TRAVEL ALONG THE HIGHWAY. EACH PORTION INCLUDES 32 PANELS, EACH 3' X 4'. THE INDIVIDUAL PANEL INCLUDES A PAIR OF 0.025" COLD-ROLLED-STEEL PLATES FORMED AND WELDED TOGETHER TO FORM CHANNELS, 0.05" X 0.25" IN CROSS SECTION, AND 2" APART ON CENTERS. THE SELECTIVE BLACK COATING IS BLACK NICKEL ON BRIGHT NICKEL. DOUBLE GLAZING IS USED; THE OUTER LAYER IS 3/16" OF IRON-FREE TEMPERED GLASS, AND THE INNER LAYER, 1.2" FROM IT, IS OF 0.004" TEDLAR. THE SPACE BETWEEN THE TEDLAR AND THE BLACK ABSORBING SHEET IS 1-1/2". THE BACKING INCLUDES TWO 3" FIBERGLAS LAYERS SEPARATED BY A VAPOR BARRIER AND 1/2" EXTERIOR-TYPE PLYWOOD. THE ENCLOSING STEEL FRAMES ARE 6" DEEP. THE FLUID IS WATER-AND-ETHYLENE-GLYCOL. THE COLLECTION EFFICIENCY IS 55%. THE COLLECTOR COMPRISES 8 ROWS OF PANELS. THE 8 ROWS ARE PARALLEL-CONNECTED. THE STORAGE SYSTEM INCLUDES TWO 450 GALLON TANKS. THE TOTAL AMOUNT OF FLUID USED IS ABOUT 1,000 GALLONS. TWO SYSTEMS ARE USED FOR COOLING IN SUMMER: ONE IS AN ARKLA LIBR ABSORPTION SYSTEM, OPERATED BY HEAT FROM THE SOLAR-HEATED WATER AT ABOUT 200 DEGREES F, AND THE OTHER IS A STANDARD COMPRESSION-TYPE AIR CONDITIONER DRIVEN BY A TURBINE, THE WORKING FLUID OF WHICH IS FREON THAT HAS BEEN HEATED TO ABOUT 200 DEGREES F BY THE SOLAR HEATED WATER AT 210 DEGREES F. BEYOND THE TURBINE THE FREON IS COOLED TO 95 DEGREES F. ALTERNATIVELY, THIS AIR CONDITIONER CAN BE POWERED CONVENTIO-NALLY.
MATERIAL	0.025" COLD-ROLLED-STEEL PLATES; BLACK NICKEL ON BRIGHT NICKEL; 3/16" IRON-FREE TEMPERED GLASS; 0.004" TEDLAR; 3" FIBERGLAS LAYERS; 1/2" EXTERIOR-TYPE PLYWOOD; 6" DEEP STEEL FRAMES; ETHYLENE-GLYCOL; TWO 450 GALLON TANKS; ARKLA LIBR ABSORPTION SYSTEM; STANDARD COMPRESSION-TYPE AIR CONDITIONER; FREON
AVAILABILITY	ADDITIONAL INFORMATION MAY BE OBTAINED FROM HONEYWELL, INC., MINNEAPOLIS, MINNESOTA.
REFERENCE	SHURCLIFF, W.A. SOLAR HEATED BUILDINGS, A BRIEF SURVEY CAMBRIDGE, MASSACHUSETTS, W.A. SHURCLIFF, MARCH 1976, 212 P

TITLE	HOOVER SCHOOL
CODE	
KEYWORDS	BERM RECESSED UNDERGROUND
AUTHOR	RALPH ALLEN OF ALLEN AND MILLER ARCHITECTS
DATE & STATUS	THIS SCHOOL ENTERED ITS THIRD YEAR OF OPERATION IN SEPTEMBER, 1976.
LOCATION	SANTA ANA, CALIFORNIA
SCOPE	THIS SCHOOL INCORPORATED THREE EXISTING BUILDINGS ON THE 4-ACRE SITE INTO THE NEW STRUCTURE. THE EXISTING TWO-STORY HAD TO BE TORN DOWN IN COMPLIANCE WITH 1968 CALIFORNIA EARTHQUAKE LAWS. OLD BUILDINGS WERE CONVERTED INTO KINDERGARTEN SPACE AND MAJOR CLASSROOMS WERE ACCOMODATED IN COMPLETELY NEW ONE-STORY STRUCTURES, RECESSED 5' INTO THE EARTH. SCHOOL OFFICIALS PRE-DICTED THAT HEATING AND COOLING COSTS SHOULD BE REDUCED CON-SIDERABLY OVER THOSE OF CONVENTIONAL SCHOOLS. THE INITIAL COST FOR THE BUILDING WAS $850,000.
TECHNIQUE	BUILDING INSULATION IS OBTAINED BY SINKING THE MAJOR STRUCTURE 5' INTO THE GROUND. WELL LANDSCAPED EARTH BERMS NEXT TO THE EXTERIOR WALLS CREATE A PARK-LIKE ENVIRONMENT FOR THE NEIGH-BORHOOD.
MATERIAL	EARTH BERMS, RECESSED CONSTRUCTION
AVAILABILITY	ADDITIONAL INFORMATION MAY BE OBTAINED FROM ALLEN AND MILLER ARCHITECTS, 1606 BUSH STREET, SANTA ANA, CALIFORNIA 92701.
REFERENCE	PROGRESSIVE ARCHITECTURE THREE UNDERGROUND SCHOOLS PROGRESSIVE ARCHITECTURE, OCTOBER 1975, P 30

Main entry to Hoover School.

Interiors of underground schools designed on 5 ft flexible modules.

TITLE	HOWARD BELL ENTERPRISE HOUSE

TITLE   HOWARD BELL ENTERPRISE HOUSE

CODE

KEYWORDS   HOUSE WINDOW AREA INSULATION COLLECTOR AIR TYPE REFLECTOR STORAGE

AUTHOR   HOWARD BELL ENTERPRISES, INC.

DATE & STATUS THE HOWARD BELL ENTERPRISE HOUSE WAS COMPLETED IN JULY 1975.

LOCATION   VALLEY CITY, OHIO

SCOPE   THE BUILDING IS A 1-STORY, 2,600 SQ. FT., 3-BEDROOM, U-SHAPED,
               WOOD-FRAME HOUSE WITH A SMALL ATTIC BUT NO BASEMENT. ALONG
               THE SOUTH SIDE OF THE HOUSE IS A 66' LONG, 4' TO 8' WIDE PORCH.
               THE NORTHEAST WING IS A 2-CAR GARAGE. THE OVERALL DIMENSIONS
               ARE 66' X 50'.

TECHNIQUE   THE BUILDING FACES DUE SOUTH. THERE IS A LARGE AREA OF WINDOWS,
               240 SQ. FT. AND SINGLE GLAZED, ON THE SOUTH SIDE OF THE HOUSE,
               WHICH OPENS ONTO THE PORCH. THE PORCH ITSELF HAS A LARGE
               WINDOW AREA. THE WINDOWS ON THE EAST, NORTH AND WEST SIDES
               OF THE HOUSE ARE FEW, SMALL, AND TRIPLE GLAZED. THE INSULATION
               IS EXCELLENT: 3-1/2" IN THE WALLS, 14" IN THE ATTIC FLOOR,
               AND PLASTIC FOAM SHEATHING ON THE EXTERIOR WALLS. THE ENTIRE
               GARAGE IS INSULATED. THERE IS EXTENSIVE USE OF WEATHERSTRIPPING
               AND VAPOR BARRIERS.

               THE COLLECTOR IS 180 SQ. FT. GROSS, OF THE AIR TYPE, MEASURING
               60' X 3', AND SLOPING 50 DEGREES FROM THE HORIZONTAL. IT IS
               LOCATED ALONG THE LOWER SOUTH EDGE OF THE PORCH. THE ALUMINUM
               SHEET HAS A NON-SELECTIVE BLACK COATING. THE DOUBLE GLAZING IS
               OF TEMPERED GLASS. THE COLLECTOR INCLUDES TEN PANELS, ARRANGED
               END TO END. BLOWER-DRIVEN AIR CIRCULATES IN THE 1-1/2" SPACE
               BEHIND THE ALUMINUM SHEET. TWO BLOWERS, EACH 1/2 HP, ARE USED.
               THEY SERVE TWO DIFFERENT REGIONS OF THE HOUSE. HOT AIR FROM
               THE COLLECTOR CAN BE SENT TO THE ROOMS, TO THE STORAGE SYSTEM,
               OR TO THE HEAT PUMPS.

               JUST TO THE SOUTH OF THE COLLECTOR THERE IS A PLANAR REFLECTOR,
               60' X 3', OF COATED, PLYWOOD-BACKED ALUMINUM. IN SUMMER, IT
               CAN BE SWUNG TOWARD THE COLLECTOR, SHIELDING IT FROM THE SUN.
               IN WINTER, IT CAN BE SWUNG DOWN TO ANY EXTENT DESIRED, EXCEPT
               THAT IT CANNOT SWING QUITE AS FAR AS HORIZONTAL, AND CANNOT
               BE TILTED TOWARD THE SOUTH.

               THE STORAGE SYSTEM CONTAINS 10 TONS OF STONES AND 10 TONS OF
               CONCRETE SLAB. THE STONES, 5" IN DIAMETER, ARE IN A 60' X 8' X 1'
               BIN. 4" OF POURED CONCRETE PORCH-FLOOR-SLAB SERVES AS THE
               TOP OF THE BIN. THERE ARE 1' DIAMETER HEADER-CONDUITS ALONG
               THE NORTH AND SOUTH EDGES OF THE SLAB. AIR FLOWS HORIZONTALLY
               NORTHWARD THROUGH THE BIN OF STONES. THE NOMINAL PATHLENGTH OF
               THE AIR IN THE BIN IS LESS THAN 8'. ROOM AIR IS CIRCULATED
               THROUGH THE BIN OF STONES BY THE SAME TWO BLOWERS MENTIONED
               ABOVE.

               TWO G.E. OUTDOOR AIR-TO-AIR HEAT-PUMPS ARE USED, EACH SERVING
               A DIFFERENT REGION OF THE HOUSE. EACH DEVICE IS OF 2-TON
               CAPACITY. AUXILIARY HEAT IS PROVIDED BY ELECTRIC HEATERS
               ASSOCIATED WITH THE HEAT PUMPS, AND TWO CAST-IRON FRANKLIN
               FIREPLACES WITH STACK HEATERS, ONE FIREPLACE AT EACH END OF
               THE PORCH. EACH FIREPLACE IS SERVED BY A 1/8 HP BLOWER WHICH
               CAN DIRECT THE HOT AIR TO THE ROOMS OR TO THE STORAGE.

MATERIAL   SINGLE GLAZED AND TRIPLE GLAZED WINDOWS; 3-1/2" AND 14" INSU-
               LATION; PLASTIC FOAM SHEATHING; WEATHERSTRIPPING; VAPOR BARRIERS;
               ALUMINUM SHEET WITH NON-SELECTIVE BLACK COATING; DOUBLE
               GLAZING; TEMPERED GLASS; TWO 1/2 HP BLOWERS; HEAT PUMPS; PLANAR
               REFLECTOR; COATED PLYWOOD-BACKED ALUMINUM; 10 TONS OF STONES;
               10 TONS OF CONCRETE SLAB; 4" OF POURED CONCRETE; TWO G.E.
               OUTDOOR AIR-TO-AIR HEAT PUMPS OF 2-TON CAPACITY; ELECTRIC
               HEATERS; TWO FRANKLIN FIREPLACES; 1/8 HP BLOWER

AVAILABILITY   ADDITIONAL INFORMATION MAY BE OBTAINED FROM HOWARD BELL
               ENTERPRISES, INC., P.O.BOX 413, VALLEY CITY, OHIO 44280.

REFERENCE   SHURCLIFF, W.A.
               SOLAR HEATED BUILDINGS, A BRIEF SURVEY
               CAMBRIDGE, MASSACHUSETTS, W.A. SHURCLIFF, MARCH 1976, 212 P

459

TITLE	HYATT REGENCY HOUSE HOTEL AND PLAZA
CODE	
KEYWORDS	HELIOSTAT MIRRORED REFLECTING GLASS ROOF XENON ARC SPOTLIGHT SOLAR LIGHTING
AUTHOR	METROPOLITAN STRUCTURES, INC. AND CARSON ASTRONOMICAL INSTRUMENTS
DATE & STATUS	THE ADDITION OF HELIOSTATS TO THE HYATT REGENCY HOUSE HOTEL WAS COMPLETED IN 1975.
LOCATION	CHICAGO, ILLINOIS
SCOPE	THE HYATT REGENCY HOUSE HOTEL IS A THIRTYSIX STORY HOTEL IN DOWNTOWN CHICAGO. BECAUSE OF ITS CLOSE PROXIMITY TO TWO NEIGH-BORING OFFICE BUILDINGS, IT GOT VIRTUALLY NO SUNLIGHT. BUT NOW, WITH THE USE OF HELIOSTATS, THE RESTAURANT, AN ART GALLERY, PART OF THE HOTEL LOBBY, AND A SCULPTURE IN THE OUTDOORS PLAZA WILL ALL HAVE SUNLIGHT.
TECHNIQUE	A HELIOSTAT IS A MIRRORED REFLECTING INSTRUMENT, WHICH TRACKS THE SUN THROUGH THE SKY, CATCHING THE RAYS AND TRANSFERRING THEM TO ANOTHER POINT. THE HELIOSTAT SYSTEM WILL PRODUCE FOUR SUNLET CIRCLES OF LIGHT, ROUGHLY 7 FEET IN DIAMETER. THREE OF THE BEAMS WILL BE REFLECTED THROUGH A GLASS-ROOFED SECTION OF THE HOTEL TO ILLUMINATE A RESTAURANT, AN ART GALLERY, AND PART OF THE HOTEL LOBBY. THE FOURTH BEAM WILL SPOTLIGHT A SCULPTURE IN THE OUTDOOR PLAZA. AT NIGHT, OR ON CLOUDY DAYS, THE MIRRORS WILL ASSUME A FIXED POSITION AND REFLECT LIGHT FROM FOUR HIGH POWERED XENON ARC SPOTLIGHTS TO PRODUCE A LIGHTING EFFECT SIMILAR TO DAYLIGHT.
MATERIAL	4 HELIOSTATS, GLASS ROOF, HIGH POWERED XENON ARC SPOTLIGHTS
AVAILABILITY	ADDITIONAL INFORMATION ON THE HELIOSTAT LIGHTING SYSTEM USED IN THE HYATT REGENCY HOUSE HOTEL MAY BE OBTAINED FROM METROPOLITAN STRUCTURES, INC., 111 E. WALKER DR., CHICAGO, ILLINOIS 60601.
REFERENCE	BUILDING DESIGN & CONSTRUCTION SUNSHINE: DOING IT WITH MIRRORS BUILDING DESIGN & CONSTRUCTION, DECEMBER 1974, P 16

TITLE	HYDE HOUSE
CODE	
KEYWORDS	HOUSE WINDOW AREAS WATER TYPE SOLAR COLLECTOR STORAGE CONTROL MICROPROCESSOR
AUTHOR	RAYTHEON CO.
DATE & STATUS	THE HYDE HOUSE WAS COMPLETED IN FEBRUARY 1976.
LOCATION	STOW, MASSACUSETTS
SCOPE	THE BUILDING IS A 4-LEVEL, WOOD-FRAME HOUSE WITH A LOFT AND A BASEMENT. THE AREA INCLUDES 2,950 SQ. FT., PLUS A 1,000 SQ. FT. GARAGE AND STORAGE AREA IN THE BASEMENT. THE LOWER LEVEL INCLUDES THE CHILDREN'S BEDROOMS AND THE FAMILY ROOM. THE NEXT, MAIN, LEVEL INCLUDES A LIVING-DINING-KITCHEN AREA AND THE MASTER BEDROOM. THE LOFT INCLUDES SPACE FOR AN OFFICE OR A BEDROOM.
TECHNIQUE	THE WINDOW AREAS ON THE EAST, NORTH, WEST, AND SOUTH ARE 311 SQ. FT., 58 SQ. FT., 67 SQ. FT., AND 0 SQ. FT., 436 SQ. FT. IN ALL. DOUBLE GLAZING IS USED. THE WALLS AND ROOF ARE INSULATED TO R-14 AND R-21, RESPECTIVELY. THE BASEMENT IS INSULATED WITH 2" OF STYROFOAM. THE HOUSE AIMS 10 DEGREES EAST OF SOUTH.
	THE COLLECTOR OCCUPIES 720 SQ. FT. GROSS, 652 SQ. FT. NET. IT IS A WATER TYPE COLLECTOR, AND COVERS A 32' X 24' SOUTH ROOF, SLOPING 45 DEGREES FROM THE HORIZONTAL. THE COLLECTOR HAS FOUR ROWS OF PANELS WITH 12 PANELS PER ROW, 48 PANELS IN ALL. EACH PANEL IS MADE BY REVERE COPPER AND BRASS CO., INC. EACH IS 7'8" X 2', AND EMPLOYS A COPPER ABSORBER PLATE WITH COPPER TUBES ATTACHED BY EPOXY CEMENT AND CLIPS. EACH PANEL IS SINGLE GLAZED WITH 3/16" OF TEMPERED GLASS. THE BACKING CONSISTS OF 6" OF FIBERGLAS. THE LIQUID IS WATER AND 45% PROPYLENE GLYCOL, AND IS CIRCULATED AT 10 GPM BY A 1/8 HP CENTRIFUGAL PUMP.
	THE STORAGE SYSTEM CONSISTS OF A 2,500 GALLON CYLINDRICAL HORIZONTAL STEEL TANK, 6' IN DIAMETER AND 13' LONG, CONTAINING WATER. THE INSULATION IS 4" OF URETHANE AND 1/2" OF PLASTER. THE ROOMS ARE HEATED BY A THREE-ZONE FAN-COIL SYSTEM.
	AUXILIARY HEAT IS PROVIDED BY A HEAT PUMP AND ELECTRICAL RESISTANCE HEATERS IN THE DUCTS. DOMESTIC HOT WATER IS PRE-HEATED BY THE STORAGE TANK AND/OR THE COLLECTOR, VIA A HEAT EXCHANGER. THERE IS ALSO AN ELECTRIC IMMERSION HEATER.
MATERIAL	DOUBLE GLAZED WINDOWS; 2" STYROFOAM; 48 REVERE COPPER & BRASS, INC. PANELS; COPPER ABSORBER PLATE WITH COPPER TUBES ATTACHED BY EPOXY CEMENT AND CLIPS; 3/16" TEMPERED GLASS; 6" FIBERGLAS; WATER AND 45% PROPYLENE GLYCOL; 1/8 HP CENTRIFUGAL PUMP; 2,500 GALLON CYLINDRICAL HORIZONTAL STEEL TANK; 4" URETHANE; 1/2" PLASTER; HEAT PUMP
AVAILABILITY	THE INSTALLATION IS PROTOTYPICAL, AND WILL NOT BE MARKETED. PERFORMANCE INFORMATION SHOULD BE AVAILABLE TENTATIVELY BY JUNE, 1977. ADDITIONAL INFORMATION MAY BE OBTAINED FROM THE VICE PRESIDENT, PLANNING DEPARTMENT, RAYTHEON CO., LEXINGTON, MASSACHUSETTS 02173.
REFERENCE	TIME MAGAZINE GIFT FROM THE SUN, THE TIME MAGAZINE, NOVEMBER 29, 1976, PP 69-72

TITLE	HYMAN HOUSE
CODE	
KEYWORDS	SOLAR PANEL COLLECTOR HEAT STORAGE TANK PUMP WATER HOUSE
AUTHOR	DR. MARK HYMAN
DATE & STATUS	THIS HOUSE WAS COMPLETED IN MARCH, 1976.
LOCATION	WALTHAM, MASSACHUSETTS
SCOPE	THIS 9-ROOM HOME IS DESIGNED FOR 100% SOLAR HEATING FOR AN AVERAGE WINTER IN WALTHAM, MASSACHUSETTS.
TECHNIQUE	THIS HOUSE USES 1,200 SQ. FT. OF ROOF MOUNTED SOLAR PANELS TO COLLECT HEAT. AN UNDERGROUND STORAGE TANK STORES WATER THAT IS WARMED WHEN IT PASSES THROUGH THE SOLAR PANELS. THIS SYSTEM IS DESIGNED TO PROVIDE 100% OF THE HEATING NEEDS FOR THIS HOUSE WITH AN ELECTRIC BASE BOARD AS BACKUP SYSTEM.
MATERIAL	SOLAR PANELS, ONE 16,000 GALLON STORAGE TANK
AVAILABILITY	ADDITIONAL INFORMATION MAY BE OBTAINED FROM DR. MARK HYMAN, PRESIDENT, SOLAR HEAT CORP., 108 SUMNER STREET, ARLINGTON, MASSACHUSETTS.
REFERENCE	ACKERMAN, JERRY SOLAR HEAT: AN INDUSTRY ABOUT TO TAKE OFF OR A FUTURISTIC PIPE DREAM BOSTON GLOBE, NOVEMBER 5, 1976, P 39 SLADE, MURMA SOLAR HEATED HOUSES, ARCHITECTURALLY NOT "FAR OUT" BOSTON HERALD AMERICAN, JANUARY 1, 1977, PP 6-7

TITLE	INA TOWER
CODE	
KEYWORDS	VARIEGATION ORIENTATION FACADES RECESSED SOFFITS OFFICE
AUTHOR	MITCHELL/GIURGOLA, ARCHITECTS
DATE & STATUS	THE INA TOWER WAS COMPLETED IN 1976.
LOCATION	PHILADELPHIA, PENNSYLVANIA
SCOPE	NO MONUMENT IN TERMS OF COST, THE INA TOWER IS MONUMENTALLY COMPELLING AS A WORK OF BOTH TECHNOLOGICAL SKILL AND ENVIRON- MENTAL DESIGN. FULLY SPRINKLERED INSIDE, WITH EQUIPMENT FOR EARLY SMOKE DETECTION, ITS ALTERNATE FACADE TREATMENTS, MODU- LATING THE ADMISSION OF SUNLIGHT IN ACCORD WITH ORIENTATION, HELPS KEEP OPERATING COSTS LOW DUE TO MINIMIZED ENERGY WASTE, ACCORDING TO THE CLIENT. IN LINE WITH STRINGENT BUDGET, THE COLUMNLESS INTERIORS ARE SPARE. BUT IN THE MORE PUBLIC SPACES, LIKE THE BLUE AND WHITE LOBBY, THE SIMPLICITY IS ELEGANT DUE TO THE CAREFUL HANDLING OF LIGHTING, THE PRACTICAL GESTURE OF THE CIRCULAR STAIR, AND THE INTRIGUING JUXTAPOSITION OF THE FIRST FLOOR AND BALCONY LEVELS, WHERE THE K-BRACED STRUCTURAL BENTS CUT DIAGONALLY THROUGH AT THE PERIMETER OF THE CORE.
TECHNIQUE	THE VARIEGATION OF THE CURTAIN WALL IS NOT CAPRICE, BUT CLIMATIC IN NATURE, RESPONDING TO THE ORIENTATION OF THE FACADES AND DENOTING THE PERIMETER RUN OF MECHANICALS. FOR EXAMPLE, ON THE UPPER PORTION OF THE SOUTH AND WEST FACADES, AS WELL AS ON THE EASTERN FACE, THE GLASS IS RECESSED WELL BEHIND THE OUTER WALL PLANE, THIS SETBACK POINTED UP BY CURVING SOFFITS AND SILLS. NATURALLY, THIS DETAIL REDUCES HEAT AND GLARE DURING SUMMERS. ON THE NORTH FACADE, HOWEVER, AND ON THE LOWER PORTION OF THE SOUTHERN ONE, WHICH IS SHADED BY A NEARBY BUILDING, THE GLASS IS FLUSH, FLOOR TO CEILING. WHERE THE RECESSED DETAIL APPEARS, THOSE CURVED SILLS CARRY THE INDUCTION UNITS; ELSEWHERE, THE UNITS STAND FREELY.
MATERIAL	COMPOSITE CONCRETE DECK, STEEL FRAME, GLASS
AVAILABILITY	ADDITIONAL INFORMATION MAY BE OBTAINED FROM MITCHELL/GIURGOLA, 12 SOUTH 12TH STREET, PHILADELPHIA, PA 19107.
REFERENCE	ARCHITECTURAL RECORD INA TOWER, INSURANCE COMPANY OF NORTH AMERICA, PHILADELPHIA ARCHITECTURAL RECORD, APRIL 1976, PP 114-116

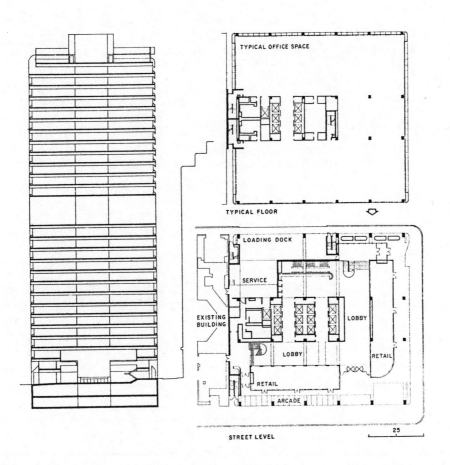

TYPICAL OFFICE SPACE

TYPICAL FLOOR

LOADING DOCK

SERVICE

EXISTING
BUILDING

LOBBY

LOBBY

RETAIL

RETAIL

ARCADE

STREET LEVEL

25

TITLE	INDIA VILLAGE ARCOLOGY
CODE	
KEYWORDS	SOLAR ENERGIZED CITY GREENHOUSE CHIMNEY APSE EFFECT ARCOLOGY
AUTHOR	PAOLO SOLERI
DATE & STATUS	INDIA VILLAGE HAS BEEN DESIGNED, BUT NOT BUILT, AS OF 1976.
LOCATION	THE THAR DESERT, INDIA
SCOPE	THE INDIA VILLAGE UTILIZES SIMPLE CONSTRUCTION TECHNOLOGY. IT IS DESIGNED FOR A COMMUNITY OF ARTISANS (3,000 TO 4,000 PEOPLE).
TECHNIQUE	INDIA VILLAGE IS BASED ON SEVERAL EFFECTS OF THE SUN, THE GREENHOUSE, CHIMNEY, AND APSE EFFECTS. AN ARCOLOGY IS SITUATED DIRECTLY ABOVE A LARGE TERRACED GREENHOUSE WHICH SENDS UP WARM, MOIST AIR THROUGH THE CHIMNEY EFFECT INTO THE CITY WHERE IT MAY BE USED OR STORED. DOME SHAPED ELEMENTS CATCH SUNLIGHT IN THE WINTER PROVIDING NATURAL HEAT.
MATERIAL	TRANSPARENT SHIELD
AVAILABILITY	ADDITIONAL INFORMATION MAY BE OBTAINED FROM PAOLO SOLERI, COSANTI FOUNDATION, 6433 DOUBLETREE ROAD, SCOTTSDALE, ARIZONA 85253.
REFERENCE	ARCOSANTI NEWSLETTER TWO SUNS EXHIBITION OPENS ARCOSANTI NEWSLETTER, SUMMER 1976, PP 1, 4-5

TITLE	INTEGRATED LIFE-SUPPORT SYSTEMS LABORATORIES
CODE	
KEYWORDS	HEMISPHERICAL DOME HOUSE GROUND LEVEL LOFT SPACE SUN TEMPERATURE STORAGE CAPACITY WIND ELECTRIC SYSTEM ELECTRICITY GENERATED HYDROPONIC HORTICULTURE PLASTIC FOAM INSULATION SOLAR HEATED BEDS HEATING FLAT PLATE DOUBLE GLAZED COLLECTOR TRANSFER MEDIUM WATER GLYCOL SOLUTION GALLON POLYURETHANE DOMESTIC HOT RADIATORS CONVECTION THETFORD TOILET LEAD ACID BATTERIES AIRLOCK PORTHOLE VENT
AUTHOR	ROBERT REINES
DATE & STATUS	THE INTEGRATED LIFE-SUPPORT SYSTEMS LABORATORIES WERE BUILT IN 1973.
LOCATION	TIJERAS, ALBUQUERQUE, NEW MEXICO

SCOPE

THE INTEGRATED LIFE-SUPPORT SYSTEMS LABORATORIES IS A HEMIS-PHERICAL DOME HOUSE WITH A 31'6" DIAMETER, A USABLE AREA OF 655 SQUARE FEET AT GROUND LEVEL, AND 200 SQUARE FEET IN LOFT SPACE.

ONE HUNDRED PERCENT OF THE SPACE AND WATER HEAT COMES FROM THE SUN. INDOOR TEMPERATURES HAVE BEEN MAINTAINED IN THE RANGE OF 65 - 85 DEGREES F., WITH OUTDOOR TEMPERATURES RANGING FROM -5 TO 100+ DEGREES F. THERE IS A STORAGE CAPACITY FOR SEVEN SUNLESS DAYS.

THE TOTAL COST OF THE DOME HOUSE (INCLUDING WIND ELECTRIC POWER SYSTEM) WAS $12,000.

ALL OF THE ELECTRICITY USED BY THE DOME HOUSE IS GENERATED BY THE WIND.

THE PARTS FOR TWO FURTHER DOMES HAVE BEEN FABRICATED AND AWAIT ERECTION. EXPERIMENTS HAVE BEGUN WITH HYDROPONIC HORTICULTURE, USING PLASTIC FOAM INSULATED, SOLAR HEATED BEDS.

TECHNIQUE

THE SOLAR HEATING SYSTEM CONSISTS OF A CONVENTIONAL FLAT-PLATE, DOUBLE GLAZED SOLAR COLLECTOR. THE HEAT TRANSFER MEDIUM IS A WATER-GLYCOL SOLUTION. A 3,000 GALLON WATER STORAGE TANK IS USED TO STORE HEAT. THE TANK IS HEAVILY INSULATED WITH POLYURETHANE FOAM. THERE IS ALSO A SMALL TANK FOR DOMESTIC HOT WATER STORAGE. THE HEATED WATER FROM THE TANK IS RUN THROUGH A RING OF RADIATORS AT THE BASE OF THE DOME, THUS GIVING A TORUS PATTERN OF HOT AIR CIRCULATION THROUGHOUT THE SPACE BY CONVECTION.

THE ELECTRIC POWER FOR RUNNING LIGHTS (150 WATTS TOTAL), STEREO, RADIO, TV, POWER TOOLS, THETFORD ELECTRIC TOILET AND SOLAR SYSTEM PUMPS IS SUPPLIED BY RECONDITIONED WIND MACHINES, OF WHICH THERE ARE PRESENTLY THREE IN OPERATION, GENERATING A TOTAL OF 4 KW. ELECTRICITY STORAGE IS IN 16 LEAD-ACID BATTERIES, WITH A CAPACITY SUFFICIENT FOR SEVEN WINDLESS DAYS; THE HOUSE CIRCUIT IS AT 110 V AC.

AN AIRLOCK IS FORMED BY THE 7' LONG ENTRANCE HALLWAY. THE DOME HAS 19 PORTHOLES OF 8-1/2" DIAMETER, AND A 6' SKYLIGHT AT THE TOP, WITH 11 CIRCULAR FIXED LIGHTS OF VARYING SIZES, AND ONE RECTANGULAR OPENING TO ACT AS A VENT.

MATERIAL

THE DOME IS BUILT FROM PRESSED STEEL SEGMENTAL SECTIONS, WITH 3" INTERNAL FLAME RESISTANT POLYURETHANE FOAM INSULATION. CONVENTIONAL DOUBLE GLAZED FLAT PLATE SOLAR COLLECTORS ARE USED. THE HEAT TRANSFER MEDIUM IS WATER-GLYCOL SOLUTION. THE WATER IS STORED IN A 3,000 GALLON TANK, AND CIRCULATED THROUGH THE HOUSE BY A RING OF RADIATORS. A THETFORD ELECTRIC TOILET IS USED. THERE ARE 4 RECONDITIONED WIND MACHINES. ELECTRICITY IS STORED IN 16 LEAD-ACID BATTERIES.

AVAILABILITY

ADDITIONAL INFORMATION ON THE INTEGRATED LIFE-SUPPORT SYSTEMS LABORATORIES CAN BE OBTAINED FROM THE OCCUPANTS, AT TIJERAS, ALBUQUERQUE, NEW MEXICO.

REFERENCE

STEADMAN, PHILIP
ENERGY, ENVIRONMENT AND BUILDING
CAMBRIDGE, ENGLAND, CAMBRIDGE UNIVERSITY PRESS, 1975, PP 149-150

TITLE   JACKSON HOUSE

CODE

KEYWORDS   HOUSE SOLAR COLLECTOR WATER STORAGE CONTROL

AUTHOR   ROY H. JACKSON

DATE & STATUS THE JACKSON HOUSE WAS COMPLETED IN NOVEMBER 1974. THE COLLECTOR
               IS PRESENTLY BEING CHANGED FROM A "TRICKLE" SYSTEM TO A
               "CLOSED" SYSTEM. THE TRICKLE SYSTEM PROVIDED 85% OF THE HEAT
               REQUIREMENT IN THE WINTER OF 1975-76, BUT NEEDED A NEW BLACK
               PAINT JOB. SO IT WAS DECIDED TO CHANGE IT TO A CLOSED SYSTEM FOR
               THE OBVIOUS ADVANTAGES PROVIDED. THE NEW PANEL WILL BE 432 SQ.
               FT., AND COVERED WITH TEDLAR FIBERGLAS. IT SHOULD BE IN OPERATION
               IN DECEMBER 1976.

LOCATION   COLORADO SPRINGS, COLORADO

SCOPE   THE BUILDING IS A ONE-STORY, 3-BEDROOM HOME WITH A WALK-IN
               BASEMENT. THERE ARE 4,400 SQ. FT., 4,000 SQ. FT. OF WHICH IS
               HEATED. THE OWNER BELIEVES THAT THE COLLECTOR IS SOMEWHAT
               UNDERSIZED, THAT IN COLD WEATHER THE BIN OF STONES MAY NOT
               BECOME HOT ENOUGH TO BE USEFUL. IT WOULD BE HELPFUL TO HAVE A
               LARGER COLLECTOR AND A LARGER STORAGE SYSTEM. HOWEVER, THE
               SYSTEM IS VERY SUCCESSFUL AS IS.

TECHNIQUE   THE BUILDING IS WELL INSULATED, WITH THERMOPANE WINDOWS, 6"
               BATT INSULATION IN THE CATHEDRAL CEILING, AND 3-1/2" BATT INSU-
               LATION IN THE WALLS.

               THE COLLECTOR AREA IS 380 SQ. FT. THE COLLECTOR IS THE TRICKLING
               WATER TYPE. IT IS DETACHED, AND ABOUT 20' TO THE EAST OF THE
               HOUSE. IT SLOPES 45 DEGREES FROM THE HORIZONTAL, AND FACES
               SOUTH. IT IS 16' HIGH AND 24' LONG. IT IS DOUBLE GLAZED WITH
               SALVAGED GLASS PANES, 1/4" THICK. WATER TRICKLES DOWN THE
               CORRUGATIONS OF THE BLACKENED SHEET, AND WATER COLLECTED BY THE
               GUTTER AT THE BOTTOM THEN FLOWS TO THE STORAGE SYSTEM, WHERE
               IT IS PUMPED BACK TO A DISTRIBUTION PIPE AT THE TOP OF THE
               COLLECTOR.

               THE STORAGE SYSTEM CONSISTS OF 1,000 GALLONS OF WATER IN A 1,000
               GALLON STEEL TANK WITHIN A BIN CONTAINING 29 CUBIC YARDS OF
               STONES, 3 TO 5" IN DIAMETER. THE BLOWER CIRCULATES AIR THROUGH
               THE BIN OF STONES TO THE ROOMS.

               A SET OF HONEYWELL CONTROL COMPONENTS IS USED FOR THE CONTROL
               SYSTEM. THE SET INCLUDES THREE THERMOSTATS. ONE IS AT THE INPUT
               TO THE COLLECTOR, AND ONE AT THE OUTPUT. THE PUMP IS TURNED OFF
               WHEN THE LATTER TEMPERATURE FAILS TO EXCEED THE FORMER TEMPERA-
               TURE. THE THIRD STARTS THE PUMP WHENEVER THE COLLECTOR TEMPERA-
               TURE EXCEEDS 118 DEGREES F. A CLOCK PROVIDES A DELAY THAT CAUSES
               THE SYSTEM TO TRY AGAIN AT LEAST ONCE EVERY 30 MINUTES AFTER
               CLOUDS HAVE COME OVER AND CAUSED THE COLLECTOR TEMPERATURE TO
               DROP.

MATERIAL   1/4" THICK GLASS PANES; 1,000 GALLON STEEL TANK; STONES, 3 TO 5"
               IN DIAMETER; HONEYWELL CONTROL COMPONENTS

AVAILABILITY   ADDITIONAL INFORMATION MAY BE OBTAINED FROM R.H. JACKSON,
               1066 E. WOODMAN VALLEY ROAD, COLORADO SPRINGS, COLORADO.

REFERENCE   SHURCLIFF, W.A.
               SOLAR HEATED BUILDINGS, A BRIEF SURVEY
               CAMBRIDGE, MASSACHUSETTS, W.A. SHURCLIFF, MARCH 1976, 212 P

TITLE	JANTZEN VACATION HOUSE
CODE	
KEYWORDS	VACATION HOUSE COLLECTOR TRANSPARENT ROOF REFLECTOR PASSIVE SHUTTERS CANOPIES SOLAR
AUTHOR	M. JANTZEN
DATE & STATUS	THE JANTZEN HOUSE WAS COMPLETED IN MARCH 1975.
LOCATION	CARLYLE, ILLINOIS
SCOPE	THE BUILDING IS A TWO-STORY, WEEK-END VACATION HOUSE, WITH TWO ROOMS AND A BATH, AND 650 SQ. FT. OF HEATED LIVING SPACE AND 290 SQ. FT. OF FENCED IN WOODEN DECK AT THE SOUTHERN EXPOSURE. THE SECOND STORY IS A 24' DIAMETER HALF-DOME OF ALUMINIZED STEEL. THE SIDES ARE MADE OF CORRUGATED STEEL. THE COST OF MATERIALS FOR THE ENTIRE HOUSE IS $10,000. THE INCREMENTAL COST OF THE SOLAR HEATING SYSTEM IS $350.
TECHNIQUE	THE SECOND STORY WALLS AND CEILING HAVE 2" TO 3" OF SPRAYED URE-THANE INSULATION TO WHICH A FIRE-RESISTANT PAINT IS APPLIED. THE FIRST STORY WALLS AND FLOOR HAVE 5" OF FIBERGLAS INSULATION.
	THE COLLECTOR AREA IS 96 SQ. FT. OF TRANSPARENT ROOF AREA, USED IN CONJUNCTION WITH A REFLECTOR AND A SMALL AMOUNT OF WATER STORAGE. THE SOLAR HEATING SYSTEM IS MAINLY OF THE PASSIVE TYPE. THE TRANSPARENT ROOF AREA IS AT A 30 DEGREE ANGLE FROM THE HORIZONTAL. IT CONSISTS OF A SINGLE LAYER OF CORRUGATED, 5-OZ. FILON (FIBERGLAS AND POLYESTER). TWO 12' X 4' SHEETS ARE USED.
	JUST ABOVE, THERE IS A 96 SQ. FT. REFLECTOR OF ALUMINIZED MYLAR BACKED BY CANVAS, PLYWOOD AND INSULATION, THAT HAS THE SAME SIZE AND SHAPE, SO THAT IT CAN BE CLOSED AND WILL THEN SERVE AS AN INSULATING COVER FOR THAT AREA. THE REFLECTOR, WHEN IN USE AS SUCH, DIRECTS MUCH RADIATION ALMOST STRAIGHT DOWNWARD, THROUGH THE TRANSPARENT ROOF AREA, AND TOWARD THE STORAGE TANKS. THE ANGLE OF THE REFLECTOR CAN BE ADJUSTED MANUALLY FROM WEEK TO WEEK, OR EVEN FROM HOUR TO HOUR IF THE OCCUPANT SO DESIRES. AT THE SOUTH SIDE OF THE BUILDING PROPER, THERE ARE SLIDING THERMOPANE GLASS DOORS, AND THREE PLASTIC HEMISPHERICAL WINDOWS WITH DIAMETERS OF 4', 4', AND 5', EACH WITH SHUTTERS OR CANOPIES THAT CAN BE USED TO INSULATE IN WINTER OR TO EXCLUDE THE SUN IN SUMMER.
	THE STORAGE SYSTEM CONTAINS 200 GALLONS OF LIQUID - 80% WATER, 20% ETHYLENE GLYCOL - IN TWO INSULATED, RECTANGULAR, 1' X 2' X 8' TANKS OF ANTI-CORROSION COATED, GALVANIZED STEEL. THE TANKS ARE SITUATED ON THE FLOOR, 6' BELOW THE TRANSPARENT ROOF AREA. THEY RECEIVE RADIATION TRAVELING DOWNWARD FROM THE REFLECTOR. THE VERTICAL STEEL "HONEYCOMB" INSIDE OF THE TANKS HELPS CARRY HEAT DOWNWARD INTO THE TANK AS A WHOLE. THE TANK TOP IS OF BLACK CORRUGATED METAL AND, 1-1/2" ABOVE IT, WITH INTERVENING AIR SPACE, IS A TRANSPARENT CORRUGATED PLASTIC SHEET. AT NIGHT, AN INSULATING CUSHION, SERVING ALSO AS A SEAT, CAN BE SWUNG DOWN TO CONFINE HEAT INSIDE THE TANK.
MATERIAL	ALUMINIZED STEEL; 2" TO 3" OF SPRAYED URETHANE INSULATION; 5" OF FIBERGLAS INSULATION; WATER; CORRUGATED, 5-OZ. FILON (FIBERGLAS AND POLYESTER); REFLECTOR OF ALUMINIZED MYLAR BACKED BY CANVAS, PLYWOOD AND INSULATION; THERMOPHANE GLASS; PLASTIC HEMISPHERICAL WINDOWS; SHUTTERS; CANOPIES; 200 GALLONS OF LIQUID (80% WATER, 20% ETHYLENE GLYCOL); TWO INSULATED, RECTANGULAR, 1' X 2' X 8' TANKS OF ANTI-CORROSION COATED, GALVANIZED STEEL; BLACK CORRUGATED METAL; TRANSPARENT CORRUGATED PLASTIC SHEET
AVAILABILITY	ADDITIONAL INFORMATION MAY BE OBTAINED FROM MIKE JANTZEN, BOX 172, CARLYLE, ILLINOIS 62231.
REFERENCE	HOUSE & GARDEN VACATION HOUSE WARMED BY SUN POWER HOUSE & GARDEN MAGAZINE, 1976, 4 P

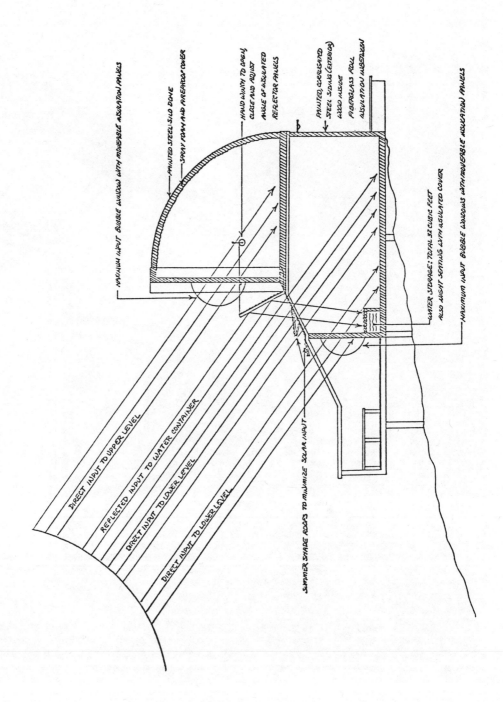

MAXIMUM INPUT BUBBLE WINDOW WITH MOVEABLE INSULATION PANELS

PAINTED STEEL SKID DOME

SPRAY FOAM AND FIREPROOF COVER

HAND WINCH TO OPEN CLOSE AND ADJUST ANGLE OF INSULATED REFLECTOR PANELS

PAINTED CORRUGATED STEEL SIDING (EXTERIOR)

WOOD INSIDE

FIBERGLASS ROLL

INSULATION SUBSTREEN

WATER STORAGE : TOTAL 32 CUBIC FEET ALSO NIGHT SEATING WITH INSULATED COVER

MAXIMUM INPUT BUBBLE WINDOWS WITH MOVEABLE INSULATION PANELS

DIRECT INPUT TO UPPER LEVEL

REFLECTED INPUT TO WATER CONTAINER

DIRECT INPUT TO LOWER LEVEL

DIRECT INPUT TO LOWER LEVEL

SUMMER SHADE ROOFS TO MINIMIZE SOLAR INPUT

TITLE   JEFFREY HOUSE

CODE

KEYWORDS   COLLECTOR WATER LIQUID SEALANTS STORAGE

AUTHOR   J.E. CLINTON

DATE & STATUS JEFFREY HOUSE WAS COMPLETED IN THE FALL OF 1975.

LOCATION   KENTFIELD HILLS, CALIFORNIA

SCOPE   THE BUILDING IS A THREE-STORY, 2,300 SQ. FT., 3-BEDROOM HOUSE-
               AND-OFFICE, ON THE SOUTHERN SLOPE OF A HILL. IT AIMS EXACTLY
               SOUTH.

TECHNIQUE   THE COLLECTOR AREA IS 900 SQ. FT. THE COLLECTOR IS THE WATER-
               TYPE. IT SLOPES 30 DEGREES FROM THE HORIZONTAL. THE HEART OF
               THE COLLECTOR IS A CORRUGATED ALUMINUM SHEET THAT, AFTER
               RECEIVING CORROSION-INHIBITING PRIMER AND A STANDARD CHEMICAL
               CONVERSION COATING, WAS COATED WITH A BLACK, NON-SELECTIVE,
               FLUOROCARBON POLYMER ENAMEL. THE CORRUGATION VALLEYS, 2.7" APART
               ON THE CENTERS, RUN UP AND DOWN THE SLOPE. THE LIQUID IS WATER
               THAT HAS BEEN SOFTENED BUT CONTAINS NO CORROSION INHIBITOR OR
               ANTIFREEZE. THE WATER IS FED TO THE CORRUGATION VALLEYS AT A
               CONTROLLABLE-RATE (0 TO 20 GPM) BY A 3/4 HP PUMP; IT IS FED BY
               A 1-1/4" DIAMETER MAIN SUPPLY PIPE OF COPPER AND A HORIZONTAL
               DISTRIBUTION PIPE ALONG THE RIDGE OF THE ROOF. IT IS FED TO
               EACH VALLEY BY TWO OR THREE 1/16" DIAMETER HOLES IN THE DISTRI-
               BUTION PIPE. THERE IS A COLLECTOR PIPE (GUTTER) AT THE BOTTOM.
               THE GLAZING CONSISTS OF A TIGHTLY PACKED PLANAR ARRAY OF GLASS
               TUBES OF A COMMON TYPE USED IN FLUORESCENT LAMPS; THE DIAMETER
               AND LENGTH ARE 1-1/2" BY 8 FT. THE TUBES CONTAIN AIR AT ATMOS-
               PHERIC PRESSURE. THE TUBES ARE SUFFICIENTLY STRONG THAT SUPPORTS
               ARE PROVIDED AT THE TUBE ENDS ONLY, THAT IS, THE SUPPORTS ARE
               8 FT. APART ON THE CENTERS. ORIENTATION OF THE TUBES IS UP-AND-
               DOWN THE ROOF; THUS RAIN HAS SOME TENDENCY TO KEEP THE TUBES
               CLEAN. THE SEALANTS USED AT TUBE ENDS AND ELSEWHERE ARE SILICONE
               AND ECH, I.E. EPICHLOROHYDRIN. THERE IS NO SEALANT BETWEEN
               ADJACENT TUBES. THE TUBES ARE EASILY REPLACED BY A PERSON ON
               A LADDER. THE TUBES ARE TOUCHING THE CORRUGATED ALUMINUM.
               COLLECTOR-BACK INSULATION CONSISTS OF AN AIRSPACE AND 1-1/2" OF
               ISOCYANURATE FOAM.

               THE STORAGE SYSTEM CONSISTS OF 3,000 GALLONS OF WATER IN A
               RECTANGULAR CONCRETE TANK IN THE CRAWL SPACE. THE TANK DIMENSIONS
               ARE 8-1/2 FT. X 8-1/2 FT. X 8 FT. HIGH. THE WALL THICKNESS IS
               8". THE TANK IS INSULATED WITH ISOCYANURATE FOAM. HOT WATER IS
               CIRCULATED BY A 1/20 HP PUMP TO TWO FAN-COIL SYSTEMS SERVING
               THE TWO HEATING ZONES.

               NO AIR COOLING SYSTEM IS PROVIDED. THE EAVES HELP EXCLUDE THE
               SUMMER SUN. THE WEST WINDOWS ARE OF REFLECTIVE GLASS. THE OPEN
               STRUCTURE OF THE FIRST TWO STORIES (WITH LOFT) AND A CLERESTORY
               PROVIDE NATURAL VENTILATION, WITH THE VENTING NEAR THE TOP OF
               THE ROOF ASSISTED BY A FAN.

MATERIAL   CORRUGATED ALUMINUM SHEET; CORROSION-INHIBITING PRIMER; STANDARD
               CHEMICAL CONVERSION COATING; BLACK, NON-SELECTIVE, FLUOROCARBON
               POLYMER ENAMEL; WATER THAT HAS BEEN SOFTENED BUT CONTAINS NO
               CORROSION INHIBITOR OR ANTIFREEZE; 3/4 HP PUMP; 1-1/4" DIAMETER
               MAIN SUPPLY PIPE OF COPPER; TIGHTLY PACKED PLANAR ARRAY OF GLASS
               TUBES; SILICONE; EPICHLOROHYDRIN; COLLECTOR-BACK INSULATION;
               1-1/2" ISOCYANURATE FOAM; 3,000 GALLONS OF WATER; RECTANGULAR
               CONCRETE TANK; 1/20 HP PUMP; REFLECTIVE GLASS

AVAILABILITY   ADDITIONAL INFORMATION MAY BE OBTAINED FROM JOHN E. CLINTON,
               INTERACTIVE RESOURCES, INC., 39 WASHINGTON AVE., POINT RICHMOND,
               CALIFORNIA 94801.

REFERENCE   SHURCLIFF, W.A.
               SOLAR HEATED BUILDINGS, A BRIEF SURVEY
               CAMBRIDGE, MASSACHUSETTS, W.A. SHURCLIFF, MARCH 1976, 212 P

TITLE	JENNIFER MASTERSON STUDIO
CODE	
KEYWORDS	POTTERY WORKSHOP WINDOW AREA SHADE AIR TYPE COLLECTOR GRAVITY CONVECTION STORAGE WOODEN FLOOR
AUTHOR	R. MASTERSON, P. VAN DRESSER, S. BAER
DATE & STATUS	THE JENNIFER MASTERSON STUDIO WAS COMPLETED IN 1973. THE BUILDING IS BASICALLY IDENTICAL IN DESIGN TO THE DAVIS HOUSE.
LOCATION	LA CIENEGA, NEW MEXICO
SCOPE	THE BUILDING IS A ONE-STORY STUDIO (POTTERY WORKSHOP), ABOUT 24' X 18'. THE FLOOR AREA IS ABOUT 500 SQ. FT.
TECHNIQUE	THE HOUSE FACES SOUTH. THE SOUTH WALL OF THE BASEMENT IS EXPOSED AND HOUSES THE COLLECTOR. THE SOUTH FACE OF THE MAIN STORY INCLUDES A WINDOW AREA, 16' X 8', SINGLE GLAZED, AND FLANKED BY TWO SLIDING COVERS, LIKE BARN DOORS, THAT CAN BE CLOSED MANUALLY BY A PERSON STANDING ON THE OUTDOOR WALKWAY. THIS WALKWAY HELPS TO SHADE THE COLLECTOR IN SUMMER. THE FLOOR IS OF WOOD. THE WALLS ARE OF 24" ADOBE. THE CEILING IS INSULATED WITH 6" FIBERGLAS.
	THE COLLECTOR IS ABOUT 100 SQ. FT., OF THE AIR TYPE, AND IS INTEGRAL WITH THE VERTICAL SOUTH WALL OF THE BASEMENT. THE COLLECTOR DIMENSIONS ARE ABOUT 22' X 4-1/2'. THE COLLECTOR IS GLAZED WITH A SINGLE LAYER OF ORDINARY WINDOW GLASS. THE PANE SIZE IS 1-1/2' X 1-1/2'. THE ABSORBER IS A SHEET OF CORRUGATED GALVANIZED STEEL THAT HAS A NON-SELECTIVE BLACK COATING. BETWEEN THE GLASS AND THE BLACK SHEET THERE ARE 4 LAYERS OF BLACK METAL LATHS WHICH INCREASE THE HEAT-TRANSFER SURFACE AREA. THE AIRFLOW IS CREATED BY GRAVITY CONVECTION. THE LARGE WINDOW AREA OF THE MAIN STORY ADMITS MUCH SOLAR RADIATION, WHICH STRIKES THE FLOOR AND WALLS.
	THE STORAGE SYSTEM CONTAINS ABOUT 20 TONS OF STONES. THE TYPICAL DIAMETER OF A STONE IS 5" TO 8". THE STONES COMPRISE A 24" THICK HORIZONTAL LAYER SITUATED BENEATH THE WOODEN FLOOR. AIR FROM THE TOP OF THE COLLECTOR FLOWS NORTH ALONG AND WITHIN THE LAYER (BIN) OF STONES, THEN FLOWS SOUTH AND DOWNWARD INTO THE BASEMENT FREE SPACE, AND EVENTUALLY REENTERS THE COLLECTOR ALONG ITS BASE. THERE ARE THREE REGISTERS IN THE FLOOR NEAR THE NORTH SIDE OF THE BUILDING. THESE ALLOW SOME WARM AIR TO PASS DIRECTLY FROM THE BIN TO THE ROOM. MUCH HEAT FLOWS TO THE ROOM VIA THE WOODEN FLOOR ALSO.
MATERIAL	24" ADOBE; 6" FIBERGLAS; SINGLE GLAZED, ORDINARY WINDOW GLASS; CORRUGATED GALVANIZED STEEL; NON-SELECTIVE BLACK COATING; 4 LAYERS OF BLACK METAL LATHS; 20 TONS OF STONES
AVAILABILITY	ADDITIONAL INFORMATION MAY BE OBTAINED FROM RICHARD MASTERSON, RT. 2, BOX 216, LA CIENEGA, NEW MEXICO 87501.
REFERENCE	SHURCLIFF, W.A. SOLAR HEATED BUILDINGS, A BRIEF SURVEY CAMBRIDGE, MASSACHUSETTS, W.A. SHURCLIFF, MARCH 1976, 212 P

TITLE	JOHNSON HOUSE
CODE	
KEYWORDS	HOUSE WINDOWS COLLECTOR WATER TYPE STORAGE
AUTHOR	TOTAL ENVIRONMENTAL ACTION, INC.
DATE & STATUS	THE JOHNSON HOUSE WAS RETROFITTED FOR SOLAR HEATING IN 1976.
LOCATION	SOUTH HADLEY, MASSACHUSETTS

SCOPE
THE BUILDING IS AN EXISTING HOUSE WITH 2-1/2 STORIES AND 4 BEDROOMS. IT IS A 2,550 SQ. FT., WOOD-FRAME HOUSE, 48' X 24'.

TECHNIQUE
THERE ARE FEW WINDOWS ON THE EAST AND NORTH. THE WINDOWS AND GLASS DOORS ARE DOUBLE GLAZED. THE THICKNESS OF THE FIBERGLAS INSULATION IS 3-1/2" IN THE WALLS AND CEILINGS.

THE COLLECTOR OCCUPIES 610 SQ. FT. GROSS, 535 SQ. FT. NET. IT IS A WATER TYPE COLLECTOR, MOUNTED ON THE SOUTH ROOF, SLOPING 56 DEGREES FROM THE HORIZONTAL. IT INCLUDES 35 SUNWORKS, INC., PANELS, OF WHICH 25 ARE 5'4" X 3', AND 10 ARE 7' X 3'. EACH PANEL CONTAINS A COPPER SHEET TO WHICH COPPER TUBES ARE SOLDERED, 5" OR 6" APART ON CENTERS. THE BLACK COATING IS SELECTIVE. IT IS SINGLE GLAZED WITH 3/16" TEMPERED GLASS. THE BACKING INCLUDES 2-1/2" OF FIBERGLAS (R-10-1/2). THE LIQUID IS WATER. THE SYSTEM IS DRAINED BEFORE FREEZE-UP CAN OCCUR. A COOLANT IS CIRCULATED AT 10 GPM BY A 1/3 HP CENTRI-FUGAL PUMP. ENERGY IS DELIVERED DIRECTLY TO THE STORAGE SYSTEM. HEAT EXCHANGER.

THE STORAGE SYSTEM CONTAINS 1,600 GALLONS OF WATER IN A 2,000 GALLON CONCRETE SEPTIC TANK INSULATED ON THE INSIDE TO R-20 WITH 5" OF POLYSTYRENE FOAM. A WATERPROOF LINER IS USED. THE TANK RESTS ON THE FLOOR OF THE BASEMENT. WHEN THE ROOMS NEED HEAT, WATER FROM THE TANK IS CIRCULATED THROUGH AN EXTRA-GREAT-LENGTH (300') OF BASEBOARD RADIATOR.

A CONVENTIONAL BOILER OPERATES IN PARALLEL WITH THE SOLAR SYSTEM. A CONTROL SYSTEM, INCORPORATING OUTDOOR TEMPERATURE MEASUREMENTS AND STORAGE TEMPERATURE MEASUREMENTS, IS USED TO MAXIMIZE THE USE OF THE RELATIVELY LOW-TEMPERATURE STORED HEAT.

MATERIAL
DOUBLE GLAZED WINDOWS AND GLASS DOORS; 3-1/2" OF FIBERGLAS INSULATION; 35 SUNWORKS, INC. PANELS; COPPER SHEET; COPPER TUBES; SELECTIVE BLACK COATING; 3/16" TEMPERED GLASS; 2-1/2" FIBER-GLAS; WATER; COOLANT; 1/3 HP CENTRIFUGAL PUMP; 2,000 GALLON CONCRETE SEPTIC TANK INSULATED WITH 5" OF POLYSTYRENE FOAM; WATERPROOF LINER; BASEBOARD RADIATOR; CONVENTIONAL BOILER

AVAILABILITY
ADDITIONAL INFORMATION MAY BE OBTAINED FROM TOTAL ENVIRONMENTAL ACTION, INC., CHURCH HILL, HARRISVILLE, NEW HAMPSHIRE 03450.

REFERENCE
SHURCLIFF, W.A.
SOLAR HEATED BUILDINGS, A BRIEF SURVEY
CAMBRIDGE, MASSACHUSETTS, W.A. SHURCLIFF, MARCH 1976, 212 P

TITLE	KELBAUGH HOUSE
CODE	
KEYWORDS	HOUSE HOME PASSIVE SOLAR THERMAL CHIMNEY FLYWHEEL THERMOGRAVITY STORAGE HEATILATOR FIREPLACE INFRARED HEATERS GAS FIRED FURNACE INSULATION TRIPLE DOUBLE GLAZE FORCED NATURAL CROSS VENTILATION REFLECTED
AUTHOR	DOUGLAS KELBAUGH
DATE & STATUS	THE KELBAUGH HOUSE WAS COMPLETED IN JULY 1975.
LOCATION	PRINCETON, NEW JERSEY
SCOPE	THE ARCHITECT WANTED A THREE-BEDROOM HOME WITH AT LEAST ONE LARGE INTERIOR ROOM AND A MAXIMUM OF USEABLE OUTDOOR SPACE. THE 60' X 100' SITE WITH A LARGE TREE NEAR THE CENTER IMPOSED TIGHT DESIGN RESTRICTIONS.
	THE ARCHITECT DECIDED EARLY IN THE DESIGN PROCESS TO USE A TROMBE PASSIVE SOLAR SYSTEM.
TECHNIQUE	A SOUTH-FACING GLASS WALL WITH A MASSIVE CONCRETE WALL BEHIND IT ACTS AS A THERMAL CHIMNEY AND FLYWHEEL. THE HOUSE WAS LOCATED ON THE NORTH SIDE OF THE SITE TO ESCAPE THE SHADOWS OF THE NEIGHBORING HOUSE TO THE SOUTH, TO MAKE USE OF THE SHADE TREES ON THE PROPERTY AND TO PROVIDE THE LARGEST POSSIBLE PRIVATE YARD. SHADOWS WERE RECORDED ON THE LOT ON DECEMBER 22, THE DAY WITH THE LOWEST SUN POSITION, AND IT WAS DETERMINED THAT A TWO-STORY VERTICAL SOLAR WALL WOULD NOT BE IN SHADOW.

THE RAYS OF THE WINTER SUN (1), LOW IN THE SKY, PENETRATE THE SOUTHERN WALL NEARLY PERPENDICULARLY, PASS THROUGH TWO SHEETS OF GLASS (2) AND HIT THE 15" CONCRETE WALL (3). THE BLACK PAINTED CONCRETE ABSORBS MUCH OF THE ENERGY.

OF THE HEAT THAT IS RADIATED BACK TOWARD THE GLASS, TWO-THIRDS IS RETAINED WITHIN THE GLASS. THE HOUSE IS HEATED IN THE WINTER BY DIRECT PASSAGE OF THE SUN'S HEAT BY THERMOGRAVITY. THE WARM WALL AND GLASS HEAT UP THE AIR WHICH THEN RISES UP THROUGH AN AIR SLOT TO THE INTERIOR SPACE (4). THIS CHIMNEY EFFECT SUCKS COOL AIR IN AT THE BOTTOM (5) AND VENTS WARM AIR AT THE TOP.(6). THE WARM AIR RELEASED AT THE TOP GRADUALLY COOLS AND FALLS (7). THE COOLER, HEAVIER AIR IS DRAWN OVER TO THE RETURN HOLE IN THE CONCRETE WALL AND BACK UP THE SOLAR CHIMNEY. THUS, HEAT IN THE FORM OF WARMED AIR IS CIRCULATED THROUGH THE ROOMS OF THE HOUSE -- ALL OF WHICH FACE ONTO THE CONCRETE WALL. THE SYSTEM CUT THE SPACE HEATING FUEL BILL BY 75% THE FIRST WINTER.

A GREENHOUSE ON THE SOUTH FACE AIDES THE SOLAR SYSTEM IN ITS COLLECTION AND STORAGE OF HEAT (8), AND IN HEATING THE CELLAR (9).

AN EXPOSED HEATILATOR FIREPLACE PROVIDES ADDITIONAL HEAT BUT ITS CONTRIBUTIONS ARE LIMITED (IF NOT COUNTERPRODUCTIVE) WHEN THE OUTSIDE TEMPERATURE FALLS 40 DEGREES BELOW THE INSIDE TEMPERA-TURE. THERE ARE INFRARED HEATERS INSTALLED IN THE BATHROOMS TO PROVIDE INSTANTANEOUS EXTRA HEAT, BUT THEY ARE SELDOM NEEDED.

A BACK-UP SYSTEM CONSISTS OF A CONVENTIONAL 58,000 BTU GAS-FIRED HOT-AIR FURNACE PLACED IN THE BASEMENT AND CONNECTED TO THE ROOMS BY SEWER PIPE CAST IN THE CONCRETE WALL.

THE FURNACE IS UNDERSIZED BECAUSE THE SKIN OF THE REST OF THE HOUSE IS WELL INSULATED WITH CELLULOSIC FIBER TO ACHIEVE A U FACTOR OF 0.06 IN THE WALLS AND 0.025 IN THE ROOF. THERE IS 4-1/2" FIBER INSULATION ON FIRST FLOOR WALLS AND 3-1/2" INSULA-TION ON SECOND FLOOR WALLS. THE ROOF HAS 9-1/2" OF INSULATION. KELBAUGH USED NEWSPAPER PULP BECAUSE IT HAS THE HIGHEST INSU-LATING PROPERTIES FOR THE LOWEST COST AND WAS LOCALLY AVAILABLE. INSULATING GLASS WAS USED IN ALL WINDOWS. THE ARCHITECT PLANS TO TRIPLE-GLAZE THE NORTH WINDOWS AND DOUBLE-GLAZE THE GREENHOUSE BEFORE NEXT WINTER.

THE HOUSE IS COOLED IN THE SUMMER BY FORCED AND NATURAL CROSS-VENTILATION. THIS PROCESS COOLS THE MASSIVE CONCRETE WALL AT NIGHT AND, BY ACTING AS A THERMAL FLYWHEEL, IT COOLS THE HOUSE THE FOLLOWING DAY. HEAT BUILDUP IS EXHAUSTED BY FOUR FANS AT THE TOP OF THE WALL. THERE IS MARKEDLY REDUCED SOLAR COLLECTION IN THE SUMMER BECAUSE THE SUN IS HIGH ENOUGH IN THE SKY TO BE REFLECTED BY THE VERTICAL GLASS WALL. THE GREENHOUSE, HOWEVER, MUST HAVE SHADES DRAWN OVER IT BECAUSE THE GLASS IS MORE PERPENDICULAR TO THE SUN'S RAYS. TWO STRATEGIC DECIDUOUS TREES ALSO PROVIDE SHADE FOR THE HOUSE.

THE HOUSE HAS A MINIMUM NUMBER OF WINDOWS, PARTICULARLY ON THE NORTH WALL WHERE THEY LOSE MORE HEAT THAN THEY COLLECT.

MATERIAL
GLASS WALL, 15" CONCRETE WALL, BLACK-PAINTED CONCRETE, WARMED AIR, GREENHOUSE, HEATILATOR FIREPLACE, INFRARED HEATERS, GAS-FIRED HOT-AIR FURNACE, SEWER PIPE, CELLULOSIC FIBER, NEWSPAPER PULP, TRIPLE-GLAZE WINDOWS, DOUBLE-GLAZED GREENHOUSE, FOUR FANS, SHADES, DECIDUOUS TREES

AVAILABILITY
FURTHER INFORMATION MAY BE OBTAINED FROM DOUGLAS KELBAUGH, ARCHITECT, 70 PINE STREET, PRINCETON, NEW JERSEY 08540. SLIDE AND INFORMATION KIT - $25; BLUEPRINTS AND DESIGN SERVICES ON REQUEST.

REFERENCE
AMERICAN INSTITUTE OF ARCHITECTS
CASE STUDY 5: KELBAUGH HOUSE
WASHINGTON, D.C., IN AIA ENERGY NOTEBOOK, PUBLISHED BY THE AMERICAN INSTITUTE OF ARCHITECTS, 1975, PP CS-19 TO CS-20
SOLAR AGE, JULY 1976

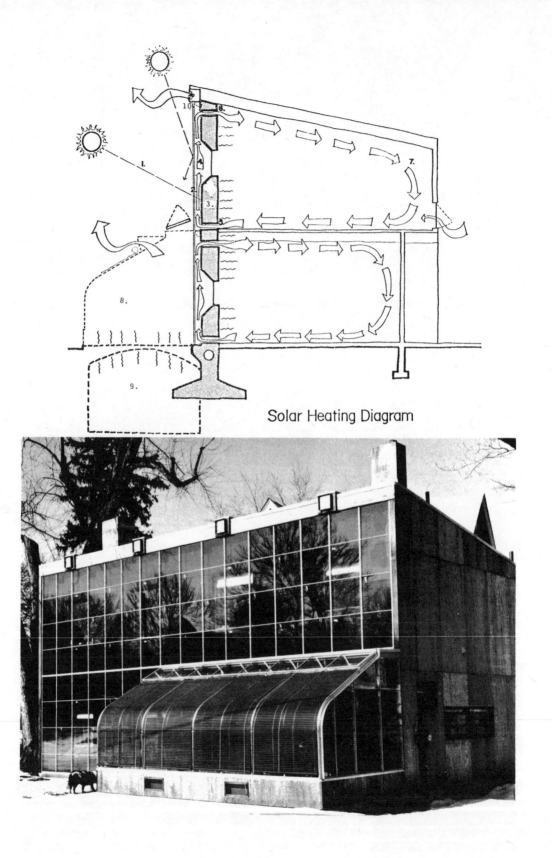

Solar Heating Diagram

TITLE	KIMURA SOLAR HOUSE
CODE	
KEYWORDS	HOUSE WATER TYPE COLLECTOR STORAGE HEAT PUMP RADIANT HEAT
AUTHOR	K. KIMURA
DATE & STATUS	THE KIMURA SOLAR HOUSE WAS CONSTRUCTED IN 1972, AND SOLAR PANELS INSTALLED IN 1973.
LOCATION	TOKOROZAWA, JAPAN
SCOPE	THE BUILDING IS A TWO-STORY, 150 SQ.M, WOOD-FRAME DWELLING WITH A HALF-BASEMENT. THE SOLAR SYSTEM HAS WORKED WELL. THE PREHEATING OF DOMESTIC HOT WATER WAS INSUFFICIENT IN WINTER, AND THE OWNER INTENDS TO INSTALL A SEPARATE SOLAR COLLECTOR FOR HEATING THE DOMESTIC HOT WATER.
TECHNIQUE	THE FIRST-STORY FLOOR CONSISTS OF A CONCRETE SLAB, CONTAINING MANY PARALLEL PIPES.
	THE COLLECTOR IS A 24 SQ.M (255 SQ. FT.), WATER TYPE COLLECTOR, MOUNTED ON THE VERTICAL SOUTH WALLS OF THE 1ST AND 2ND STORIES. IT CONSISTS OF 16 PANELS, EACH 0.9 M X 1.8 M (NET ABSORBER AREA IS 1.5 SQ.M.) THE HEART OF THE PANEL IS A 0.5 MM COPPER SHEET, TO WHICH 10 MM IN DIAMETER COPPER TUBES, 133 MM APART ON CENTERS, HAVE BEEN SOLDERED. THE BLACK COATING IS NON-SELECTIVE. THE PANEL IS SINGLE GLAZED AND HAS A FIBERGLAS BACKING. THE LIQUID IS PLAIN WATER WITH NO ANTI-FREEZE ADDED. WHEN THE OUTDOOR NIGHTTIME TEMPERATURE IS VERY LOW, THE WARM WATER IS CIRCULATED THROUGH THE COLLECTOR TO PREVENT FREEZE-UP. THE PANELS ARE CONNECTED TO THE MANIFOLDS BY RUBBER TUBINGS, AND ARE MOUNTED SO THAT THEY CAN BE SLID LATERALLY TO PERMIT SUNLIGHT TO ENTER THE ROOMS WHEN THE OCCUPANTS SO DESIRE, AND TO FACILITATE CLEANING THE INNER FACE OF THE WINDOW. THE WATER IS CIRCULATED TO THE COLLECTOR FROM THE STORAGE TANK.
	THE STORAGE SYSTEM CONSISTS OF A 1 CUBIC METER, WATER-FILLED, CONCRETE, INSULATED TANK IN THE BASEMENT. THE TANK INCLUDES A SPECIAL COIL AND AN ELECTRIC HEATING ELEMENT. WHEN THE ROOMS NEED HEAT, E.G. AT NIGHT, A 1.5 KW, COP-4, HEAT PUMP EXTRACTS ENERGY FROM THIS TANK AND DELIVERS IT TO THE WATER THAT IS CIRCULATED THROUGH THE PIPES IN THE CONCRETE SLAB, THUS PROVIDING RADIANT HEAT AND CONVECTIVE HEAT TO THE ROOMS ABOVE AND ALSO TO THE ROOMS BELOW.
MATERIAL	CONCRETE SLAB CONTAINING MANY PARALLEL PIPES; 0.5 MM COPPER SHEET; 10 MM DIAMETER COPPER TUBES; NON-SELECTIVE BLACK COATING; SINGLE GLAZED PANELS; FIBERGLAS BACKING; PLAIN WATER; RUBBER TUBINGS; 1 CUBIC METER WATER-FILLED, CONCRETE, INSULATED TANK; SPECIAL COIL; ELECTRIC HEATING ELEMENT; 1.5 KW, COP-4 HEAT PUMP
AVAILABILITY	ADDITIONAL INFORMATION MAY BE OBTAINED FROM PROF. KEN-ICHI KIMURA, DEPARTMENT OF ARCHITECTURE, WASEDA UNIVERSITY, NISHIOKUBO, SHINJUKU-KU, TOKYO, JAPAN.
REFERENCE	KIMURA, KEN-ICHI PRESENT TECHNOLOGIES OF SOLAR HEATING, COOLING AND HOT WATER SUPPLY IN JAPAN ARCHITECTURAL SCIENCE REVIEW, VOL 19, NO 2, JUNE 1976, 3 P

TITLE	KNEPSHIELD HOUSE
CODE	
KEYWORDS	HOUSE WALL MASSIVE DRAPERIES PASSIVE COLLECTOR AIR TYPE STORAGE
AUTHOR	W.H. KNEPSHIELD
DATE & STATUS	THE KNEPSHIELD HOUSE WAS COMPLETED IN FEBRUARY 1975. IN SEPTEMBER 1975, TESTS ON THE PANELS WITH 3 TO 12 LAYERS OF BRICKS WERE UNDERWAY AND WERE SAID TO SHOW PROMISE..
LOCATION	MIFFLINBURG, PENNSYLVANIA
SCOPE	THE BUILDING IS A ONE-STORY, RANCH-TYPE, 72' LONG, 1,700 SQ. FT. HOUSE. IT HAS 3 BEDROOMS, 2 BATHS, AND MANY OTHER ROOMS, PLUS A FULL BASEMENT AND A GARAGE, BUT NO ATTIC. ONLY READILY AVAILABLE MATERIALS ARE USED; NOTHING IS USED THAT WILL NOT PAY FOR ITSELF IN FUEL SAVINGS WITHIN A 3 YEAR PERIOD. THE COLLECTORS CAN BE BUILT FOR LESS THAN $2.00 PER SQ. FT.
TECHNIQUE	THE SOUTH WALL IS MASSIVE, AND INCLUDES TWO SPACED LAYERS OF BRICKS. IT HAS WINDOWS WITH AN AGGREGATE AREA OF 178 SQ. FT. THE WINDOWS ARE NON-OPENABLE, DOUBLE GLAZED, AND ARE COVERED WITH HEAVY DRAPERIES ON COLD NIGHTS. THERE ARE FEW WINDOWS ON THE EAST, NORTH AND WEST SIDES. THE LIVING AREA HAS A VERY SPECIAL FIREPLACE AT THE WEST END, AND A GARAGE AT THE EAST END.

THE PASSIVE COLLECTOR IS 178 SQ. FT. OF VERTICAL, DOUBLE-GLAZED WINDOWS ON THE SOUTH WALL. THE RADIATION ENTERS THE WINDOWS AND STRIKES THE FLOOR. THE FLOOR IS OF THE NON-MASSIVE TYPE, AND IS COVERED WITH A CARPET BENEATH WHICH IS ALUMINUM FOIL.

THE ACTIVE COLLECTOR IS A 432 SQ. FT., AIR-TYPE COLLECTOR, CONSISTING OF THREE DETACHED, SEPARATE ARRAYS, OR PANELS, 6' TO 25' FROM THE SOUTHWEST CORNER OF THE HOUSE. EACH PANEL IS TILTED 56 DEGREES FROM THE HORIZONTAL. THE HEART OF EACH PANEL IS A SHEET OF CORRUGATED ALUMINUM, THE UPPER SURFACE OF WHICH HAS A NON-SELECTIVE BLACK COATING. EACH PANEL IS SINGLE GLAZED WITH A 3.8' WIDE, UNSUPPORTED SHEET OF 0.004" TEDLAR. A 1" SPACE BETWEEN THE GLAZING AND ALUMINUM SHEET CONTAINS STILL AIR. IN A 3" SPACE BEHIND THE ALUMINUM SHEET, AN AIR-FLOW IS MAINTAINED BY A BLOWER, SITUATED ADJACENT TO THE STORAGE SYSTEM. THE REAR OF THIS LATTER SPACE IS FACED WITH ORDINARY 2" THICK BRICKS. ON DAYS WITH INTERMITTENT SUN, THE THERMAL CAPACITY OF THE ARRAY OF BRICKS INCREASES THE FLOW OF ENERGY TO THE STORAGE SYSTEM BY SMOOTHING THE FLOW. BEHIND THE BRICKS ARE ALUMINUM FOIL, 3-1/2" OF FIBERGLAS, AND PLYWOOD. THE FORCED AIR, TRAVELING IN A 6" DIAMETER GALVANIZED IRON DUCT, INSULATED WITH 3-1/2" FIBERGLAS AND PLASTIC COVER, TRAVELS THROUGH THE THREE PANELS SEQUENTIALLY. THE BRICK-LADEN PANELS ARE MOUNTED ON THE GROUND BECAUSE THEY ARE TOO HEAVY FOR THE ROOF.

THE STORAGE SYSTEM CONSISTS OF 30 TONS OF 1" TO 2" DIAMETER STONES IN A 10' X 6' X 8' BIN IN THE BASEMENT. THE STONES REST ON A WIRE MESH, FORMING THE TOP OF THE HORIZONTAL, 4" HIGH PLENUM CONTAINING, FOR SUPPORT, MANY SPACED BRICKS LYING ON THEIR 2" X 8" EDGES. 2" SPACES ARE LEFT BETWEEN THEM FOR AIRFLOW. THE INSULATION ON THE TOP OF THE BIN IS 7" OF FIBERGLAS, 1" FOAM, 1/2" SHEETROCK, AND ALUMINUM FOIL. THE INSULATION ON THE BOTTOM AND ON THE SIDES IS SIMILAR, EXCEPT THAT ONLY 3-1/2" FIBERGLAS IS USED. HOT AIR FROM THE COLLECTOR ENTERS THE BIN-PLENUM AT ONE END OF THE BIN AND LEAVES AT THE BOTTOM OF THE OTHER END, AND THEN FLOWS BACK TO THE COLLECTOR. WHEN THE ROOMS NEED HEAT, A FRACTIONAL-HP BLOWER SENDS AIR FROM THE BIN, VIA A SINGLE 6" DIAMETER EXIT PIPE AT THE TOP OF THE BIN, TO A 12-PIPE SYSTEM SERVING 12 BENEATH-WINDOW LOCATIONS IN THE HOUSE.

AUXILIARY HEAT IS PROVIDED BY A SPECIAL FIREPLACE-STOVE AT THE WEST END OF THE LIVING ROOM. THE FIRE BURNS WITHIN A 20" DIAMETER. THERE IS A FIVE-TURN COIL OF 3" DIAMETER FLEXIBLE STEEL TUBING, THROUGH WHICH AIR IS CIRCULATED TO VARIOUS ROOMS BY A SMALL BLOWER. THE COIL AXIS IS HORIZONTAL. ALTERNATIVELY, HOT AIR CAN BE SENT TO THE BIN OF STONES. ELECTRIC HEATERS ARE ALSO USED.

COOLING IN SUMMER IS ACCOMPLISHED BY A 1/2 HP BLOWER BLOWING
ROOM AIR THROUGH A TRENCH, CONSISTING OF 100 TONS OF STONES,
AND BACK TO THE ROOMS. THE STONES ARE KEPT COOL BY CONTACT
WITH THE EARTH. THE TRENCH IS LOCATED JUST TO THE SOUTH OF THE
HOUSE.

MATERIAL   DOUBLE GLAZED WINDOWS; HEAVY DRAPERIES; BRICKS; CARPET; ALUMINUM
               FOIL; NON-SELECTIVE BLACK COATING; 3.8' WIDE SHEET OF 0.004"
               TEDLAR; BLOWER; ALUMINUM FOIL; 3-1/2" FIBERGLAS; PLYWOOD;
               6" DIAMETER GALVANIZED IRON DUCT; 3-1/2" FIBERGLAS; PLASTIC
               COVER; 30 TONS OF 1" TO 2" DIAMETER STONES; WIRE MESH; 7" FOAM;
               1/2" SHEETROCK; ALUMINUM FOIL; FRACTIONAL-HP BLOWER; FIREPLACE-
               STOVE; 3" DIAMETER FLEXIBLE STEEL TUBING

AVAILABILITY   ADDITIONAL INFORMATION MAY BE OBTAINED FROM W.H. KNEPSHIELD,
               84 LANEY STREET, MIFFLINBURG, PENNSYLVANIA 17844.

REFERENCE   SHURCLIFF, W.A.
               SOLAR HEATED BUILDINGS, A BRIEF SURVEY
               CAMBRIDGE, MASSACHUSETTS, W.A. SHURCLIFF, MARCH 1976, 212 P

486

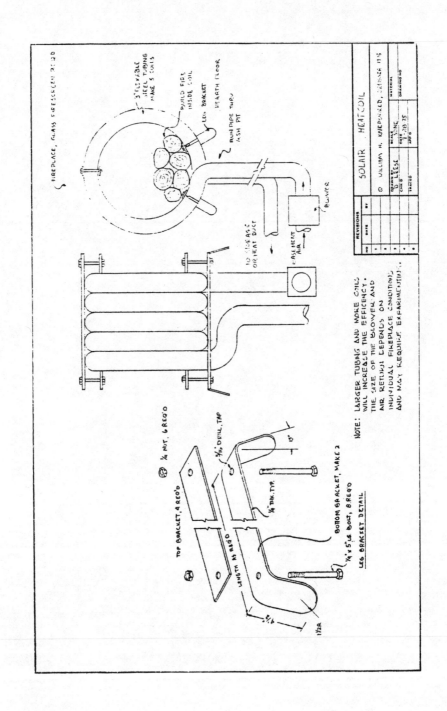

FIREPLACE, GLASS FIRESCREEN REQ'D

3" FLEXIBLE STEEL TUBING MAKE 5 COILS

BUILD FIRE INSIDE COIL

LEG BRACKET

HEARTH FLOOR

RUN PIPE THRU ASH PIT

BLOWER

TO FURNACE OR HEAT DUCT

BASEMENT AIR

NOTE: LARGER TUBING AND MORE COIL WILL INCREASE THE EFFICENCY. THE SIZE OF THE BLOWER AND AIR RETURN DEPENDS ON INDIVIDUAL FIREPLACE CONDITIONS AND MAY REQUIRE EXPERIMENTING.

¼ NUT, 6 REQ'D

³⁄₁₆ DRILL, TAP

¼ THK. TYP.

5"

TOP BRACKET, 4 REQ'D

LENGTH AS REQ'D

¼ x 5"LG BOLT, 8 REQ'D

BOTTOM BRACKET, MAKE 2

LEG BRACKET DETAIL

2¼"

1½R

SOLAIR HEATCOIL

© WILLIAM N. KNOPENGELD, DESIGNER 1975

DRAWN BY
D. LEESE

SCALE
NONE

CH'K'D

DATE
1.28.75

MATERIAL

DRAWING NO.

APP'D

TRACED

REVISIONS
NO   DATE   BY

TITLE	KORMAN CORPORATION SOLAR HOUSE
CODE	
KEYWORDS	SOLAR HEAT HOT WATER PANEL HOUSE HUD STORAGE TANK COLLECTOR
AUTHOR	THIS HOUSE IS PART OF THE HUD RESIDENTIAL SOLAR HEATING AND COOLING DEMONSTRATION PROGRAM. THE KORMAN CORP. BUILT THE HOUSE.
DATE & STATUS	THIS HOUSE WAS BUILT IN 1976.
LOCATION	GLOUCESTER TOWNSHIP, NEW JERSEY
SCOPE	SOLAR POWER IS EXPECTED TO SUPPLY ABOUT 63% OF THE PROJECTED HEATING NEEDS, AND 90% OF THE DOMESTIC HOT WATER NEEDS FOR THIS HOUSE. THE KORMAN CORP., WHICH BUILT THE HOUSE, RECEIVED A $32,000 GRANT FROM HUD FOR THE PURPOSE OF EQUIPPING TWO HOUSES WITH SOLAR HEATING SYSTEMS.
TECHNIQUE	ROOF MOUNTED COLLECTORS PROVIDE HEAT FOR THIS HOUSE. HEATED WATER IS STORED IN TWO UNDERGROUND 500 GALLON STORAGE TANKS. BLOWERS DIRECT WARMED AIR THROUGH THE DUCT SYSTEM WITHIN THE HOUSE.
MATERIAL	SOLAR COLLECTORS, STORAGE TANKS, BLOWERS
AVAILABILITY	ADDITIONAL INFORMATION MAY BE OBTAINED FROM THE KORMAN CORP., JENKINTOWN PLAZA, JENKINTOWN, PENNSYLVANIA.
REFERENCE	PROFESSIONAL BUILDER BUILDERS RATE SOLAR DEMO PROGRAM: FAST PROCESSING BUT SLOW DELIVERY PROFESSIONAL BUILDER, NOVEMBER 1976, P 52

490

490½

TITLE	KRUEGER HOME
CODE	
KEYWORDS	DIRECT CURRENT WINDMILL GENERATOR
AUTHOR	JACK KRUEGER
DATE & STATUS	THE KRUEGER WIND GENERATOR IS CURRENTLY PRODUCING AN AVERAGE OF 4 KWH OF ENERGY A DAY (1976).
LOCATION	GRAND FORKS, NORTH DAKOTA
SCOPE	KRUEGER'S WINDMILL WEIGHS 600 POUNDS AND TOOK 25 MAN-HOURS TO CONSTRUCT. THE PROP NEEDS AN 8-10 MPH GUST TO GET IT STARTED, AND KRUEGER SAYS A 20-FOOT DIAMETER BLADE WOULD QUADRUPLE OUTPUT WITHOUT CHANGING THE BASIC STRUCTURE - AND STILL COST ABOUT $400 TO BUILD. WITH THE INCREASED OUTPUT, SUCH A WINDMILL COULD PAY FOR ITSELF IN FOUR OR FIVE YEARS.
TECHNIQUE	THE DIRECT CURRENT PRODUCED RUNS TO SPACE HEATERS IN WINTER, AND TO THE WATER-HEATING SYSTEM IN SUMMER. ON CALM DAYS, OR WHEN EXTRA HEATING POWER IS NEEDED, A MANUAL SWITCHOVER TO UTILITY POWER IS POSSIBLE. ON A NORMAL WINDY DAY, WIND VELOCITY AVERAGES 15 MPH OVER THE 24-HOUR PERIOD, AND A TOTAL OF 4 KWH OF ENERGY IS EXTRACTED BY THE WINDMILL. THE WINDMILL SITS ATOP A FIFTY-FOOT TOWER, AND CONSISTS OF A USED 3-KW GENERATOR AND A JUNKED CAR'S FRONT-WHEEL SPINDLE ASSEMBLY. THE GENERATOR IS A 300-VOLT DC UNIT RATED AT 1800 RPM THAT PRODUCES 125 VOLTS AND 3,000 WATTS AT 600 RPM. THE GENERATOR CASE IS SEALED AROUND ALL BOLTS AND JOINTS WITH A HIGH-GRADE CAULKING MATERIAL, AND THE VENTI-LATING COVERS ARE REPLACED WITH SOLID COVERS.

THE 10-FOOT PROPELLER USES FIXED-PITCH BLADES ENGINEERED TO SPIN AT 600 RPM IN A 40-MPH WIND. THE BLADES WERE CARVED OUT OF FIVE 1X8 PINE BOARDS LAMINATED TOGETHER WITH RESORCINOL-RESIN WATERPROOF GLUE, AND CONTOURED WITH A COARSE (16-GRIT) SANDING DISK. THE ENTIRE SURFACE IS COVERED WITH FIBERGLAS AND POLYESTER RESIN. THE TAIL IS OF 1/2" EXTERIOR-GRADE PLYWOOD SIMILARLY PROTECTED. |
MATERIAL	10-FOOT PROPELLER, PINE BOARDS FIBERGLAS POLYESTER RESIN, PLYWOOD
AVAILABILITY	ADDITIONAL INFORMATION MAY BE OBTAINED FROM PROF. JACK KRUEGER, UNIVERSITY OF NORTH DAKOTA, GRAND FORKS, ND 58201.
REFERENCE	WILKE, JOHN JACK KRUEGER: WIND POWER HELPS HEAT HIS HOME POPULAR SCIENCE, JANUARY 1976, PP 103, 131

Turntable beneath generator is rebuilt auto brake-spindle unit—ideal housing for commutator (⅛" copper slip rings on Micarta disk inside brake drum.)

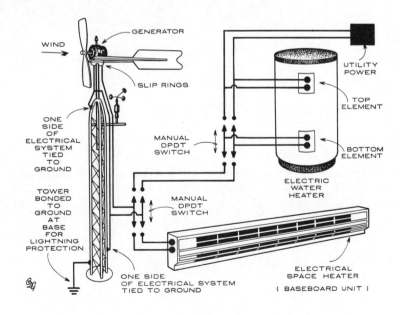

492

TITLE	KRUSCHKE HOUSE
CODE	
KEYWORDS	HOUSE INSULATION WINDOW AREA GREENHOUSE PASSIVE STORAGE
AUTHOR	D. KRUSCHKE
DATE & STATUS	THE KRUSCHKE HOUSE WAS COMPLETED IN 1975.
LOCATION	WILD ROSE, WISCONSIN
SCOPE	THE BUILDING IS A ONE-STORY, 1,000 SQ. FT., 64' X 15', HOUSE WITH NO ATTIC OR BASEMENT.
TECHNIQUE	THE HOUSE AIMS EXACTLY SOUTH. THE WALL AND ROOF INSULATION IS 6" OF FIBERGLAS. THE WINDOW AREAS ON THE EAST, NORTH, AND WEST ARE SMALL, 10 SQ. FT. IN ALL, AND DOUBLE GLAZED WITH 0.004" POLYETHYLENE FILM. THE SOUTH WINDOW AREA IS LARGE. ALONG THE SOUTH PORTION OF THE HOUSE, THERE IS A GREENHOUSE AREA, 60' X 6'. THE EARTH IS 16" DEEP, AND RESTS ON A 2" LAYER OF STYROFOAM. THE SOLAR RADIATION ENTERS THE GREENHOUSE VIA A WINDOW AREA, 56' X 5-1/2', SLOPING 60 DEGREES FROM THE HORIZONTAL; THUS THE BUILDING IS 3' WIDER AT THE FLOOR LEVEL THAN AT HEAD HEIGHT. THE WINDOW AREA IS DOUBLE GLAZED WITH 0.004" POLYETHYLENE FILM HELD BY STAPLES AND, MORE IMPORTANT, 1/4" THICK WOODEN BATTENS. THE POLYETHYLENE FILMS ARE REPLACED EACH YEAR, IF NECESSARY. THE COST IS SMALL, 2¢/SQ. FT.
	THE REMAINDER OF THE HOUSE HAS 3-1/2" OF POURED CONCRETE FLOOR SLAB RESTING ON 2" OF STYROFOAM. THE EDGES OF THE SLAB ARE INSULATED WITH 2" TO 4" OF STYROFOAM. THE GREENHOUSE WINDOW AREA IS PROVIDED, AT NIGHT, WITH STYROFOAM PANELS, 5-1/2' X 4' X 2", STORED, DURING THE DAY, JUST ABOVE THE WINDOW REGION.
	THE SOLAR HEATING SYSTEM IS OF THE PASSIVE TYPE. RADIATION ENTERS VIA THE GREENHOUSE WINDOW AREA AND STRIKES THE GREENHOUSE EARTH, THE BARRELS, THE CONCRETE FLOOR SLAB, ETC. IT PROVIDES ALSO MOST OF THE DAYLIGHT FOR THE ROOMS. 1,500 GALLONS OF WATER IS CONTAINED IN 28 STEEL DRUMS, OR BARRELS, OF 55 GALLON CAPACITY EACH. THE BARRELS FORM A SINGLE ROW, AND ARE PARALLEL TO ONE ANOTHER. THE BARREL AXIS IS NORTH-SOUTH. THE BARRELS ARE JUST ABOVE THE FLOOR LEVEL, BENEATH THEM IS EARTH. THE BARREL ENDS ARE SUPPORTED BY THE CONCRETE FLOOR SLAB. NO CORROSION INHIBITOR IS USED, AND THERE ARE NO INSULATING COVERS FOR THE BARRELS.
	AUXILIARY HEAT IS PROVIDED BY A SIMPLE, OLD-FASHIONED, SMALL SHEET-METAL, WOOD-BURNING STOVE NEAR THE CENTER OF THE BUILDING.
MATERIAL	6" OF FIBERGLAS; DOUBLE GLAZED WITH 0.004" POLYETHYLENE FILM; 2" LAYER OF STYROFOAM; 1/4" THICK WOODEN BATTENS; STYROFOAM PANELS; EARTH; CONCRETE FLOOR SLAB; 1,500 GALLONS OF WATER; 28 STEEL DRUMS OF 55-GALLON CAPACITY EACH; SHEET-METAL, WOOD-BURNING STOVE
AVAILABILITY	ADDITIONAL INFORMATION MAY BE OBTAINED FROM DAVID KRUSCHKE, ROUTE 2, BOX 34D, WILD ROSE, WISCONSIN 54984.
REFERENCE	SHURCLIFF, W.A. SOLAR HEATED BUILDINGS, A BRIEF SURVEY CAMBRIDGE, MASSACHUSETTS, W.A. SHURCLIFF, MARCH 1976, 212 P

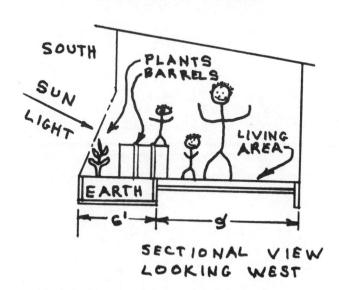

SECTIONAL VIEW
LOOKING WEST

TITLE	LAGOON STREET URBAN RENEWAL
CODE	
KEYWORDS	COMMERCIAL RESIDENTIAL PLAZA LOUVERED BRIDGE PEDESTRIAN ARCADE VENTILATION
AUTHOR	KRAMER & KRAMER
LOCATION	FREDERIKSTED, ST. CROIX
DATE & STATUS	
SCOPE	THE LAGOON STREET URBAN RENEWAL PROJECT IS A COMMERCIAL RESIDENTIAL DEVELOPMENT OCCUPYING TWO SQUARE BLOCKS. OF THE SEVEN BUILDINGS IN THE SUPERBLOCK, FOUR ARE RESIDENTIAL AND THREE ARE COMMERCIAL. THE PLAZA IS SURROUNDED BY TWO-STORY COMMERCIAL BUILDINGS, WHICH WILL HOUSE GOVERNMENT OFFICES AND SHOPS. THE OFFICES ON THE SECOND-STORY LEVEL ARE LOUVERED AND ARE REACHED THROUGH EXTERIOR STAIRS LEADING TO OPEN BRIDGES, WHICH CONNECT THE THREE BUILDINGS. THE SHOPS BELOW ARE RECESSED, FORMING PEDESTRIAN ARCADES.
TECHNIQUE	THE EXTERIOR WALLS OF THE THREE STORY APARTMENT BLOCKS ARE ALSO ENTIRELY LOUVERED, PROVIDING MAXIMUM VENTILATION AND VIEW, EVEN DURING PERIODS OF RAIN. ALL APARTMENTS ARE FLOOR-THROUGH, PERMITTING EXCELLENT CROSS BREEZES.
MATERIAL	LOUVERS
AVAILABILITY	REFER TO THE FOLLOWING REFERENCE FOR ADDITIONAL INFORMATION.
REFERENCE	ARCHITECTURE PLUS VIRGIN ISLAND BREEZY ARCHITECTURE PLUS, NOVEMBER/DECEMBER 1974, P 113

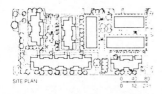

SITE PLAN

0 40 80
0 12 24

TITLE	LASAR HOUSE
CODE	
KEYWORDS	HOUSE UNDERGROUND WINDOW SOLAR PASSIVE STORAGE INSULATION RADIATION NATURAL CONVECTION
AUTHOR	STEPHEN LASAR
DATE & STATUS	THE LASAR HOUSE WAS COMPLETED IN 1976.
LOCATION	NEW MILFORD, CONNECTICUT
SCOPE	THE BUILDING IS A 2-1/2 STORY HOUSE, 22' X 55', WITH A FLOOR AREA OF 2,460 SQ. FT. IT INCLUDES A FAMILY ROOM, A GREENHOUSE, AND A STORAGE ROOM. THE SECOND, MAIN, STORY INCLUDES A LIVING ROOM, A KITCHEN, A DINING AREA, AND A STUDIO-BEDROOM. THE DORMER LOFT IS AT THE TOP CENTER OF THE HOUSE.
TECHNIQUE	THE HOUSE AIMS 5 DEGREES WEST OF SOUTH, ON A 15 DEGREE WEST-SLOPING HILL. THE FIRST STORY IS LARGELY UNDERGROUND EXCEPT AT THE WEST END. THE INSULATION IS EXCELLENT: THE CEILING HAS 9" OF FIBERGLAS, AND THE WALLS ARE INSULATED ON THE OUTSIDE WITH 3" OF POLYSTYRENE FOAM AND WATERPROOFED WITH WHITE STUCCO CEMENT. THERE IS A LARGE WINDOW AREA ON THE SOUTH SIDE. MOST OF THIS AREA IS DOUBLE GLAZED AND EQUIPPED WITH SHUTTERS OR HEAVY CURTAINS THAT ARE CLOSED ON COLD NIGHTS.

THE SOLAR HEATING SYSTEM IS OF THE PASSIVE TYPE. SOLAR COLLECTION IS ACHIEVED BY THE SOUTH-SIDE WINDOWS, WHOSE TOTAL AREA IS 575 SQ. FT. MOST OF THESE WINDOWS (400 SQ. FT.) ARE VERTICAL, THOSE SERVING THE GREENHOUSE (175 SQ. FT.) SLOPE 60 DEGREES FROM THE HORIZONTAL. ALL ARE DOUBLE GLAZED.

STORAGE IS ACCOMPLISHED BY MASSIVE CONCRETE WALLS AND FLOORS. AT THE WEST END OF THE FIRST FLOOR, THERE IS A 5' HIGH CONCRETE-BLOCK WALL JUST SOUTH OF THE 5' WIDE GREENHOUSE. AT THE EAST END, THERE IS A MASSIVE SECOND STORY WALL, 2' BEHIND THE ARRAY OF WINDOWS. THIS LEAVES A 2' WALKWAY BETWEEN THE WALL AND THE WINDOWS. THE EAST, WEST, AND NORTH WALLS ARE OF 10" POURED CONCRETE FOR THE FIRST STORY, AND 10" CONCRETE BLOCKS FOR THE SECOND STORY. ALL WALLS ARE EXPOSED ON THE INNER FACE FOR EASY INTAKE OR OUTPUT OF HEAT.

THE INSULATION IS ON THE OUTSIDE. AT THE WEST END, THE FLOOR OF THE FIRST STORY IS OF POURED CONCRETE RESTING ON AN ARRAY OF CHANNELED CONCRETE BLOCKS. THE CHANNELS CONSTITUTE NORTH-SOUTH DUCTS, 12" APART ON CENTERS. THEY ARE SERVED BY HEADER-DUCTS RUNNING EAST-WEST. AIR CAN CIRCULATE IN THE CHANNELS (NATURALLY, OR ASSISTED BY A FAN) TO DELIVER OR REMOVE HEAT. THE FLOOR OF THE SECOND STORY IS OF PRECAST CONCRETE PANELS. THE VERTICAL CONCRETE WALLS NEAR THE SOUTH SIDE OF THE HOUSE ARE PAINTED WITH NON-SELECTIVE BLACK PAINT. IN SUMMARY, SOLAR RADIATION ENTERS THE 575 SQ. FT. SOUTH WINDOW AREAS AND STRIKES THE MASSIVE CONCRETE FLOORS OR THE MASSIVE VERTICAL BLACK CONCRETE WALLS. THESE MASSIVE COMPONENTS STORE HEAT DURING SUNNY DAYS, AND GIVE OUT HEAT ON COLD NIGHTS. THE DISTRIBUTION OF HEAT IS MAINLY BY NATURAL CONVECTION, WITH SOME USE OF DUCTS AND FANS.

THE DOMESTIC HOT WATER IS PREHEATED BY PASSING ALONG SEVERAL PARALLEL, 3/4" COPPER PIPES, WHOSE TOTAL LENGTH IS 100 SQ. FT. THEY ARE IN CLOSE CONTACT WITH, AND RECESSED IN 1-1/2" X 9-1/2" DEPRESSIONS IN, THE VERTICAL BLACK CONCRETE WALL NEAR THE EAST END OF THE HOUSE. THE WATER IS STORED IN A SMALL TANK PARTLY RECESSED IN THAT SAME WALL. THE FINAL HEATING IS BY ELECTRIC COIL IN AN INSULATED TANK CLOSER TO THE FIRST TAP. |
MATERIAL	POLYSTYRENE FOAM; WHITE STUCCO CEMENT; SHUTTERS; HEAVY CURTAINS; CONCRETE WALLS AND FLOORS; DOUBLE GLAZED WINDOWS; 10" POURED CONCRETE; 10" CONCRETE BLOCKS; PRECAST CONCRETE PANELS; NON-SELECTIVE BLACK PAINT; 3/4" COPPER PIPES
AVAILABILITY	ADDITIONAL INFORMATION MAY BE OBTAINED FROM STEPHEN LASAR, 110 SAWYER HILL ROAD, RFD 3, NEW MILFORD, CONNECTICUT 06776.
REFERENCE	SHURCLIFF, W.A. SOLAR HEATED BUILDINGS, A BRIEF SURVEY CAMBRIDGE, MASSACHUSETTS, W.A. SHURCLIFF, MARCH 1976, 212 P

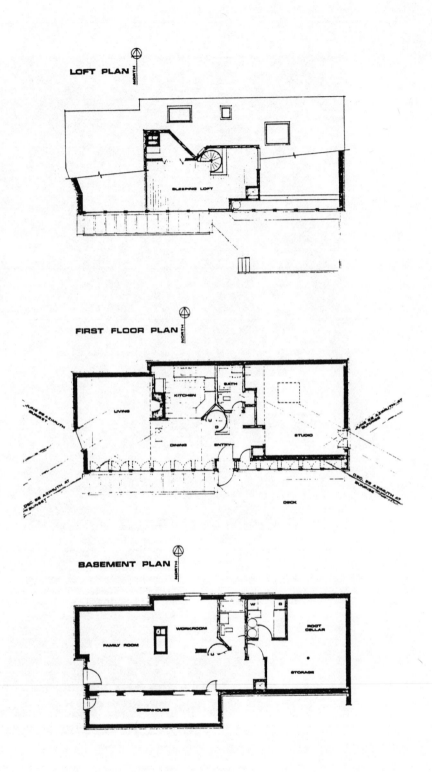

LOFT PLAN

SLEEPING LOFT

FIRST FLOOR PLAN

JUNE 21 AZIMUTH AT SUNSET

DEC. 22 AZIMUTH AT SUNSET

LIVING

KITCHEN

BATH

DINING

ENTRY

STUDIO

JUNE 21 AZIMUTH AT SUNRISE

DEC. 22 AZIMUTH AT SUNRISE

DECK

BASEMENT PLAN

FAMILY ROOM

WORKROOM

ROOT CELLAR

STORAGE

GREENHOUSE

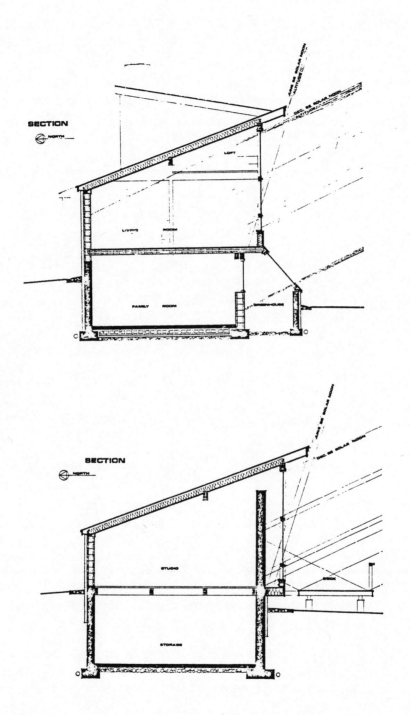

SECTION

NORTH

LOFT

LIVING ROOM

FAMILY ROOM

GREENHOUSE

SECTION

NORTH

STUDIO

DECK

STORAGE

TITLE	LAWRANCE HOUSE
CODE	
KEYWORDS	HOUSE VERTICAL WALL EAVES
AUTHOR	R. LAWRANCE
DATE & STATUS	THE LAWRANCE HOUSE WAS COMPLETED IN LATE 1975.
LOCATION	WILLETON, AUSTRALIA
SCOPE	LAWRANCE HOUSE IS A 1-1/2-STORY, BRICK-WALLED BUILDING. THE PLAN DIMENSIONS ARE 40' X 30'. THE FLOOR AREA IS 1,800 SQ. FT. THE HOUSE HAS 2 BEDROOMS, BUT NO BASEMENT OR ATTIC. THE CARPORTS ARE CLOSE TO THE SHADY SIDE OF THE HOUSE. THE BUILDING RESTS ON A CONCRETE SLAB. THE SECOND STORY INCLUDES A STUDIO AND A SMALL BEDROOM.
TECHNIQUE	THE HOUSE AIMS 12 DEGREES WEST OF NORTH, FAVORING RADIATION COLLECTION IN THE P.M.
	THE SOLAR HEATING SYSTEM IS OF THE PASSIVE TYPE, CONSISTING OF A SPECIAL SOLAR WALL, WHICH IS VERTICAL, FACES 12 DEGREES WEST OF NORTH, AND IS 22' LONG AND 11' HIGH. THE NET COLLECTION AREA IS 230 SQ. FT. THE OUTERMOST LAYER OF THE WALL IS A 0.005" FILM OF MELINEX TYPE 0 (LIKE MYLAR). 1" FROM THE FILM THERE IS A 3-1/4" THICK BRICK WALL. THE BRICKS ARE LAID ON EDGE, AND PAINTED A NON-SELECTIVE BLACK ON THE SIDE TOWARD THE SUN. BEHIND THIS THERE IS A 2-1/2" AIRSPACE, THEN A THICKER, 4-1/4", BRICK WALL, SLIGHTLY INSULATED ON THE SIDE TOWARD THIS AIRSPACE. THE INSULATION CONSISTS OF SISAL PAPER AND ALUMINUM FOIL, THE FOIL BEING IN DIRECT CONTACT WITH THE BRICKS. THE PURPOSE OF THIS INSULATION IS TO SLIGHTLY REDUCE THE AMOUNT OF ENERGY CONDUCTED DIRECTLY INTO THE ROOMS BEHIND THE WALL, AND SLIGHTLY INCREASE THE AMOUNT OF ENERGY DELIVERED TO THE UPWARD-FLOWING AIR IN THIS AIRSPACE. THE INSULATION IS EVEN MORE HELPFUL IN SUMMER. THERE ARE VENTS, EACH 2' X 1/2', AT THE TOP AND BOTTOM OF THE SOLAR WALL; THE 3 AT THE TOP ALLOW WARM AIR TO ENTER THE SECOND FLOOR ROOMS, AND THE 3 AT THE BOTTOM ALLOW COOL GROUND-FLOOR AIR TO ENTER THE BASE OF THE AIRSPACE. WARM AIR IN THE SUNNY-SIDE ROOMS OF THE SECOND FLOOR TRAVELS FREELY, HORIZONTALLY, TO THE SHADY-SIDE ROOMS VIA HUGE OPENINGS IN THE PARTITION WALL OF THE STUDIO. THE THERMAL CAPACITY OF THE BRICKS SUFFICES TO KEEP THE ROOMS WARM IN THE EVENING.
	TO PROVIDE COOLING IN SUMMER, THE UPPER VENTS TO THE INDOORS ARE CLOSED, AND THE CORRESPONDING VENTS TO THE OUTDOORS ARE OPENED. THE SOLAR-RADIATION-INDUCED UPDRAFT IN THE 2-1/2" AIR-SPACE HELPS VENT HOT AIR FROM THE GROUND-FLOOR ROOMS. NEAR THE SUMMER SOLSTICE, HOWEVER, WHEN LITTLE RADIATION STRIKES THE SOLAR WALL, EAVES SHADE THE WALL.
MATERIAL	0.005" FILM OF MELINEX TYPE 0; 3-1/4" THICK BRICK WALL, PAINTED NON-SELECTIVE BLACK; 4-1/4" BRICK WALL; SISAL PAPER; ALUMINUM FOIL
AVAILABILITY	ADDITIONAL INFORMATION MAY BE OBTAINED FROM R. LAWRANCE, 7 BROGLA PROMENADE, BURRENDAH ESTATE, WILLETON, AUSTRALIA.
REFERENCE	SHURCLIFF, W.A. SOLAR HEATED BUILDINGS, A BRIEF SURVEY CAMBRIDGE, MASSACHUSETTS, W.A. SHURCLIFF, MARCH 1976, 212 P

TITLE	LEARNING MODEL OF TALL OFFICE BUILDING

CODE

KEYWORDS   LEARNING MODEL INSULATION OVERHANG WATER TYPE COLLECTOR STORAGE

AUTHOR   ENVIRONMENTAL SYSTEMS DESIGN INC.

DATE & STATUS   THE LEARNING MODEL OF A TALL OFFICE BUILDING WAS COMPLETED
IN MAY 1975.

LOCATION   HARMARVILLE, PENNSYLVANIA

SCOPE   THE BUILDING IS A 1-1/2-STORY, OPEN-INTERIOR, LEARNING MODEL
BUILDING, 21' WIDE, 14' DEEP, AND 21' HIGH, WITH A MONOSLOPE
45 DEGREE ROOF. THE BUILDING CONTAINS MANY FEATURES PERTINENT
TO LATER-PHASE BUILDINGS.

TECHNIQUE   MANY ENERGY CONSERVING FEATURES ARE USED, INCLUDING HIGH-
QUALITY INSULATION AND DOUBLE GLAZING. THE WINDOWS AND SLIDING
DOORS, FACING STRAIGHT SOUTH, ARE SHADED IN SUMMER BY A LARGE
OVERHANG OF COLLECTOR-ROOF. THE WINDOWS ON THE EAST AND WEST
ENDS OF THE BUILDING FACE NORTHEAST AND NORTHWEST RESPECTIVELY,
AND THUS RECEIVE NO DIRECT SOLAR RADIATION DURING THE MIDDLE-OF-
THE-DAY 5-HOUR PERIOD IN SUMMER. THE BUILDING'S WOOD FRAME
RESTS ON A CONCRETE SLAB. THE WINDOW TYPE IS PPG TWINDOW WITH
INSULATING GLASS. THE WINDOW AREA IS 640 SQ. FT.

THE COLLECTOR IS A 490 SQ. FT. GROSS, 450 SQ. FT. NET, WATER-
TYPE COLLECTOR, MOUNTED ON THE 45 DEGREE ROOF AND ALSO ON THE
VERTICAL ENDS OF THE BUILDING. THE COLLECTOR CONSISTS OF 27
PANELS, OF WHICH 15 ARE ON THE ROOF, 6 ON THE EAST END OF THE
BUILDING, AND 6 ON THE WEST END OF THE BUILDING. THOSE ON THE
BUILDING ENDS FACE SOUTHEAST AND SOUTHWEST. EACH PANEL IS A
PPG ASSEMBLY, 34-3/16" X 76-3/16" X 1-1/16". THE HEART OF THE
PANEL IS A ROLL-BOND ALUMINUM SHEET, MADE BY OLIN BRASS CO.,
WITH INTEGRAL PASSAGES FOR A LIQUID, WHICH IS 80% DISTILLED
WATER AND 20% ETHYLENE GLYCOL. THE ALUMINUM SHEET HAS AN
ALCOA #655 SELECTIVE BLACK COATING. EACH PANEL HAS A 3" BACKING
OF FIBERGLAS, AND EACH IS DOUBLE GLAZED WITH PPG HERCULITE
K 1/8" TEMPERED GLASS. THE FLUID IS CIRCULATED TO THE STORAGE
SYSTEM BY A 1/2 HP, 27 GPM CENTRIFUGAL PUMP.

THE STORAGE SYSTEM CONSISTS OF 500 GALLONS OF WATER-AND-GLYCOL
SOLUTION IN AN INSULATED STEEL TANK, 3-1/2' IN DIAMETER AND 8'
LONG. THE TANK IS ORIENTED HORIZONTALLY, AND SITUATED ABOVE
GROUND JUST OUTSIDE THE BUILDING ON THE NORTH SIDE. HEAT IS
DELIVERED TO THE BUILDING INTERIOR BY MEANS OF A FAN-COIL SYSTEM.

COOLING MAY BE PROVIDED AND MAY EMPLOY, AS A POWER SOURCE, HEAT
FROM THE HOT WATER, AT ABOUT 200 DEGREES F, FROM THE COLLECTOR.

MATERIAL   DOUBLE GLAZING; PPG TWINDOW; INSULATING GLASS; PPG ASSEMBLY;
ALUMINUM SHEET; 80% DISTILLED WATER; 20% ETHYLENE GLYCOL;
#655 SELECTIVE BLACK COATING; 3" FIBERGLAS; PPG HERCULITE
K 1/8" TEMPERED GLASS; 1/2 HP, 27 GPM CENTRIFUGAL PUMP; INSULATED
STEEL TANK; FAN-COIL SYSTEM

AVAILABILITY   ADDITIONAL INFORMATION MAY BE OBTAINED FROM ENVIRONMENTAL SYSTEMS
DESIGN INC., 35 EAST WACKER DRIVE, CHICAGO, ILLINOIS 60601.

REFERENCE   SHURCLIFF, W.A.
SOLAR HEATED BUILDINGS, A BRIEF SURVEY
CAMBRIDGE, MASSACHUSETTS, W.A. SHURCLIFF, MARCH 1976, 212 P

TITLE	LEFEVER HOUSE
CODE	
KEYWORDS	HOUSE VISOR CANOPY COLLECTOR STORAGE THERMAL CAPACITY OVERHANG
AUTHOR	H.R. LEFEVER
DATE & STATUS	THE LEFEVER HOUSE WAS COMPLETED IN 1954.
LOCATION	STOVERSTOWN, PENNSYLVANIA
SCOPE	THE BUILDING IS A 1-1/2-STORY, 3-BEDROOM DWELLING. THERE IS NO BASEMENT. THE SECOND STORY IS UNHEATED. IT SUPPORTS THE COLLECTOR AND SERVES AS AN ATTIC. THE HEATED AREA, I.E. THE FIRST STORY WHICH IS 65' X 20', IS 1,250 SQ. FT.
TECHNIQUE	AN 18" WIDE VISOR OR CANOPY SHADES THE FIRST STORY WINDOWS IN SUMMER.
	THE COLLECTOR HAS A GROSS AREA OF 520 SQ. FT. THE NET AREA IS 450 SQ. FT. IT EMPLOYS 15 DOUBLE GLAZED AREAS, EACH 8' X 4', COVERING THE ENTIRE VERTICAL SOUTH FACE OF THE SECOND STORY. THE THICKNESS OF THE GLASS SHEETS IS 7/32". RADIATION PASSING THROUGH THE GLASS STRIKES A NON-SELECTIVE BLACK SURFACE, AND FAN-DRIVEN AIR CARRIES HEAT FROM THAT SURFACE, VIA DUCTS, TO SEVERAL 4' WIDE CLOSETS OR BINS BETWEEN THE ROOMS, AND THEN TO THE ROOMS. THE SYSTEM EMPLOYS ONE 1/6 HP SQUIRREL-CAGE BLOWER, AND TWO 20" DIAMETER, 800 RPM FANS. THE FIRST-STORY WINDOWS OF 80 SQ. FT. ARE DOUBLE GLAZED, AND COLLECT SOME SOLAR RADIATION.
	THERE IS NO STORAGE SYSTEM OTHER THAN THE GENERAL THERMAL CAPACITY OF THE HOUSE ITSELF. AUXILIARY HEAT IS PROVIDED BY A GAS-FIRED FURNACE. THERE IS NO COOLING IN SUMMER. THE ROOF OVERHANG KEEPS THE COLLECTOR FROM OVERHEATING IN SUMMER. THE FIRST-STORY VISOR BLOCKS MOST OF THE RADIATION APPROACHING THE FIRST-STORY SOUTH WINDOWS.
MATERIAL	7/32" THICK GLASS SHEETS; NON-SELECTIVE BLACK SURFACE; 1/6 HP SQUIRREL-CAGE BLOWER; TWO 20" DIAMETER, 800 RPM FANS
AVAILABILITY	ADDITIONAL INFORMATION MAY BE OBTAINED FROM THE SONNEWALD SERVICE, RD 1, BOX 457, SPRING GROVE, PENNSYLVANIA 17362.
REFERENCE	SHURCLIFF, W.A. SOLAR HEATED BUILDINGS, A BRIEF SURVEY CAMBRIDGE, MASSACHUSETTS, W.A. SHURCLIFF, MARCH 1976, 212 P

TITLE	LIBERTY ELEMENTARY SCHOOL
CODE	
KEYWORDS	ELEMENTARY SCHOOL REPETITIVE MODULES STEEL FRAME PORCELAIN INSULATED PANELS
AUTHOR	DON M. HISAKA & ASSOCIATES, ARCHITECTS, INC.
LOCATION	COLUMBUS, OHIO
DATE & STATUS	THE LIBERTY ELEMENTARY SCHOOL WAS COMPLETED IN APRIL 1976.
SCOPE	THE LIBERTY ELEMENTARY SCHOOL IS A ONE STORY BUILDING WITH A MEZZANINE FOR ADMINISTRATIVE OFFICES; TO THE REAR THERE IS A LOFTLIKE SPACE CONSISTING OF REPETITIVE MODULES WHICH CAN BE EASILY ADDED ONTO FOR EXPANSION.
	THE SCHOOL IS FOR KINDERGARTEN THROUGH GRADE SIX. IT IS SUITABLE FOR TEAM TEACHING AND INCLUDES A MULTIPURPOSE ROOM AND LIBRARY LEARNING CENTER.
TECHNIQUE	THE STEEL FRAME CONSTRUCTION WITH YELLOW PORCELAIN INSULATED PANELS ENHANCES ENERGY CONSERVATION.
MATERIAL	STEEL FRAME, YELLOW PORCELAIN INSULATED PANELS
AVAILABILITY	ADDITIONAL INFORMATION MAY BE OBTAINED FROM DON M. HISAKA, 237 THE ARCADE, CLEVELAND, OHIO 44114.
REFERENCE	PROGRESSIVE ARCHITECTURE DON M. HISAKA & ASSOCIATES, ARCHITECTS, INC. PROGRESSIVE ARCHITECTURE, JANUARY, 1976, PP 48-49

505

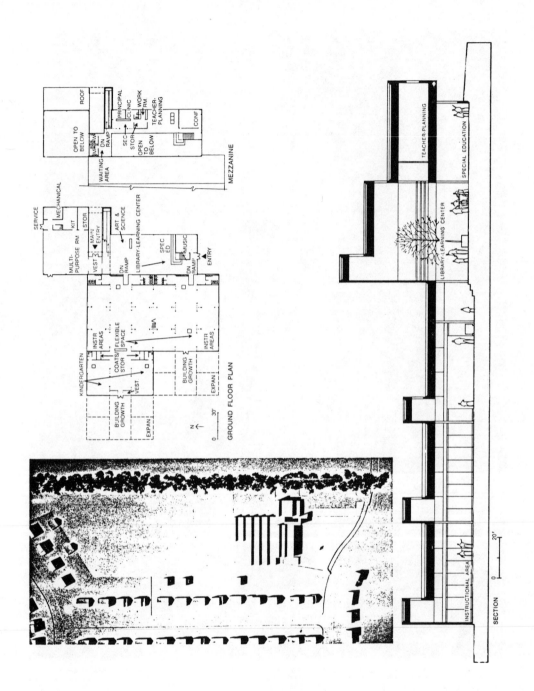

ROOF

OPEN TO BELOW

PRINCIPAL

CLINIC

WORK RM.

TEACHER-PLANNING

DN RAMP

SEC STOR

OPEN TO BELOW

CONF

WAITING AREA

MEZZANINE

SERVICE

MECHANICAL

KIT STOR

MULTI-PURPOSE RM.

ART & SCIENCE

MAIN ENTRY

VEST

LIBRARY-LEARNING CENTER

DN RAMP

SPEC ED

MUSIC

DN RAMP

ENTRY

INSTR AREAS

FLEXIBLE SPACE

COATS/ STOR

INSTR AREAS

KINDERGARTEN

VEST

BUILDING GROWTH

BUILDING GROWTH

EXPAN

EXPAN

N ←

0   30'

GROUND FLOOR PLAN

TEACHER-PLANNING

SPECIAL EDUCATION

LIBRARY-LEARNING CENTER

INSTRUCTIONAL AREA

0   20'

SECTION

TITLE	LINDE-GROTH OFFICE BUILDING
CODE	
KEYWORDS	AIR SUNTRAP COLLECTOR
AUTHOR	LINDE-GROTH
DATE & STATUS	THE LINDE-GROTH OFFICE BUILDING WAS RETROFITTED WITH THE SUN STONE SYSTEM IN 1975.
LOCATION	SHEBOYGAN, WISCONSIN
SCOPE	THE DEVELOPMENT OF AN AIR-TYPE SOLAR ENERGY SYSTEM CALLED SUN STONE IS A SIMPLE, WELL-DESIGNED SYSTEM THAT COULD BE PRODUCED AND INSTALLED BY A NON-PROFESSIONAL BUILDER. IT IS A COLLECTOR SYSTEM THAT FITS ANY TYPE OF ROOF, IS LIGHTWEIGHT, AND AS TROUBLE-FREE AS POSSIBLE. IT IS A SYSTEM THAT COULD BE SOLD BY MAIL ORDER, PERHAPS THROUGH A MAJOR CATALOG HOUSE, A REALLY UNCOMPLICATED SYSTEM, THAT CAN BE ASSEMBLED WITH A SMALL DRILL, SCREWS, AND TAPE.

THE SYSTEM, UTILIZING 14 COLLECTOR PANELS, PROVIDED 27% OF THE HEAT REQUIRED FOR THE OFFICE BUILDING DURING THE FIRST YEAR. ADDING FOUR PANELS HAS INCREASED THE FIGURE TO 45%.

THE SYSTEM AS ORIGINALLY INSTALLED COST ABOUT $2,000 FOR MATERIALS ONLY. A SIMILAR INSTALLATION, WITH APPROPRIATE LABOR COSTS, WOULD BE PRICED AT $4,000. (THE USE OF CANNED WATER IN PLACE OF ROCK ADDS $600 TO THE MATERIAL COST, BUT THE AMOUNT OF SPACE REQUIRED FOR THE STORAGE MEDIUM IS ONLY A THIRD OF THAT REQUIRED FOR STONE STORAGE.)

TECHNIQUE	A FLUID COLLECTOR WAS ENGINEERED BUT THEN DISCARDED BECAUSE OF FREEZING PROBLEMS. THE COLLECTOR UNIT IS A TRIANGULAR DEVICE 6 FEET HIGH BY 40 FEET LONG, MADE UP OF A DOUBLE LAYER OF GLASS, BACKED BY 3 INCHES OF FIBERGLAS INSULATION. FOR MAXIMUM EFFICIENCY AT SHEBOYGAN'S LATITUDE, THE COLLECTOR IS ANGLED 50 DEGREES.

AFTER THE COLLECTOR WAS MOUNTED, A 6-BY-12 FOOT STORAGE ROOM IN THE BASEMENT WAS FILLED WITH ORDINARY STONES TO CREATE A STORAGE MEDIUM. GETTING THE STONES INTO POSITION INVOLVED THREE OR FOUR DAYS OF AIR-HAMMERING AND SHOVELING 18 TONS OF GRAVEL AROUND IN THE AREA. (AN ALTERNATIVE STORAGE MEDIUM HAS SINCE BEEN SUBSTITUTED, INVOLVING TIN CANS FILLED WITH PASTEURIZED WATER AND PLACED IN A SHEET METAL STORAGE BOX 2 BY 4 BY 5 FEET, WHICH CAN BE EASILY MANEUVERED DOWN A FLIGHT OF BASEMENT STEPS.)

THE COLLECTOR AND THE STORAGE MEDIUM ARE CONNECTED BY 12-INCH DUCTS, WHILE A THERMOSTATICALLY CONTROLLED BLOWER KEEPS AIR MOVING THROUGH THE SYSTEM AT A STEADY RATE OF JUST OVER 10 MPH. BESIDES THE BLOWER, THE ONLY MOVING PARTS ARE A SET OF AUTOMATIC DAMPERS. ANNUAL MAINTENANCE CONSISTS OF REPLACING A STANDARD THROWAWAY FILTER, INSPECTING THE BLOWER AND DAMPERS, AND WASHING THE COLLECTOR'S GLASS SURFACE WHEN NECESSARY.

MATERIAL	DOUBLE LAYER OF GLASS, 3 INCHES OF FIBERGLAS INSULATION, STONES, THERMOSTATICALLY CONTROLLED BLOWER, AUTOMATIC DAMPERS
AVAILABILITY	ADDITIONAL INFORMATION MAY BE OBTAINED FROM SUN STONE SOLAR ENERGY EQUIPMENT, P.O. BOX 941, SHEBOYGAN, WISCONSIN 53081
REFERENCE	KOEHLER, ROBERT E. NEW MARKETS FOR ARCHITECTS' SERVICES AIA JOURNAL, JULY 30, 1976, MEMO INSERT, 4 P

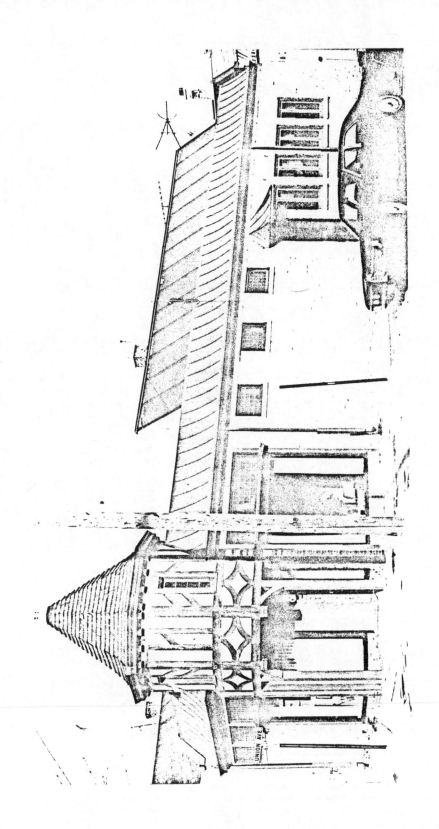

508

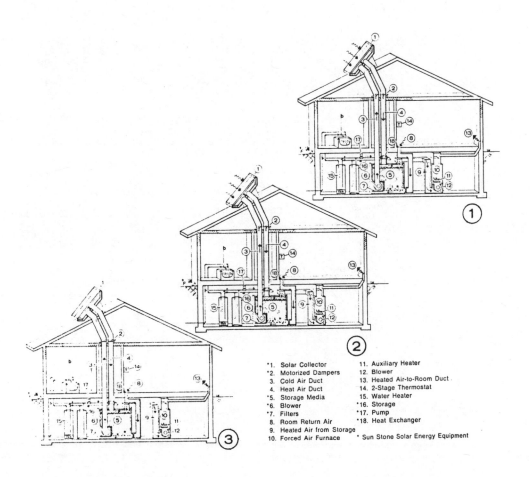

*1. Solar Collector
*2. Motorized Dampers
3. Cold Air Duct
4. Heat Air Duct
*5. Storage Media
*6. Blower
*7. Filters
8. Room Return Air
9. Heated Air from Storage
10. Forced Air Furnace

11. Auxiliary Heater
12. Blower
13. Heated Air-to-Room Duct
14. 2-Stage Thermostat
15. Water Heater
*16. Storage
*17. Pump
*18. Heat Exchanger

* Sun Stone Solar Energy Equipment

509

TITLE	LIVONIA OFFICE BUILDING
CODE	
KEYWORDS	U FORM PANEL OFFICE CONCRETE REINFORCING STEEL CIRCULAR VOIDS INTEGRATED SYSTEM
AUTHOR	MELVIN H. SACHS, AIA, NCARB
DATE & STATUS	THIS BUILDING WAS FINISHED IN 1973.
LOCATION	LIVONIA, MICHIGAN
SCOPE	THIS 119,000 SQ. FT., 8-STORY, OFFICE BUILDING WAS ERECTED IN LESS THAN 7 MONTHS, USING SACHS' INNOVATIVE U-FORM WALL PANEL SYSTEM. AFTER TESTS, THIS BUILDING WAS FOUND TO HAVE USED 22% LESS ENERGY THAN CURRENT ADMINISTRATION GUIDELINES CALL FOR.
TECHNIQUE	THE TECHNIQUE INVOLVED IN SAVING ENERGY IN THIS BUILDING IS THE UNIQUE WALL SYSTEM INVENTED BY MR. SACHS. A BASIC UNIT INVOLVES INSULATING PANELS, WHICH ARE 6' WIDE, 32" HIGH AND 8" THICK. VERTICAL CIRCULAR VOIDS IN THE PANEL ARE FILLED WITH CONCRETE AND STEEL REINFORCEMENT AT THE SITE TO PROVIDE A SOLID INTERLOCKING STRUCTURAL SYSTEM, NOT UNLIKE A SCHEME DEVELOPED BY R. BUCKMINISTER FULLER SOME FIFTY YEARS AGO.

MORE RECENT DEVELOPMENTS INCLUDE INCREASING THE PANEL SIZE TO 13' WIDE, 8'10" (A FULL STORY) HIGH, AND 12" THICK. THE PANELS CAN ALSO INCLUDE SHOP INSTALLED WINDOWS, SPANDRELS AND RELATED CAULKING AS ARE BEING USED ON A TWIN, 12-STORY, 232 UNIT (EACH BUILDING 116 UNITS), HUD APPROVED APARTMENT DEVELOPMENT FOR THE ELDERLY NOW BEING COMPLETED IN DETROIT, MICHIGAN. A HUD STRUCTURAL ENGINEERING BULLETIN (NO. 869) HAS BEEN ISSUED. CHOICE OF EXTERIOR VENEERS IS ALMOST UNLIMITED. |
MATERIAL	U-FORM WALL PANELS, CONCRETE, REINFORCING STEEL, U-FACTOR 0.032
AVAILABILITY	ADDITIONAL INFORMATION MAY BE OBTAINED FROM U-FORM SYSTEMS AND TECHNOLOGY, INC., 29200 VASSAR AVENUE, SUITE 700, LIVONIA, MICHIGAN 48152.
REFERENCE	AMERICAN INSTITUTE OF ARCHITECTS FOCUS ON PRODUCTS: OLD IDEAS AND ENERGY EFFICIENCY ENERGY, DECEMBER 1976, #13, 4 P

This boiler heats this building (with only 8 BTU's per square foot) Total energy consumption for all purposes (heating, air conditioning, lighting, hot water, elevators, misc. power, etc.) is only 43,000 BTU's per square foot per year, which is 22% lower than the new U.S. General Services Administration (G.S.A.) recommendations.

Sachs-Freed Associates, Architects, Engineers, Planners

512